Springer Series in Translational Stroke Research

Series Editor
John Zhang
Neurosurgery, Physiology, Loma Linda University, Loma Linda, USA

The Springer Series in Translational Stroke Research will cover basic, translational and clinical research, emphasizing novel approaches to help translate scientific discoveries from basic stroke research into the development of new strategies for the prevention, assessment, treatment, and enhancement of central nervous system repair after stroke and other forms of neurotrauma.

More information about this series at http://www.springer.com/series/10064

Li-Ru Zhao • John H. Zhang
Editors

Cellular Therapy for Stroke and CNS Injuries

Springer

Editors
Li-Ru Zhao
Department of Neurosurgery
SUNY Upstate Medical College
Syracuse
New York
USA

John H. Zhang
Loma Linda University School of Medicine
Depts of Physiology
Loma Linda
California
USA

ISBN 978-3-319-11480-4 ISBN 978-3-319-11481-1 (eBook)
DOI 10.1007/978-3-319-11481-1
Springer Cham Heidelberg New York Dordrecht London

Library of Congress Control Number: 2014955546

Printed on acid-free paper

Springer is part of Springer Science+Business Media (www.springer.com)

The Preface

Stroke and injuries in the central nervous system (CNS) such as traumatic brain injury and spinal cord injury are the leading causes of long-term disability worldwide. During the past several decades, great efforts have been made to understand the pathological process and develop therapeutic strategies for reducing neuron loss and improving functional restoration after stroke or CNS injuries. Accumulating evidence has revealed that the pathological process after stroke and CNS injuries is orchestrated in a very complex manner throughout the acute, subacute and chronic phases. A large number of molecules, proteins and cells both inside and outside the CNS are crucially involved in the events occurring from initiating stroke and CNS injuries to recovery from the injuries; these events include oxidative stress, inflammatory, neuroprotection, dead tissue removal, neurogenesis, angiogenesis, and neural network rewiring. Learning from the lessons that thousands of drugs targeting a signal molecule, protein or cell failed in clinical trials, combinational therapies targeting multiple molecules, proteins or cells have recently been proposed to be the new direction for developing treatment for stroke and CNS injuries.

The Cellular Therapy for Stroke and CNS Injuries was chosen as the theme of this book because a large body of evidence suggests that stem cell therapy would be a good candidate to meet the new mission of searching for treatment for stroke and CNS injuries. Numerous studies have demonstrated that stem cells can release neuron trophic factors, growth factors, and anti-inflammatory cytokines in situ, which protect neurons from injuries, reduce inflammation, and enhance neurogenesis, angiogenesis, and neuronal network remodeling in the recipient CNS after transplantation. Importantly, recent studies have demonstrated that stem cells also release abundant exosomes and microvesicles (EMVs), which contain membrane receptors, proteins, mRNAs, miroRNAs and organelles. Stem cell-derived EMVs have been shown to rescue cells from injury, reprogram terminally differentiated cells to re-enter the cycle, and promote tissue regeneration and repair. Although transplantation of neural stem cells or bone marrow stem cells has moved to clinical trials, many open questions in basic stem cell research still remain to be addressed in future. Further studies needed in future include mechanistic elaboration of stem cells in rebuilding the neurovascular unit and in neuroprotection and neurorepair in the settings of stroke and CNS injuries, limitation of stem cell passages before transplantation, refined methods for culturing stem cells ex vivo or in vitro, the optimal timing, route and dose of stem cell delivery, the frequency of stem cell therapy, and the combination of stem cell therapy with other treatments.

Many investigators in the world, especially those who have given important contributions to the field of stem cell research and stem cell therapy, were invited to create the chapters of this book. In these chapters, the development of stem cell therapy and new advances in stem cell research and therapies for stroke and CNS injuries have been summarized, and challenging questions and future directions have been discussed. This book is organized by several sections divided by stem cell types. We believe that the readers of this book will obtain new knowledge on the progress in stem cell research and therapy for stroke and CNS injuries, will get new insights into how the current approaches can be improved, and will find the gaps to advance this research field to a deeper level.

We would greatly appreciate all the contributors for each of the chapters. Obviously without their support and contributions, the creation of this book would not have been possible.

Li-Ru Zhao
John H. Zhang

Foreword

The concept of cell-based therapy can be traced back to the nineteenth century when it was discovered that cells were the building blocks of life and that some cells had the ability to produce other types of cells such as blood cells. The term "stem cell" appeared in the scientific literature as early as the middle of the nineteenth century in the works of the eminent German biologist Ernst Haeckel. Inspired by the German word "stammbäume" meaning family trees or "stem trees", Haeckel used the term "stammzelle" (German for stem cell) to describe the ancestor unicellular organism from which he presumed all multicellular organisms evolved. In 1968, the first bone marrow transplant was performed to successfully treat two siblings with severe combined immunodeficiency. It was 10 years later when multipotent stem cells were discovered in human cord blood. Groundbreaking discoveries in recent history include the creation of the sheep Dolly, which was the first mammal cloned from an adult somatic cell, the identification of hematopoietic cancer stem cells, isolation of embryonic stem cells and, more recently, the creation of induced pluripotent stem (iPS) cells from adult mouse and human cells. These discoveries were exciting for the fields such as cell biology, embryology, and carcinogenesis, and they provided broad tools for drug discovery using in vitro tests. Meanwhile, cell-based transplantation therapy has been investigated for several decades. In addition to bone marrow transplantation, most early transplantation studies used mature cells and generated some enthusiastic results. This approach, however, was hampered by very poor cell survival, unreliable efficacy, insufficient cell sources, and adverse side effects. In this respect, the discovery of embryonic stem cells and iPS cells opened the door for more effective and clinically feasible cell-based transplantation therapies. Today, stem cell transplantation therapy has been studied in many animal models and tested in a number of clinical trials for the treatment of stroke, traumatic brain or spinal cord injury, neurodegenerative diseases, heart ischemia, and many other disorders.

Stem cell research is a rapidly progressing, multifaceted research area. Nowadays, there are fewer doubts that cell-based therapy possesses the promising potential for reducing neuronal injury and promoting regenerative effects following brain injuries and disorders. Transplantation of various types of stem cells, including embryonic stem cells, neural stem cells, endothelial progenitor cells, mesenchymal stem cells, hematopoietic stem cells, iPS cell-derived neural progenitor cells, and many others has been extensively tested. In animal models of ischemic stroke, stem cell transplantation therapy shows consistent functional benefits after intracerebral and systemic administration of cells. Transplanted stem cells are able to differentiate into neurons and glial cells in the injured brain and spinal cord, and provide growth and trophic supports for neural reconstruction and blood flow recovery in the ischemic region. As a trend in recent years, stem cell transplantation therapy has been combined with other protective and/or regenerative treatments to enhance the efficacy and efficiency of the therapy. This is an area that will be benefited by the accumulating knowledge in basic research of cellular and molecular mechanisms of cell death, proliferation, differentiation, migration, and neural development. A good example of the combination therapy is the preconditioning strategy that shows multiple benefits such as enhancing cell viability, trophic supports, and functional recovery after transplantation.

Compared to previous neuroprotective treatments, cell-based therapy in the treatment of stroke and CNS injuries are thought to have longer windows of administration and fewer exclusion criteria for clinical indications. However, issues regarding cell sources, long-term safety and efficacy, delivery route, homing to the lesion site, cell tracking, and tissue engineering/reconstruction are the examples of some bottlenecks in the translational process. Although pre-clinical and clinical trials for cell-based therapies have accumulated compelling results on it's therapeutic potential, the mechanisms of the beneficial effects are not well understood. The information presented in this book is our collective effort to address some of these concerns and provide information for future research. The collection of research summaries and research perspectives should be useful to basic scientists, physicians, medical or graduate students, and biologists who are interested in stem cells biology and translational research focused on stroke, traumatic injuries, and other CNS disorders.

The time is ripe for many of the ideas presented in this book. It is expected that in a few years, preclinical and clinical studies will demonstrate some successes in resolving the aforementioned bottlenecks in the development of stem cell therapy. At this exciting period when we are entering a more advanced phase of regenerative medicine, this book seeks to illustrate critical points relevant to the important concerns. Part I of this book introduces cellular therapies as a promising treatment for stroke. Part II summarizes applications of cellular therapies on traumatic brain injury and spinal cord injury. The cell types tested for treating stroke and CNS injuries are summarized and the issue of motivating endogenous stem cells is discussed. Several chapters address the need for understanding the microenvironment following stroke and CNS injuries. The functional and morphological integrations with endogenous cells and neural networks are discussed by several authors. Based on their own research, authors provide current understanding on the trophic support from transplanted cells, which may hold the key for the success of transplantation therapy. The regenerative and restorative benefits, as well as their dependence on cell types and delivery routes are the focus of some chapters in this book. The selection of distinct cell types and monitoring of migration patterns post-transplantation are discussed as the critical questions in translational research. The book also provides useful information on the collection methods and cell characterization. In this regard, optimized protocols and industry standards are emerging in recent years that will facilitate clinical application of cell-based therapies.

The Editors have done a great job attracting the leading scientists on the research frontier to bring forth novel ideas and present comprehensive summaries of their research achievements. The book offers new updates on many research areas and criticisms on the current status of stem cell transplantation therapy, which can be regarded as useful references and guidelines for improving future pre-clinical and clinical investigations.

Ling Wei
Professor and John E. Steinhaus Endowed
Chair for Stem Cell Research
Emory University School of Medicine

Contents

Contributors

Sandra Acosta Center of Excellence for Aging and Brain Repair, Department of Neurosurgery and Brain Repair, University of South Florida College of Medicine, Tampa, FL, USA

Daniela Aguirre Center of Excellence for Aging and Brain Repair, Department of Neurosurgery and Brain Repair, University of South Florida College of Medicine, Tampa, FL, USA

Marina Bastawrous Center of Excellence for Aging and Brain Repair, Department of Neurosurgery and Brain Repair, University of South Florida College of Medicine, Tampa, FL, USA

Cesar V. Borlongan Center of Excellence for Aging and Brain Repair, Department of Neurosurgery and Brain Repair, University of South Florida College of Medicine, Tampa, FL, USA

Francisco Campos Department of Neurology, Clinical Neurosciences Research Laboratory, Hospital Clínico Universitario, c/ Travesa da Choupana s/n, Health Research Institute of Santiago de Compostela (IDIS), University of Santiago de Compostela, Santiago de Compostela, Spain

Maddalena Caruso Centro di Ricerca E. Menni, Fondazione Poliambulanza-Istituto Ospedaliero, Brescia, Italy

José Castillo Department of Neurology, Clinical Neurosciences Research Laboratory, Hospital Clínico Universitario, c/ Travesa da Choupana s/n, Health Research Institute of Santiago de Compostela (IDIS), University of Santiago de Compostela, Santiago de Compostela, Spain

Jieli Chen Department of Neurology, Henry Ford Hospital, Detroit, MI, USA

Michael Chopp Department of Neurology, Henry Ford Hospital, Detroit, MI, USA

Department of Physics, Oakland University, Rochester, MI, USA

Isis Cristina do Nascimento Departamento de Bioquímica, Instituto de Química, Universidade de São Paulo, São Paulo, Brazil

Maria-Grazia De Simoni Department of Neuroscience, IRCCS-Istituto di Ricerche Farmacologiche Mario Negri, Milan, Italy

Ike dela Peña Center of Excellence for Aging and Brain Repair, Department of Neurosurgery and Brain Repair, University of South Florida College of Medicine, Tampa, FL, USA

Jenny A. K. Ekberg Eskitis Institute for Drug Discovery, Griffith University, Brisbane, Australia

School of Biomedical Sciences, Queensland University of Technology, Brisbane, Australia

Vaughan Feisst Dunbar Laboratory, School of Biological Sciences, The University of Auckland, Auckland, New Zealand

E. A. Foran Center for Excellence in Aging and Brain Repair, Department of Neurosurgery and Brain Repair, Morsani College of Medicine MDC 78, University of South Florida, Tampa, FL, USA

Diana Hernandez Center of Excellence for Aging and Brain Repair, Department of Neurosurgery and Brain Repair, University of South Florida College of Medicine, Tampa, FL, USA

Kunlin Jin Zhejiang Provincial Key Laboratory of Aging and Neurological Disorder Research, First Affiliated Hospital, Wenzhou Medical University, Wenzhou, China

Department of Pharmacology and Neuroscience, University of North Texas Health Science Center, TX, USA

Yuji Kaneko Center of Excellence for Aging and Brain Repair, Department of Neurosurgery and Brain Repair, University of South Florida College of Medicine, Tampa, FL, USA

Yaning Li Neuroscience and Neuroengineering Research Center, Med-X Research Institute and School of Biomedical Engineering, Jiao Tong University, Shanghai, China

Daniel J. Liebl The Miami Project to Cure Paralysis and Department of Neurological Surgery, University of Miami, Miami, FL, USA

The Miami Project to Cure Paralysis, Department of Neurological Surgery, University of Miami Miller School of Medicine, Miami, FL, USA

Michelle B. Locke Faculty of Medicine and Health Sciences, South Auckland Clinical School, C/- Middlemore Hospital, The University of Auckland, Auckland, New Zealand

Diego Lozano Center of Excellence for Aging and Brain Repair, Department of Neurosurgery and Brain Repair, University of South Florida College of Medicine, Tampa, FL, USA

Yuanyuan Ma Department of Neurology, School of Medicine, Ruijin Hospital, Jiao Tong University, Shanghai, China

Marek Molcanyi Institute for Neurophysiology, University of Cologne, Cologne, Germany

Edmund A. M. Neugebauer Institute for Research in Operative Medicine, University of Witten/Herdecke, Köln, Germany

Mibel Pabon Center of Excellence for Aging and Brain Repair, Department of Neurosurgery and Brain Repair, University of South Florida College of Medicine, Tampa, FL, USA

Ornella Parolini Centro di Ricerca E. Menni, Fondazione Poliambulanza-Istituto Ospedaliero, Brescia, Italy

Emanuela Parotto Institute of Anesthesia and Intensive Care, University of Padova, Padova, Italy

Francesca Pischiutta Department of Neuroscience, IRCCS-Istituto di Ricerche Farmacologiche Mario Negri, Milan, Italy

Peter Riess Department of Trauma and Orthopedic Surgery, HELIOS Klinikum Bad Berleburg, Teaching Hospital University Marburg, Bad Berleburg, Germany

Linhui Ruan Zhejiang Provincial Key Laboratory of Aging and Neurological Disorder Research, First Affiliated Hospital, Wenzhou Medical University, Wenzhou, China

Department of Pharmacology and Neuroscience, University of North Texas Health Science Center, TX, USA

Tomás Sobrino Department of Neurology, Clinical Neurosciences Research Laboratory, Hospital Clínico Universitario, c/ Travesa da Choupana s/n, Health Research Institute of Santiago de Compostela (IDIS), University of Santiago de Compostela, Santiago de Compostela, Spain

James A. St John Eskitis Institute for Drug Discovery, Griffith University, Brisbane, Australia

Ewa K. Stuermer Institute for Research in Operative Medicine, University of Witten/Herdecke, Köln, Germany

Cheuk-Kwan Sun Department of Emergency Medicine, E-Da Hospital, I-Shou University, Kaohsiung City, Taiwan, Republic of China

Naoki Tajiri Center of Excellence for Aging and Brain Repair, Department of Neurosurgery and Brain Repair, University of South Florida College of Medicine, Tampa, FL, USA

Johana Tello Velasquez Eskitis Institute for Drug Discovery, Griffith University, Brisbane, Australia

Michelle H. Theus The Department of Biomedical Sciences and Pathobiology, Virginia-Maryland Regional College of Veterinary Medicine, Blacksburg, VA, USA

Henning Ulrich Departamento de Bioquímica, Instituto de Química, Universidade de São Paulo, São Paulo, Brazil

Poornima Venkat Department of Neurology, Henry Ford Hospital, Detroit, MI, USA

Department of Physics, Oakland University, Rochester, MI, USA

Yongting Wang Neuroscience and Neuroengineering Research Center, Med-X Research Institute and School of Biomedical Engineering, Jiao Tong University, Shanghai, China

Alison E. Willing Center for Excellence in Aging and Brain Repair, Department of Neurosurgery and Brain Repair, Morsani College of Medicine MDC 78, University of South Florida, Tampa, FL, USA

Guo-Yuan Yang Neuroscience and Neuroengineering Research Center, Med-X Research Institute and School of Biomedical Engineering, Jiao Tong University, Shanghai, China

Department of Neurology, School of Medicine, Ruijin Hospital, Jiao Tong University, Shanghai, China

Elisa R. Zanier Department of Neuroscience, IRCCS-Istituto di Ricerche Farmacologiche Mario Negri, Milan, Italy

Li-Ru Zhao Department of Neurosurgery, State University of New York, Upstate Medical University, Institute for Human Performance, Syracuse, NY, USA

Qichuan ZhuGe Zhejiang Provincial Key Laboratory of Aging and Neurological Disorder Research, First Affiliated Hospital, Wenzhou Medical University, Wenzhou, China

Department of Neurosurgery, The First Affiliated Hospital, Wenzhou Medical University, Wenzhou, China

Part I
Stem Cell Research in Stroke

Chapter 1
Basic Studies on Neural Stem Cells in the Brain

Isis Cristina do Nascimento and Henning Ulrich

Contents

Abstract Neural stem cells self-renew and differentiate into neurons, astrocytes and oligodendrocytes. In the developing brain, they proliferate, migrate and differentiate into neurons and glial cells, which form trillions of connections in the adult brain. Different from the developing nervous system, where neural stem cells (NSC) are widely distributed, in the adult brain the occurrence of NSC is restricted to the subventricular zone of the lateral ventricles and the subgranular zone of the dentate gyrus of the hippocampus. NSC, residing in so-called stem cell niches, promote neurogenesis throughout life for maintaining plasticity of the brain. They can also be induced to proliferate and migrate to brain injury sites upon extrinsic signals such as growth factors and other signaling cues. In this book chapter we will explain the principles of neural stem cell isolation, their culture as well as proliferation and differentiation regulation, studied *in vitro* and *in vivo*.

Abbreviations

BDNF	Brain-derived neurotrophic factor
BMP	Bone-morphogenetic protein
CNS	Central nervous system
CNTF	Ciliary neurotrophic factor

H. Ulrich (✉) · I. C. do Nascimento
Departamento de Bioquímica, Instituto de Química, Universidade de São Paulo,
Avenida Professor Lineu Prestes 748, São Paulo 05508-000, Brazil
e-mail: henning@iq.usp.br

L.-R. Zhao, J. H. Zhang (eds.), *Cellular Therapy for Stroke and CNS Injuries*,
Springer Series in Translational Stroke Research, DOI 10.1007/978-3-319-11481-1_1

EGF	Epidermal growth factor
FGF	Fibroblast growth factor
GDNF	Glial cell line-derived neurotrophic factor
GFAP	Glial fibrillary acidic protein
IFN- γ	Interferon-γ
IGF	Insulin-like growth factor
IL	Interleukin
LIF	Leukemia-inhibitory factor
miR	Micro RNA
MMP	Metaloproteinases
nAChRs	Nicotinic acetylcholine receptors
NPC	Neural progenitor cells
NSC	Neural stem cells
PDGF	Platelet-derived growth factor
SGZ	Subgranular zone
Shh	Sonic hedgehog
SVZ	Subventricular zone
T3	Triiodothyronine
TNF	Tumor necrosis factor
VEGF	Vascular endothelial growth factor

1.1 Neural Stem Cells

The growing interest on stem cells is justified by the diversity of applications of these cells in basic research as well as in the regenerative medicine seeking functional recovery of injured tissues. Stem cells are unspecialized cells that can self-renew, i.e. are able to generate cells identical to themselves indefinitely by cell division and in certain physiological or experimental conditions, can differentiate into specialized cells of one or more tissues. Based on their potency stem cells can be classified into at least four types: *Totipotent* cells capable of generating cells of all tissues of the body and the extra embryonic structures, *pluripotent* cells originating all cells of organisms, but not extra embryonic cells; *multipotent* cells committed to originate only a limited number of cells of a particular tissue, and *unipotent* cells capable to differentiate into only one cell type. Stem cells can also be classified according to their origin as embryonic, fetal and adult ones.

For a long time it has been believed that the final structure of the central nervous system (CNS) is defined during development, and that after this process the tissue structure would be unchanged and the formation of new cellular components impossible. According to this dogma, it was postulated that the replacement of old cells or neuronal reconstitution after injury would not occur. The discovery of neural stem cells (NSC) broke down this dogma, demonstrated brain plasticity and allowed advances in the knowledge of neurogenesis in the adult brain, and subsequently the development of novel therapeutic strategies (Gage and Temple 2013).

NSC are multipotent, possess the ability of self-renewal and may give rise to the three major cell types of neural tissue (neurons, astrocytes and oligodendrocytes). They can be obtained from embryonic, fetal or adult neural tissue (Reynolds and Weiss 1992; Reynolds et al. 1992). Their identification is usually performed by the expression of markers CD133, Vimentin, Nestin, SOX1 & 2, Musashi, Notch1, GD3, and CXCR4 (Bottai et al. 2003; Hemmati et al. 2003; Nakatani et al. 2010; Tárnok et al. 2010; Li et al. 2011; Oliveira et al. 2013). A detailed list of NSC markers is given in Table 1.1.

1.2 Neural Stem Cells in Developing and Adult Brain

CNS development starts with neural tube formation from the ectoderm. During early stage of CNS development, neuroepithelial progenitor cells through symmetric divisions expand the neuroepithelium. Later these cells by asymmetric division give rise to a stem cell and a more differentiated neural progenitor cell (NPC) cell characterizing neurogenesis initiation (Fig. 1.1). Overall, temporally and spatially distinct multipotent cell populations are widely present in the CNS tissue and respond to specific microenvironmental cues regulating proliferation and differentiation of these cells and being crucial for acquisition of the final cell phenotype and regionalization of the brain.

In the mouse, initial primitive NSC from E5.5 to 8.5 are responsive to leukemia-inhibitory factor (LIF) or fibroblast growth factor (FGF). Then, NSC expand in the ventricular zone and respond to epidermal growth factor (EGF) and FGF on E13. Peaks of neurogenesis occur on E14.5 involving Notch signaling and platelet-derived growth factor (PDGF) receptor activation. With the progression of development, the number of NSC begins to decrease and represents only a small percentage of cells present in specific regions of the brain (reviewed by Ramasamy et al. 2013). In addition to the above-mentioned cues, one of the crucial factors for forebrain patterning is the bone-morphogenetic protein (BMP), which in conjunction with Wnt, FGF, and Sonic hedgehog (Shh) signaling determines dorsal-ventral forebrain patterning (Harrison-Uy and Pleasure 2012). Inibition of BMP-induced signaling is necessary for neural induction (Pera et al. 2014). Some cells from the neuroepithelium, known as radial glial cells, become bipolar and extend processes, which provide support for the migration of newborn neurons. These radial glial cells, regarded also as stem cells in neocortical development, are capable of self-renewal and originate differentiated neural cells, including neurons (Pinto and Gotz 2007; Gan et al. 2014). These cells expressed markers such as CXCR4, Vimentin, Nestin, and Sox2, characteristic for the NSC phenotype (Li et al. 2011).

Table 1.1 Phenotypic markers expressed by adult and embryonic neural stem cells

Marker profile	Cell type	Brain tissue	Reference
Toll-like receptors 2 and 4	Adult NSC/NPC	Mouse dentate gyrus	Rolls et al. 2007
CD133, *Musashi-1*, Sox2, melk, PSP, bmi-1, *and nestin*	Adult NSC	Human brain	Hemmati et al. 2003
GD3	Adult NSC	Mouse striata and subventricular zone	Nakatani et al. 2010
$HRD1^+/Nestin^+/GFAP^+$	Adult NSC	Mouse subventricular zone	Kawada et al. 2011
EGF receptor	Adult NSC/NPC	Mouse subventricular zone	Doetsch et al. 2002
Ki-67	Adult NSC/NPC	Pig subventricular zone	Liard et al. 2009
Querkopf	Adult NSC	Mouse subventricular zone	Sheikh et al. 2012
$Vimentin^+$, $Nestin^+$, $Sox2^+$	Radial glial cells	Rat germinal zone E 16.5	Li et al. 2011
CXCR4	Radial glial cells	Embryonic spinal cord	Mithal et al. 2013
$CD133^+/CD15^+$	Embryonic NSC	Human fetal brain E50–55	Sun et al. 2009
E-PHA binding N-glycans	Embryonic NSC	Mouse fetal brain E12, E14, E16	Hamanoue et al. 2009
Musashi1 and Musashi2	Embryonic NSC	Mouse embryonic forebrain	Sakakibara et al. 2002
Syndecan-1, Integrin-b1, and Notch-1	Embryonic NSC	Mouse embryonic forebrain E14.5	Nagato et al. 2005
Nestin, GFAP	Embryonic NSC and NPC	Rat embryonic forebrain E14.5	Martins et al. 2008
Toll-like receptor 3	Embryonic NSC	Mouse embryonic forebrain E14	Yaddanapudi et al. 2011
Pax-6	Embryonic NSC and NPC	Mouse embryonic forebrain	Estivill-Torrus et al. 2002
Sox 1, Sox 2, CD133	Embryonic and adult NSC	Mouse embryonic forebrain E12.5 Mouse subventricular zone	Corti et al. 2007

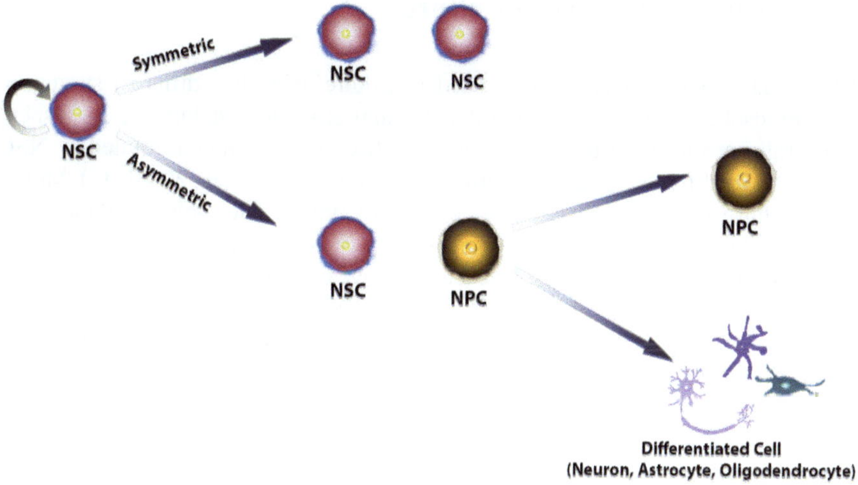

Fig. 1.1 Scheme of cell division promoting self-renewal and differentiation of neural stem cells

In adult brain, NSC are found in the subgranular zone (SGZ) of the dentate gyrus of the hippocampus and the subventricular zone (SVZ) of the lateral ventricles (Yamashima et al. 2007). Some molecules, such as BMP, Notch, and Shh, known to be involved in embryonic development of the CNS, are also presents in neurogenic regions of adult brain, and are also important for adult neurogenesis. Adult NSC respond to stimuli from the microenvironment and can proliferate, migrate to other regions of the brain and give rise to different neural phenotypes and functionally integrate into local circuits (Van Praag et al. 2002; Carleton et al. 2003; Gage and Temple 2013; Sequerra et al. 2013). Although NSC derived from embryonic and adult CNS are considered as multipotent, it is believed that there are temporal limitations for the potency of adult stem cells (Frantz and McConnell 1996; Nguyen et al. 2013).

The establishment of protocols for isolation and culture of NSC has been of great value, both for basic research in order to understand mechanisms of neural development and adult neurogenesis as well as therapeutic applications. NSC can be isolated from the adult brain tissue only in small amounts, making difficult their use in cell replacement therapy. A therapeutic concept could be to isolate these cells during biopsy and then replicate them *in vitro* to increase the number of cells prior to using them for transplantation purposes.

Alternatively, NSC obtained from fetal CNS or can be derived from other stem cell types, such as embryonic stem cells (Garzón-Muvdi and Quiñones-Hinojosa 2009). More recently, NSC are being obtained by reprogramming mature somatic cells into induced pluripotent stem cells or direct-induced NSC (Li et al. 2013). We will focus on embryonic/fetal and adult brain NSC.

1.3 Neural Stem Cells in Culture

NSC usually are isolated and expanded in culture using two different strategies: (1) Neurospheres, which are three-dimensional structures maintained in suspension originated from a single cell by asymmetric division and constituted by NSC and NPC at different stages of differentiation (Suslov et al. 2000). (2) Monolayer culture using adhesive substrate such as poly-lysine, poly-ornithine, laminin and fibronectin. The proliferative capacity of NSC is very similar using both methods. However, the neurosphere culture reveals a higher potential for generating neurons compared to NSC cultured in monolayers (Meyer et al. 2012). In fact, neurospheres and other cultures of neuroepithelial cells contain low numbers of NSC and a much higher percentage of NPC with restricted differentiation potential.

The neural tissue of the developing or adult CNS tissue is dissociated *in vitro* providing single cells that are cultured as neurospheres or adherent monolayer at low density in growth medium containing hormonal supplements (B27 and N2) and the growth factors EGF and basic FGF (Kornblum 2007; Negraes et al. 2012). The supplemented medium with growth factors maintains the ability of neurospheres to proliferate and prevents spontaneous differentiation of the cells. The removal of growth factors induces the differentiation of these cells into neurons, astrocytes and oligodendrocytes (Reynolds and Weiss 1996; Meyer et al. 2012). Using embryonic telencephalon rat E14.5 neurospheres, following 7 days of differentiation induction, a radial pattern of cell migration is observed together with an increase in neuron-specific β-3 Tubulin and glial specific GFAP (glial fibrillary acidic protein) expression from 25 to 75% and 20 to 35% of the cell population, respectively, while the percentage of Nestin-positive cells declines following induction of NSC to differentiation and the occurrence of oligodendrocytes remains below 10% of the cell population. This cell system provides simplified conditions for studying the effects of extrinsic and intrinsic parameters on embryonic cortex development (Trujillo et al. 2012). Induction of differentiation results yet in a further increase of the number of neurons in the differentiated neurosphere population (Schwindt et al. 2011), while astroglial cell populations are enriched in long term cultures of neurospheres (Chiang et al. 1996).

1.4 Neural Stem Cell Proliferation, Migration and Differentiation

Neurogenesis is a highly regulated process in both development and the adult brain, where bioactive compounds are produced and released by the proper stem cells of the neurogenic niche. These extracellular messengers act together with intrinsic cues on proliferation, migration and differentiation of newborn cells. Growth factors, hormones, extracellular matrix molecules, neurotransmitters, cytokines, small non-coding RNAs have been related with regulation of NSC mobilization (Trujillo et al. 2009; Trujillo et al. 2012; Schouten et al. 2012; Oliveira et al. 2013; Bellenchi et al. 2013).

Neurogenesis is also modulated by the activity of ion channels. The concept of depolarization-driven progress of neural differentiation and phenotype determination goes back to the fundamental studies of Nicholas Spitzer (Ben-Ari and Spitzer 2010). Different checkpoints need to be passed for putting control over proliferation, migration and neurotransmitter and receptor specification. In view that, acetylcholine receptor expression patterns guide the progress of neurogenesis (reviewed by Trujillo et al. 2009; Berg et al. 2013). Activation of α7 nicotinic acetylcholine receptors inhibits proliferation and promotes neuronal differentiation of NSC (Narla et al. 2013). Moreover, the selective lesion of cholinergic input decreases neurogenesis in the dentate gyrus (Mohapel et al. 2005). The activation of muscarinic M1 receptors in the hippocampus induces neurogenesis (Ma et al. 2004; Van Kampen and Eckman 2010), as described as well for M2 receptors in the ventricular zone of the embryonic rat cortex (Resende and Adhikari 2009).

Dopaminergic receptors are involved in regulation of proliferation and differentiation of NPC in the SVZ (O'Keeffe et al. 2009). Stimulation of NPC proliferation by dopamine can be mediated by EGF (O'Keeffe et al. 2009). Moreover activation of dopamine receptors in neurospheres from the SVZ stimulate brain-derived neurotrophic factor (BDNF) release, increase cell proliferation and the number of differentiating cells (Merlo et al. 2011).

Involvement of glutamate receptors in proliferation of precursor cells depends on the expressed subtype and involves expression of neurotrophins (Mackowiak 2002). It has been shown that inhibited NMDA-glutamate receptor activity results in increased neurogenesis (Nacher et al. 2003), while activation of the metabotropic mGluR5 receptor augments NPC proliferation (Di Giorgi-Gerevini et al. 2005). SVZ progenitor activities depend on reactive astrocytes, which release glutamate and ATP resulting in activation of glutamate and purinergic receptors (Guo et al. 2013; Boccazzi et al. 2014).

Furthermore, some studies suggest that trophic factors such as BDNF, glial cell line-derived neurotrophic factor (GDNF) and vascular endothelial growth factor (VEGF) act in conjunction with neurotransmitter systems on neurogenesis (Berg et al. 2013). BDNF promotes differentiation and maturation of progenitor cells by enhancing GABA release in the SGZ (Waterhouse et al. 2012). This neurotrophic factor promotes neuronal migration via p75 neurotrophin receptor (NTR) activation in the adult SVZ (Snapyan et al. 2009). It has been suggested that adult neurogenesis augments by environmental enrichment and voluntary physical exercise. Bekinschtein et al. (2011) suggest that this effect results from BDNF released by newborn neurons.

Platelet-derived growth factor-BB (PDGF-BB) leads to increased neuronal differentiation and maintains the proliferation and migration of NPC. Such actions were observed in the embryonic ventricular zone and in post-natal brain development (Smits et al. 1991; Sasahara et al. 1992). Promotion of neurogenesis by PDGF-

BB was also observed in an animal model of impaired hippocampal neurogenic proliferation inflicted by HIV Tat & cocaine (Yao et al. 2012).

Bradykinin, a biologically active peptide classically involved in inflammation and blood pressure regulation, performs its functions by binding to G protein coupled kinin-B2 receptors. Activation of B2 receptors results in proliferation block and induction of NPC migration and neurogenesis. Inhibition experiments in the presence of the selective B2 receptor blocker HOE-140 showed an increase in gliogenesis over neurogenesis as well as down-regulation of muscarinic and purinergic receptor activity and expression, indicating that bradykinin participated in neural fate determination (Trujillo et al. 2012).

Metaloproteinases (MMP) have been attributed with functions in proliferation and differentiation of NPC in the embryonic development of the mouse brain (Tonti et al. 2009). As extensively reviewed by Tonti and co-workers, their participation is suggested in embryonic and postnatal developmental neurogenesis by degrading extracellular matrix and allowing for neural migration. While MMP-9 is highly expressed in early developmental stages together with migrating external granular cell precursors, MMP-2 expression is observed during postnatal development. MMP-2 also seems to be involved in synaptic plasticity, as increasing activity of the enzyme was related to the recovery of plasticity along formation of cortex projection to the hippocampal dentate gyrus (Dzwonek et al. 2004).

Previous reports have shown the involvement of immune mediators in the control of neurogenesis in the adult brain (Gonzalez-Perez et al. 2010; 2012). Some studies suggest that inflammatory cytokines promote stimulating effects on neurogenesis especially in brain damage situations (McPherson et al. 2011). The brain is considered an immune-privileged environment due to the selectivity of blood-brain barrier, which limits the entrance of molecules and cells of the peripheral immune system and the presence of microglia responsible for the identification of signs of injury and inflammation. However, in the healthy brain, both during development and in the adult brain, the microglia is involved in phagocytosis of apoptotic cells. It is believed that microglia releases cytokines and free radicals acting on surrounding cells, thereby in dependence on the environmental situation promoting survival or death of such cells (Marín-Teva et al. 2011). Microglia is present in neurogenic niches and suggested to regulate neurogenesis by interacting with NSC.

Several cells in the brain including microglia, astrocytes and neurons produce and release cytokines, chemokines and express receptors for these factors, providing autocrine and paracrine loops for modulation of migration, fate choice and viability of NPC of the SVZ NPC (Gordon et al. 2012). Cytokines may have pro- or anti-inflammatory actions. While interferon-γ (IFN- γ), interleukin (IL)-1β, 6 and 18 and tumor necrosis factor (TNF)- α promote inflammatory reactions, insulin-like growth factor (IGF)-1 and IL-15 have anti-inflammatory properties, and IL-4 exerts opposing effects under different cellular conditions. However, their properties in augmenting or reducing inflammatory events cannot be directly linked to their inflammation-promoting or -inhibiting capabilities.

The pro-inflammatory cytokine TNFα leads to increased proliferation and differentiation into astrocytes and decreased the number of neurons. IFN-γ augments differentiation and neurite outgrowth in NSC, while it is also involved in reduction of proliferation and survival of NPC. On the other side, IL-1β inhibits proliferation and differentiation of NSC, such as has been observed for IL-6.

The anti-inflammatory cytokine IGF1, produced by activated microglia, is related to the promotion of neurogenesis. IL-4 leads differentiation to oligodendrocytes. In addition, NPC decreases proliferation and promotes the migration of these cells. IL-15 inhibits differentiation and self-renewal capacity of NSC. IL-18 produced by astrocytes and microglia acts by decreasing differentiation and promoting neuronal apoptosis. Leukemia inhibiting factor (LIF) and ciliary neurotrophic factor (CNTF), known to maintain pluripotency in embryonic stem cells during replication (Wolf et al. 1994), are also important for the self-renewal capacity of NSC (Deverman and Patterson 2009; Gonzalez-Perez et al. 2010, 2012; Yoneyama et al. 2011).

The above-discussed modulators are extrinsically acting, either by receptor ligand interaction with induction of proliferation-/ differentiation signal transduction in the target cells or by protease secretion for digestion of extracellular matrix compounds. However, intracellular hormone receptors, activated by lipophilic ligands, also participate in developmental and adult neurogenesis. Triiodothyronine (T3) is known to be involved in neuronal development, and deficiency of T3 affects formation of the CNS during childhood. Using cellular models, T3 has been attributed with functions in neuronal migration, cortex formation and cerebellar differentiation as well as with glial differentiation and proliferation (Ausó et al. 2004; Dezonne et al. 2009; Portella et al. 2010; Johe et al. 1996; Lima et al. 1997; see for a complete review Dezonne et al. 2013).

The action of small noncoding RNAs, denominated also micro RNAs (miR), such as also epigenetic transcription regulation, is intrinsic. Post-transcriptional control by such RNAs has been described for proliferation, differentiation and apoptosis (Stefani and Slack 2008), being a critical event for embryonic and adult neurogenesis (Visvanathan et al. 2007; Cheng et al. 2009). Small non-coding RNAs are needed for specific cellular fate for maintenance of the adult CNS, i.e. for oligodendrocyte lineage specification and differentiation by miR-219 and miR-338, thereby providing key points for myelin repair (Zhao et al. 2010). The here reported mechanisms are summarized in Fig. 1.2.

Conclusions

Since the first evidence of the existence of neural stem cells around the 1960s, many studies have been conducted to understand mechanisms involved in their formation, proliferation, differentiation, migration and integration into the local neural network. In this chapter, we summarized some of the advances made in this field of knowledge. However, despite the growing number of studies conducted it seems that we are still far away from fully understanding the biology of NSC. The success

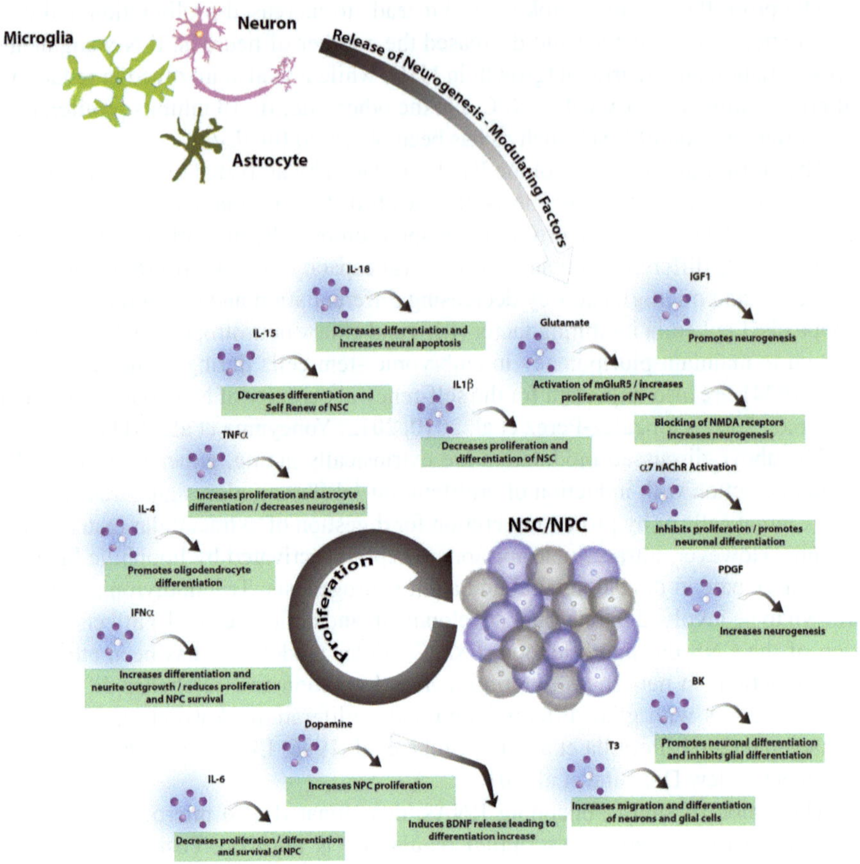

Fig. 1.2 Autocrine and paracrine secretion of growth factors, hormones, neurotransmitters and cytokine as control mechanisms for embryonic and adult NSC proliferation, differentiation and phenotype specification. Abbreviations: *BK* bradykinin, *α7 nAchR, α7* subtype of the nicotinic acetylcholine receptor

in the clinical application of NSC/NPC depends on the methodologies used for their isolation and enrichment as well as in directing these cells to the target location by endogenous factors. The latter should be a microenvironment that favors the survival and functional integration of transplanted NSC/NPC and/or the recruitment of endogenous stem cells to the site of injury.

Effects of NSC transplantation of various origins have been tested in animal models of diseases, such as amyotrophic lateral sclerosis, Parkinson's disease, Alzheimer's disease, multiple sclerosis, Huntington's disease and glioblastoma; however, in most applications there were large limitations due to poor cell survival and engraftment into functional networks.

References

Ausó E, Lavado-Autric R, Cuevas E, Del Rey FE, Morreale De Escobar G, Berbel P (2004) A moderate and transient deficiency of maternal thyroid function at the beginning of fetal neocorticogenesis alters neuronal migration. Endocrinology 145:4037–4047

Bekinschtein P, Oomen CA, Saksida LM, Bussey TJ (2011) Effects of environmental enrichment and voluntary exercise on neurogenesis, learning and memory, and pattern separation: BDNF as a critical variable? Semin Cell Dev Biol 22:536–542

Bellenchi GC, Volpicelli F, Piscopo V, Perrone-Capano C, di Porzio U (2013) Adult neural stem cells: an endogenous tool to repair brain injury? J Neurochem 124:159–167

Ben-Ari Y, Spitzer NC (2010) Phenotypic checkpoints regulate neuronal development. Trends Neurosci 33:485–492

Berg DA, Belnoue L, Song H, Simon A (2013) Neurotransmitter-mediated control of neurogenesis in the adult vertebrate brain. Development 140:2548–2561

Boccazzi M, Rolando C, Abbracchio MP, Buffo A, Ceruti S (2014) Purines regulate adult brain subventricular zone cell functions: contribution of reactive astrocytes. Glia 62:428–439

Bottai D, Fiocco R, Gelain F, Defilippis L, Galli R, Gritti A, Vescovi LA (2003) Neural stem cells in the adult nervous system. J Hematother Stem Cell Res 12:655–670

Carleton A, Petreanu LT, Lansford R, Alvarez-Buylla A, Lledo PM (2003) Becoming a new neuron in the adult olfactory bulb. Nat Neurosci 6:507–518

Cheng LC, Pastrana E, Tavazoie M, Doetsch F (2009) miR-124 regulates adult neurogenesis in the subventricular zone stem cell niche. Nat Neurosci 12:399–408

Chiang YH, Silani V, Zhou FC (1996) Morphological differentiation of astroglial progenitor cells from EGF-responsive neurospheres in response to fetal calf serum, basic fibroblast growth factor, and retinol. Cell Transplant 5:179–189

Corti S, Nizzardo M, Nardini M, Donadoni C, Locatelli F, Papadimitriou D, Salani S, Del Bo R, Ghezzi S, Strazzer S, Bresolin N, Comi GP (2007) Isolation and characterization of murine neural stem/progenitor cells based on Prominin-1 expression. Exp Neurol 205:547–562

Deverman BE, Patterson PH (2009) Cytokines and CNS development. Neuron 64:61–78

Dezonne RS, Stipursky J, Gomes FC (2009) Effect of thyroid hormone depletion on cultured murine cerebral cortex astrocytes. Neurosci Lett 467:58–62

Dezonne RS, Stipursky J, Araujo AP, Nones J, Pavão MS, Porcionatto M, Gomes FC (2013) Thyroid hormone treated astrocytes induce maturation of cerebral cortical neurons through modulation of proteoglycan levels. Front Cell Neurosci 7:125

Di Giorgi-Gerevini V, Melchiorri D, Battaglia G, Ricci-Vitiani L, Ciceroni C, Busceti CL, Biagioni F, Iacovelli L, Canudas AM, Parati E, De Maria R, Nicoletti F (2005) Endogenous activation of metabotropic glutamate receptors supports the proliferation and survival of neural progenitor cells. Cell Death Differ 12:1124–1133

Doetsch F, Petreanu L, Caille I, Garcia-Verdugo JM, Alvarez-Buylla A (2002) EGF converts transit-amplifying neurogenic precursors in the adult brain into multipotent stem cells. Neuron 36:1021–1034

Dzwonek J, Rylski M, Kaczmarek L (2004) Matrix metalloproteinases and their endogenous inhibitors in neuronal physiology of the adult brain. FEBS Lett 567:129–135

Estivill-Torrus G, Pearson H, van Heyningen V, Price DJ, Rashbass P (2002) Pax6 is required to regulate the cell cycle and the rate of progression from symmetrical to asymmetrical division in mammalian cortical progenitors. Development 129:455–466

Frantz GD, McConnell SK (1996) Restriction of late cerebral cortical progenitors to an upper-layer fate. Neuron 17:55–61

Gage FH, Temple S (2013) Neural stem cells: generating and regenerating the brain. Neuron 80:588–601

Gan Q, Lee A, Suzuki R, Yamagami T, Stokes A, Nguyen BC, Pleasure D, Wang J, Chen HW, Zhou CJ (2014) Pax6 mediates ß-catenin signaling for self-renewal and neurogenesis by neocortical radial glial stem cells. Stem Cells 32:45–58

Garzón-Muvdi T, Quiñones-Hinojosa A (2009) Neural stem cell niches and homing: recruitment and integration into functional tissues. ILAR J 51:3–23
Gonzalez-Perez O, Quiñones-Hinojosa A, Garcia-Verdugo JM (2010) Immunological control of adult neural stem cells. J Stem Cells 5:23–31
Gonzalez-Perez O, Gutierrez-Fernandez F, Lopez-Virgen V, Collas-Aguilar J, Quinones-Hinojosa A, Garcia-Verdugo JM (2012) Immunological regulation of neurogenic niches in the adult brain. Neuroscience 226:270–281
Gordon RJ, Mehrabi NF, Maucksch C, Connor B (2012) Chemokines influence the migration and fate of neural precursor cells from the young adult and middle-aged rat subventricular zone. Exp Neurol 233:587–594
Guo Y, Wei Q, Huang Y, Xia W, Zhou Y, Wang S (2013) The effects of astrocytes on differentiation of neural stem cells are influenced by knock-down of the glutamate transporter, GLT-1. Neurochem Int 63:498–506
Hamanoue M, Matsuzaki Y, Sato K, Okano HJ, Shibata S, Sato I, Suzuki S, Ogawara M, Takamatsu K, Okano H (2009) Cell surface N-glycans mediated isolation of mouse neural stem cells. J Neurochem 110:1575–1584
Harrison-Uy SJ, Pleasure SJ (2012) Wnt signaling and forebrain development. Cold Spring Harb Perspect Biol 4:a008094
Hemmati HD, Nakano I, Lazareff JA, Masterman-Smith M, Geschwind DH, Bronner-Fraser M, Kornblum HI (2003) Cancerous stem cells can arise from pediatric brain tumors. Proc Natl Acad Sci USA 100:15178–15183
Johe KK, Hazel TG, Muller T, Dugich-Djordjevic MM, McKay RD (1996) Single factors direct the differentiation of stem cells from the fetal and adult central nervous system. Genes Dev 10:3129–3140
Kawada K, Kaneko M, Nomura Y, Mimori S, Hamana H, Ogita K, Murayama T, Fujino H, Okuma Y (2011) Expression of the ubiquitin ligase HRD1 in neural stem/progenitor cells of the adult mouse brain. J Pharmacol Sci 117:208–212
Kornblum HI (2007) Introduction to neural stem cells. Stroke 38:810–816
Li H, Jin G, Qin J, Tian M, Shi J, Yang W, Tan X, Zhang X, Zou L (2011) Characterization and identification of Sox2 + radial glia cells derived from rat embryonic cerebral cortex. Histochem Cell Biol 136:515–526
Li R, Bai Y, Liu T, Wang X, Wu Q (2013) Induced pluripotency and direct reprogramming: a new window for treatment of neurodegenerative diseases. Protein Cell 4:415–424
Liard O, Segura S, Pascual A, Gaudreau P, Fusai T, Moyse E (2009) In vitro isolation of neural precursor cells from the adult pig subventricular zone. J Neurosci Methods 182:172–179
Lima FR, Trentin AG, Rosenthal D, Chagas C, Moura Neto V (1997) Thyroid hormone induces protein secretion and morphological changes in astroglial cells with an increase in expression of glial fibrillary acidic protein. J Endocrinol 154:167–175
Ma W, Li BS, Zhang L, Pant HC (2004) Signaling cascades implicated in muscarinic regulation of proliferation of neural stem and progenitor cells. Drug News Perspect 17:258–266
Mackowiak M, O'Neill MJ, Hicks CA, Bleakman D, Skolnick P (2002) An AMPA receptor potentiator modulates hippocampal expression of BDNF: an in vivo study. Neuropharmacology 43:1–10
Marín-Teva JL, Cuadros MA, Martín-Oliva D, Navascués J (2011) Microglia and neuronal cell death. Neuron Glia Biol 7:25–40
Martins AH, Alves JM, Trujillo CA, Schwindt TT, Barnabé GF, Motta FL, Guimaraes AO, Casarini DE, Mello LE, Pesquero JB, Ulrich H (2008) Kinin-B2 receptor expression and activity during differentiation of embryonic rat neurospheres. Cytometry A 73:361–368
McPherson CA, Kraft AD, Harry GJ (2011) Injury-induced neurogenesis: consideration of resident microglia as supportive of neural progenitor cells. Neurotox Res 19:341–352
Merlo S, Canonico PL, Sortino MA (2011) Distinct effects of pramipexole on the proliferation of adult mouse sub-ventricular zone-derived cells and the appearance of a neuronal phenotype. Neuropharmacology 60:892–900
Meyer AK, Jarosch A, Schurig K, Nuesslein I, Kißenkötter S, Storch A (2012) Fetal mouse mesencephalic NPCs generate dopaminergic neurons from post-mitotic precursors and maintain long-term neural but not dopaminergic potential in vitro. Brain Res 1474:8–18

Mithal DS, Ren D, Miller RJ (2013) CXCR4 signaling regulates radial glial morphology and cell fate during embryonic spinal cord development. Glia 61:1288–1305

Mohapel P, Leanza G, Kokaia M, Lindvall O (2005) Forebrain acetylcholine regulates adult hippocampal neurogenesis and learning. Neurobiol Aging 26:939–946

Nacher J, Alonso-Llosa G, Rosell DR, McEwen BS (2003) NMDA receptor antagonist treatment increases the production of new neurons in the aged rat hippocampus. Neurobiol Aging 24:273–284

Nagato M, Heike T, Kato T, Yamanaka Y, Yoshimoto M, Shimazaki T, Okano H, Nakahata T (2005) Prospective characterization of neural stem cells by flow cytometry analysis using a combination of surface markers. J Neurosci Res 80: 456–466

Nakatani Y, Yanagisawa M, Suzuki Y, Yu RK (2010) Characterization of GD3 ganglioside as a novel biomarker of mouse neural stem cells. Glycobiology 20:78–86

Narla S, Klejbor I, Birkaya B, Lee YW, Morys J, Stachowiak EK, Terranova C, Bencherif M, Stachowiak MK. (2013) α7 nicotinic receptor agonist reactivates neurogenesis in adult brain. Biochem Pharmacol. 86:1099–1104

Negraes PD, Schwindt TT, Trujillo CA, Ulrich H (2012) Neural differentiation of P19 carcinoma cells and primary neurospheres: cell morphology, proliferation, viability, and functionality. Curr Protoc Stem Cell Biol Chapter 2:Unit 2D.9

Nguyen TD, Widera D, Greiner J, Müller J, Martin I, Slotta C, Hauser S, Kaltschmidt C, Kaltschmidt B (2013) Prolonged cultivation of hippocampal neural precursor cells shifts their differentiation potential and selects for aneuploid cells. Biol Chem 394:1623–1636

O'Keeffe GC, Barker RA, Caldwell MA (2009) Dopaminergic modulation of neurogenesis in the subventricular zone of the adult brain. Cell Cycle 8:2888–2894

Oliveira SL, Pillat MM, Cheffer A, Lameu C, Schwindt TT, Ulrich H (2013) Functions of neurotrophins and growth factors in neurogenesis and brain repair. Cytometry A 83:76–89

Pera EM, Acosta H, Gouignard N, Climent M, Arregi I (2014) Active signals, gradient formation and regional specificity in neural induction. Exp Cell Res 321:25–31

Pinto L, Götz M (2007) Radial glial cell heterogeneity-the source of diverse progeny in the CNS. Prog Neurobiol 83:2–23

Portella AC, Carvalho F, Faustino L, Wondisford FE, Ortiga-Carvalho TM, Gomes FC (2010) Thyroid hormone receptor beta mutation causes severe impairment of cerebellar development. Mol Cell Neurosci 44:68–77

Ramasamy S, Narayanan G, Sankaran S, Yu YH, Ahmed S (2013) Neural stem cell survival factors. Arch Biochem Biophys 534:71–87

Resende RR, Adhikari A (2009) Cholinergic receptor pathways involved in apoptosis, cell proliferation and neuronal differentiation. Cell Commun Signal 7:20

Reynolds BA, Weiss S (1992) Generation of neurons and astrocytes from isolated cells of the adult mammalian central nervous system. Science 255:1707–1710

Reynolds BA, Weiss S (1996) Clonal and population analyses demonstrate that an EGF responsive mammalian embryonic CNS precursor is a stem cell. Dev Biol 175:1–13

Reynolds BA, Tetzlaff W, Weiss S (1992) A multipotent EGF-responsive striatal embryonic progenitor cell produces neurons and astrocytes. J Neurosci 12:4565–4574

Rolls A, Shechter R, London A, Ziv Y, Ronen A, Levy R, Schwartz M (2007) Toll-like receptors modulate adult hippocampal neurogenesis. Nat Cell Biol 9:1081–1088

Sakakibara S, Nakamura Y, Yoshida T, Shibata S, Koike M, Takano H, Ueda S, Uchiyama Y, Noda T, Okano H (2002) RNA-binding protein Musashi family: roles for CNS stem cells and a subpopulation of ependymal cells revealed by targeted disruption and antisense ablation. Proc Natl Acad Sci USA 99:15194–15199

Sasahara A, Kott JN, Sasahara M, Raines EW, Ross R, Westrum LE (1992) Platelet-derived growth factor B-chain-like immunoreactivity in the developing and adult rat brain. Brain Res Dev Brain Res 68:41–53

Schouten M, Buijink MR, Lucassen PJ, Fitzsimons CP (2012) New neurons in aging brains: molecular control by small non-coding RNAs. Front Neurosci 6:25

Schwindt TT, Trujillo CA, Negraes PD, Lameu C, Ulrich H (2011) Directed differentiation of neural progenitors into neurons is accompanied by altered expression of P2X purinergic receptors. J Mol Neurosci 44:141–146
Sequerra EB, Costa MR, Menezes JR, Hedin-Pereira C (2013) Adult neural stem cells: plastic or restricted neuronal fates? Development 140:3303–3309
Sheikh BN, Dixon MP, Thomas T, Voss AK (2012) Querkopf is a key marker of self-renewal and multipotency of adult neural stem cells. J Cell Sci 125:295–309
Smits A, Kato M, Westermark B, Nistér M, Heldin CH, Funa K (1991) Neurotrophic activity of platelet-derived growth factor (PDGF): rat neuronal cells possess functional PDGF beta-type receptors and respond to PDGF. Proc Natl Acad Sci USA 88:8159–8163
Snapyan M, Lemasson M, Brill MS, Blais M, Massouh M, Ninkovic J, Gravel C, Berthod F, Götz M, Barker PA, Parent A, Saghatelyan A (2009) Vasculature guides migrating neuronal precursors in the adult mammalian forebrain via brain-derived neurotrophic factor signaling. J Neurosci 29:4172–4188
Stefani G, Slack FJ (2008) Small non-coding RNAs in animal development. Nat Rev Mol Cell Biol 9:219–230
Sun Y, Kong W, Falk A, Hu J, Zhou L, Pollard S, Smith A (2009) CD133 (Prominin) negative human neural stem cells are clonogenic and tripotent. PLoS ONE 4:e5498
Suslov ON, Kukekov VG, Laywell ED, Scheffler B, Steindler DA (2000) RT-PCR amplification of mRNA from single brain neurospheres. J Neurosci Methods 96:57–61
Tárnok A, Ulrich H, Bocsi J (2010) Phenotypes of stem cells from diverse origin. Cytometry A 77:6–10
Tonti GA, Mannello F, Cacci E, Biagioni S (2009) Neural stem cells at the crossroads: MMPs may tell the way. Int J Dev Biol 53:1–17
Trujillo CA, Schwindt TT, Martins AH, Alves JM, Mello LE, Ulrich H (2009) Novel perspectives of neural stem cell differentiation: from neurotransmitters to therapeutics. Cytometry A 75:38–53
Trujillo CA, Negraes PD, Schwindt TT, Lameu C, Carromeu C, Muotri AR, Pesquero JB, Cerqueira DM, Pillat MM, de Souza HD, Turaça LT, Abreu JG, Ulrich H (2012) Kinin-B2 receptor activity determines the differentiation fate of neural stem cells. J Biol Chem 287:44046–44061
Van Kampen JM, Eckman CB (2010) Agonist-induced restoration of hippocampal neurogenesis and cognitive improvement in a model of cholinergic denervation. Neuropharmacology 58:921–929
Van Praag H, Schinder AF, Christie BR, Toni N, Palmer TD, Gage FH (2002) Functional neurogenesis in the adult hippocampus. Nature 415:1030–1034
Visvanathan J, Lee S, Lee B, Lee JW, Lee SK (2007) The microRNA miR-124 antagonizes the anti-neural REST/SCP1 pathway during embryonic CNS development. Genes Dev 21:744–749
Waterhouse EG, An JJ, Orefice LL, Baydyuk M, Liao GY, Zheng K, Lu B, Xu B (2012) BDNF promotes differentiation and maturation of adult-born neurons through GABAergic transmission. J Neurosci 32:14318–14330
Wolf E, Kramer R, Polejaeva I, Thoenen H, Brem G (1994) Efficient generation of chimaeric mice using embryonic stem cells after long-term culture in the presence of ciliary neurotrophic factor. Transgenic Res 3:152–158
Yaddanapudi K, De Miranda J, Hornig M, Lipkin WI (2011) Toll-like receptor 3 regulates neural stem cell proliferation by modulating the Sonic Hedgehog pathway. PLoS ONE 6:e26766
Yamashima T, Tonchev AB, Yukie M (2007) Adult hippocampal neurogenesis in rodents and primates: endogenous, enhanced, and engrafted. Rev Neurosci 18:67–82
Yao H, Duan M, Yang L, Buch S (2012) Platelet derived growth factor-BB restores human immunodeficiency virus Tat-cocaine-mediated impairment of neurogenesis: role of TRPC1 channels. J Neurosci 32:9835–9847
Yoneyama M, Shiba T, Hasebe S, Ogita K (2011) Adult neurogenesis is regulated by endogenous factors produced during neurodegeneration. J Pharmacol Sci 115:425–432
Zhao X, He X, Han X, Yu Y, Ye F, Chen Y, Hoang T, Xu X, Mi QS, Xin M, Wang F, Appel B, Lu QR (2010) MicroRNA-mediated control of oligodendrocyte differentiation. Neuron 65:612–626

Chapter 2
Neural Stem Cells in Response to Microenvironment Changes Inside and Outside of the Brain

Li-Ru Zhao

Contents

Abstract Neural stem cells/neural progenitor cells (NSCs/NPCs) residing in the neurogenic regions of the brain play a crucial role in brain development, brain plasticity, and brain repair. Recent progress in NSC/NPC research has advanced our understanding of the NSC/NPC niche and signaling networks controlling NSC/NPC proliferation and differentiation. However, the natural behavioral of NSCs/NPCs in response to physiological and pathological changes inside and outside the brain remains poorly understood. This chapter has summarized recent findings concerning NSC/NPC activity in the early stage of cortical brain ischemia and the hematopoietic stem cell growth factors in regulating NSC/NPC proliferation and differentiation. Moreover, future directions for NSC/NPC research are also discussed in this chapter.

L.-R. Zhao (✉)
Department of Neurosurgery, State University of New York, Upstate Medical University, Institute for Human Performance, Room 4110, 750 E. Adams Street, Syracuse, NY 13210, USA
e-mail: ZHAOL@upstate.edu; ZLR.LUND@gmail.com

L.-R. Zhao, J. H. Zhang (eds.), *Cellular Therapy for Stroke and CNS Injuries*, Springer Series in Translational Stroke Research, DOI 10.1007/978-3-319-11481-1_2

Abbreviations

Ara-C	Cytosine-β-d-Arabinofuranoside
bFGF	Basic fibroblast growth factor
bHLH	Basic helix-loop-helix
BDNF	Brain-derived neurotrophic factor
BrdU	Bromodeoxyuridine
EGF	Epidermal growth factor
EMVs	Exosomes and microvesicles
G-CSF	Granulocyte-colony stimulating factor
GFAP	Glial fibrillary acidic protein
GFP	Green fluorescent protein
LTP	Long-term potentiation
Ngn1	Neurogenin 1
NSCs/NPCs	Neural stem cells/neural progenitor cells
OB	Olfactory bulb
RMS	Rostral migratory stream
RT-PCR	Reverse transcription polymerase chain reaction
SCF	Stem cell factor
SGZ	Subgranular zone
SHRs	Spontaneously hypertensive rats
SVZ	Subventricular zone
TuJ1	Neuron-specific class iii beta-tubulin
VEGF	Vascular endothelial growth factor

2.1 Introduction

Neural stem cells/neural progenitor cells (NSCs/NPCs) have the capacity for self-expansion (symmetrical division), self-renewal (asymmetrical dividing) and differentiation into neurons and glial cells. During embryonic brain development, NSCs/NPCs are located in the neuroepithelium lining the lumen of the lateral ventricle (the ventricular zone) to generate neurons and glial cells throughout the brain (Farkas and Huttner 2008; Egger et al. 2011). It has been widely accepted that there are two neurogenic regions in which NSCs/NPCs reside in the postnatal or adult mammalian brain: the subgranular zone (SGZ) in the dentate gyrus of the hippocampus and the subventricular zone (SVZ) in the forebrain (Gage 2000; Curtis et al. 2007). Convincing evidence has shown that the NSCs/NPCs in the postnatal brain or adult brain are involved in learning and memory (Kempermann et al. 2004a), and that they may also participate in brain repair in the settings of brain injury (Jin et al. 2001; Arvidsson et al. 2002).

Although much progress has been made towards understanding the NSC/NPC niche and signaling networks controlling NSC/NPC proliferation and differentiation, the natural behavioral of NSCs/NPCs in physiological and pathological conditions remains poorly understood. Greater knowledge also needs to be gained regarding the role of NSCs/NPCs in self-protection and brain protection in the settings of brain injury and the regulation of NSC/NPC proliferation and differentiation by the growth factors of other systems or organs. In this chapter, our recent findings concerning the NSC/NPC behavior in intact brain and in the brain of ischemic injury, the effects of NSCs/NPCs on brain protection, and the involvement of hematopoietic stem cell growth factors in directing NSC/NPC proliferation and differentiation will be reviewed, and future directions for NSC/NPC research will be discussed.

2.2 NSC/NPC Classification: Bridge Cells and Migrating Cells

A large body of evidence has supported the notion that NSCs/NPCs in the SGZ generate granular neurons locally in the hippocampus, whereas the NSCs/NPCs in the SVZ travel a relatively long distance along the rostral migratory stream (RMS) to the olfactory bulb (OB) where they give rise to the interneurons (Luskin 1993; Lois and Alvarez-Buylla 1994; Gage 2000; Lie et al. 2004) in the postnatal or adult brain. During the migration along the RMS, NSCs/NPCs form chains providing supportive substrates to help NSCs/NPCs move forward to their destination, the OB (Lois et al. 1996; Chazal et al. 2000).

In addition to the findings collected from the fixed brain tissue as stated above, a live cell imaging technique enabled us to extend our knowledge on the natural status of NSC/NPC migration. Using multiphoton microscopy, we recorded NSC/NPC migration in the RMS in the transgenic mice carrying a report gene, green fluorescent protein (GFP) driven by nestin promoter (nestin-GFP mice). The nestin-GFP mice were sacrificed at the age of 3–4 weeks in the conditions of intact or ischemic injury, and the sagittal slices of the brain were prepared for live imaging. We observed something very interesting: there were two different types of nestin-GFP-labeled NSCs/NPCs in the RMS: non-migrating cells and migrating cells. We named the non-migrating NSCs/NPCs as bridge NSCs/NPCs and the traveling NSCs/NPCs as migrating NSCs/NPCs (Zhao et al. 2007a). The bridge NSCs/NPCs act as a bridge to support migrating NSCs/NPCs for travel. The traveling directions of the migrating NSCs/NPCs are multiple: forward to the OB, backward to the SVZ, up or down in the RMS (Fig. 2.1 a). Interestingly, the bidirectional migration of NSCs/NPCs has also been found by other investigators in vitro (Wichterle et al. 1997) or in the neonatal rat brain (Suzuki and Goldman 2003). It remains uncertain, however, how NSCs/NPCs are divided into bridge cells and migrating cells, what molecules are involved in distinguishing these two subtypes of NSCs/NPCs and how they retain their unique biological function.

2.3 The Early Reaction of NSCs/NPCs in Response to Cortical Brain Ischemia is Self-Protective

Time-lapse images revealed that the early response to cortical brain ischemia for NSCs/NPCs was to escape away from the area of brain damage (Zhao et al. 2007a). Cortical brain ischemia was made by the permanent occlusion of the right common carotid artery and the right middle cerebral artery distal to the striatal branch. Brain slices were prepared 3 h after induction of cortical brain ischemia in the right hemisphere of nestin-GFP mice, live brain slice imaging was acquired with a multiphoton microscope, and time-lapse images were recorded for 15 h at 7 min intervals. Very interestingly, we observed that only bridge NSCs/NPCs remained in the RMS next to the infarct cortex, whereas there were no migrating NSCs/NPCs in the RMS adjacent to the infarct cortex (Fig. 2.1 b–f). The migrating NSCs/NPCs were seen in the RMS distal to the infarct area. The migrating NSCs/NPCs in this part of RMS continued touching each other for a while (Fig. 2.1 b); thereafter, some of the NSCs/NPCs immediately moved away from the lesion area and towards to the direction of OB, and some other NSCs/NPCs traveled toward to the direction of infarct area along the RMS (Fig. 2.1 c). However, once they touched the bridge NSCs/NPCs in the location closer to the infarct cortex (Fig. 2.1 d), these migrating NSCs/NPCs changed their direction 180° and started moving away from the infarct area (Fig. 2.1 e and f). This observation suggests that cell-cell communication between the NSCs/NPCs leads to changes of direction for the migrating NSCs/NPCs, and that the early reaction of the migrating NSCs/NPCs in response to ischemic damage in the cortex is to move away from the infarct area during the 3–18 h after the induction of cortical ischemia. The results of this study offer insights into the natural self-protective behavior of migrating NSCs/NPCs in response to brain ischemic injury in the early period of cortical ischemia—escape from the infarct area, and this behavior is self-protection.

The open questions to be addressed in the future studies are: how the NSCs/NPCs sense the microenvironment changes, how the signals are transported/delivered among the NSCs/NPCs, and how NSCs/NPCs escape away from "danger"– the dead zone (infarct) induced by focal brain ischemia.

2.4 Focal Brain Ischemia-Induced NSC/NPC Proliferation is Required for Brain-Self Protective

It has been well documented that focal brain ischemia is a robust trigger for stimulating NSC/NPC proliferation in both SVZ and SGZ in the brain of adult rodents (Arvidsson et al. 2001; Jin et al. 2001; Arvidsson et al. 2002). It has been proposed that focal brain ischemia-induced NSC/NPC proliferation and neurogenesis may play a role in brain self-repair. However, more than 80 % of newly formed neurons have been found dead in the setting of focal brain ischemia, and the surviving

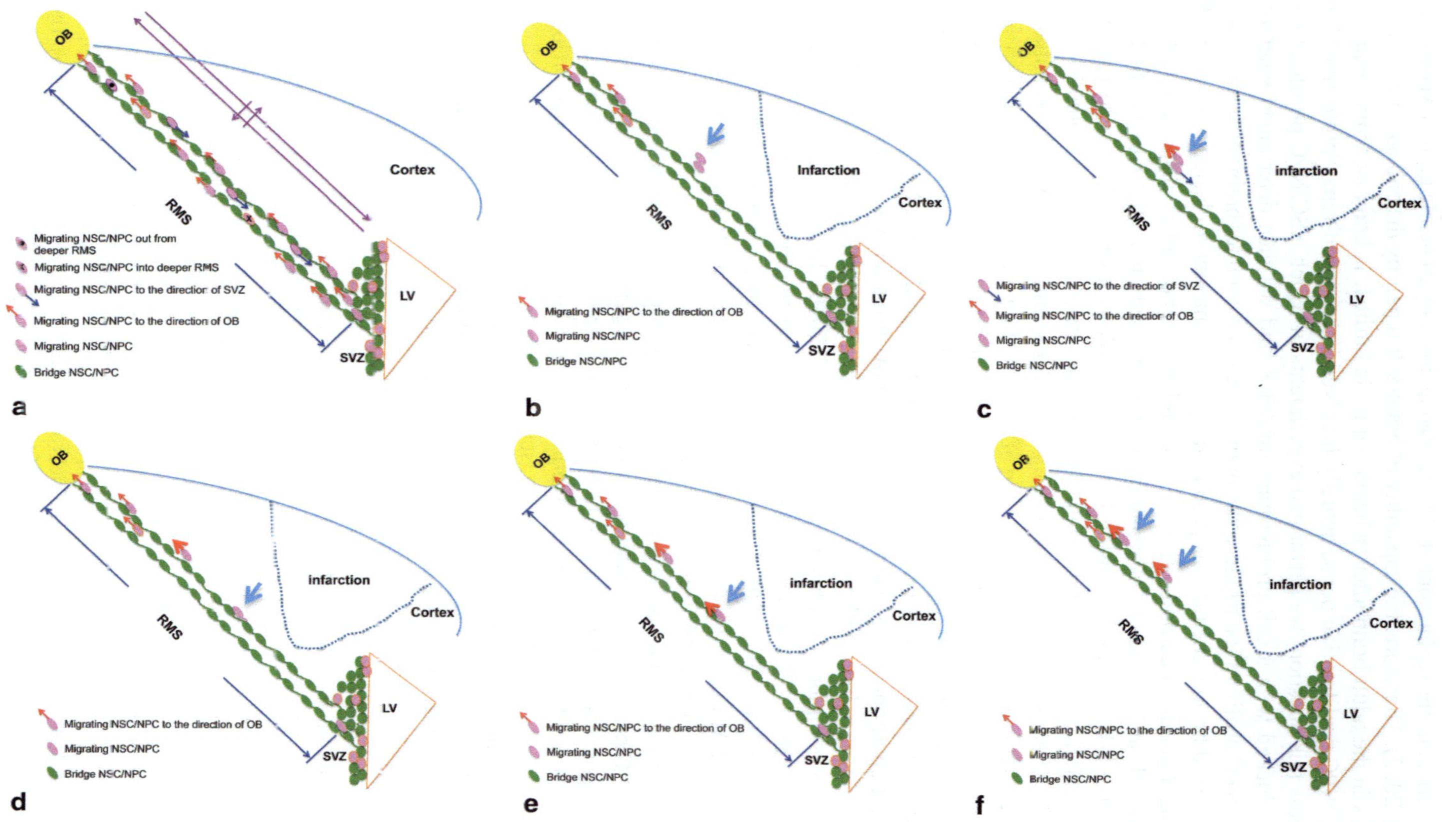

Fig. 2.1 Schematic diagram of NSC/NPC behavior in the RMS. **a** NSC/NPC behavior in the RMS of intact brain. Note that NSCs/NPCs in the RMS are divided into two sub-types: bridge NSCs/NPCs (*green*) and migrating NSCs/NPCs (*pink*). The bridge NSCs/NPCs are not traveling cells, and their biological function appears to generate a "bridge" between the olfactory bulb (OB) and subventricular zone (SVZ) for supporting NSC/NPC migration. The migrating

newborn neurons replace only 0.2 % of dead neurons by ischemic injury (Arvidsson et al. 2002). Therefore, the capability of brain-self repair by the endogenous NSCs/NPCs in the adult brain after ischemic injury is limited. Is brain ischemia-induced NSC/NPC proliferation per se actually involved in brain self-repair/protection? To address this question, we performed experiments in which NSC/NPC proliferation was blocked with a cell proliferating inhibitor, and we then determined whether NSC/NPC proliferation triggered by brain ischemia is vital for brain self-protection.

Cortical brain ischemia was produced by the permanent occlusion or ligation of the right common carotid artery and the middle cerebral artery distal to the striatal branch in male adult C57BL mice or spontaneously hypertensive rats (SHRs). Cytosine-β-D-Arabinofuranoside (Ara-C), a DNA synthesis inhibitor to prevent cell proliferation, was infused into the cerebral lateral ventricle ipsilateral to the ischemic hemisphere with a 7-day-infusion osmotic minipump. Ara-C was initially delivered one day before the induction of cortical brain ischemia or immediately after the induction of cortical brain ischemia, and the infusion of Ara-C was continued for 7 days. Bromodeoxyuridine (BrdU), a synthetic thymidine analog, can be incorporated into DNA during cell dividing. To identify proliferating NSCs/NPCs, BrdU was intraperitoneally injected daily during Ara-C infusion. We found that cortical brain ischemia-induced NSC/NPC proliferation in the bilateral SVZ and ipsilateral SGZ was completely inhibited by Ara-C (Li et al. 2010). In addition, slow dividing type −1 NSCs/NPCs in the bilateral SGZ were also significantly reduced by Ara-C infusion (Li et al. 2010). These findings indicate that cerebral lateral ventricle infusion of Ara-C is sufficient to prevent cortical brain ischemia-induced NSC/NPC proliferation in both SVZ and SGZ.

Our research data also revealed that inhibiting cortical brain ischemia-induced NSC/NPC proliferation in both SVZ and SGZ exaggerated neuron loss in both the hippocampus and infarct cortex) (Li et al. 2010). There are two types of neuron loss in the setting of focal brain ischemia: primary neuron loss and secondary neuron loss. The primary neuron loss occurs in the ischemic core within several hours after focal brain ischemia, whereas the secondary neuron loss appears in the brain tissue remote to the infarct core days and even months after focal brain ischemia (Hara et al. 1993; Touzani et al. 2001). It is worth noting that focal brain ischemia-induced NSC/NPC

NSCs/NPCs are the traveling cells in the RMS, and their traveling directions are multiple (*purple arrows*). **b–f** NSC/NPC behavior in the RMS of the brain suffering from cortical ischemic injury during 3–18 h after the induction of cortical brain ischemia. Note that there are no migrating NSCs/NPCs in the RMS adjacent to the cortical infarct area; only bridge NSCs/NPCs remain in the RMS nearby the cortical infarction. **b** Two migrating NSCs/NPCs physically interact with each other (communication) in the RMS distal to the cortical infarction. **c** The result of communication: one of the migrating NSCs/NPCs moves forward in the direction of OB (escaping away from the dead zone—cortical infarction), and other migrating NSC/NPC travels towards the infarction area along the RMS. **d** The migrating NSC/NPC "communicates" with a bridge NSC/NPC in the RMS close to the zone of dying brain cells (the cortical infarction area). **e** and **f** The results of the communication shown in d. The migrating NSC/NPC flees away from the dead zone (cortical infarction). The schematic diagram summarizes the results of time-lapse images published elsewhere (Zhao and Nam 2007). *LV* lateral ventricle. *RMS* the rostral migratory stream proliferation in the SVZ and SGZ is markedly increased during the period of 4 days to 2 weeks post-ischemia (Jin et al. 2001; Arvidsson et al. 2002) Therefore, focal brain ischemia-induced amplification of NSCs/NPCs in the SVZ and SGZ may contribute to protecting neurons from the ischemia-induced secondary death. The evidence supporting this hypothesis shows that inhibiting NSC/NPC proliferation in the SGZ causes significant neuron loss in the hippocampus 7 days after cortical brain ischemia (Li et al. 2010). In addition, the neuron loss in the hippocampus has been demonstrated as the secondary neuron loss through apoptosis (Li et al. 2010) because the stroke model used in this study only causes infarction in the ipsilateral cortex, whereas the hippocampus is spared from the infarct.

How focal brain ischemia-induced amplification of NSCs/NPCs in the SVZ and SGZ protects brain from ischemic damage remains poorly understood. One possibility may be due to releasing trophic factors as our findings have shown that adult brain-derived NSCs/NPCs can protect cortical neurons against excitotoxic damage through the NSC/NPC-released trophic factors, brain-derived neurotrophic factor (BDNF) and vascular endothelial growth factor (VEGF) (Li et al. 2010). In agreement with these observations, other investigators have also found that both BDNF and VEGF are produced by NSCs/NPCs (Leker et al. 2007; Mochizuki et al. 2008) and that administration of BDNF and VEGF in animal models of brain ischemia leads to reduction in infarction size (Schabitz et al. 1997; Ren and Finklestein 2005).

NSCs/NPCs may act as trophic factor providers to save neurons from the injury of focal brain ischemia. Several lines of evidence have shown that the NSCs/NPCs in the SVZ of adult brain can migrate long distances to the cortex adjacent to the infarct area. It has been demonstrated that the neuroblasts derived from proliferating NSCs/NPCs in the SVZ can migrate to the peri-infarct cortex through the white matter (Jin et al. 2003) and blood vessels (Ohab et al. 2006). In line with these findings, we also observed that SVZ-derived NSCs/NPCs migrated toward the cortical infarction along the corpus callosum at day 7 after induction of cortical brain ischemia (Li et al. 2010). Proliferating NSCs/NPCs in the SGZ may rescue hippocampal neurons through the delivery of BDNF via axonal projections between the dentate gyrus and hippocampal CA1 and CA3 subregions. It has been revealed that trkB, the receptor for BDNF, is expressed in CA1 and CA3 (Yan et al. 1997). BDNF has been shown to prevent glutamate-induced apoptotic cell death (Almeida et al. 2005).

Our knowledge and understanding of the precise role of focal brain ischemia-induced amplification of NSCs/NPCs in the SVZ and SGZ in adult brain is still far from complete. Is the focal brain ischemia-induced amplification of NSCs/NPCs part of a reaction from NSCs/NPCs to protect themselves in the condition of ischemic injury? It has been shown that a widespread hyperexcitability occurs in the intact brain regions including peri-infarct cortex and hippocampus, and that the hyperexcitability is most pronounced at 3–7 days and persists 1 month after cerebral cortical ischemia (Schiene et al. 1996; Qu et al. 1998; Redecker et al. 2002). NSCs/NPCs in the adult brain share many physiological features with astrocytes (Filippov et al. 2003; Kempermann et al. 2004b). Astrocytes can regulate inhibitory synapses formation in hippocampal neurons (Elmariah et al. 2005), manipulate glutamatergic transmission and uptake glutamate (Newman 2003). Can focal brain ischemia-induced amplification of NSCs/NPCs in both SVZ and SGZ contribute to preventing the hyperexcitability-induced neuron damage?

Other interesting questions that remain elusive are: whether generating and releasing trophic factors from NSCs/NPCs are cell-status dependent (proliferation vs. quiescent), and whether there are other mechanisms involved in NSC/NPC -induced neuroprotection in addition to the trophic factors released by NSCs/NPCs. Recent studies have revealed that exosomes and microvesicles (EMVs) play a key role in cell-to-cell communication. EMVs contain membrane receptors, proteins, mRNAs, miroRNAs and organelles that offer genetic regulation of cell survival, phenotype, proliferation, differentiation or death to the target cells either nearby (paracrine) or at a distance (endocrine) (Camussi et al. 2010). Stem cells have been demonstrated to release abundant EMVs (Camussi et al. 2010; Aoki et al. 2014). Stem cell-derived EMVs have been shown to rescue cells from injury, reprogram terminally differentiated cells to re-enter the cycle, and promote tissue regeneration and repair (Camussi et al. 2010). However, the role of NSC/NPC-released EMVs in supporting brain tissue survival and regeneration in the setting of stroke remains an open question to be addressed in future.

Enhancing propagation of NSCs/NPCs has been shown to protect brain from ischemic injury in animal models. Systemic administration of stem cell factor (SCF) 3h after induction of cortical brain ischemia in SHRs dramatically increases NSC/NPC proliferation in the SVZ, reduces infarction size and eliminates neurological deficits (Zhao et al. 2007a). Sodium butyrate has been reported to enhance amplification of NSCs/NPCs, reduce infarction size and improves functional outcome when systemically administered for 7 days after focal brain ischemia in rats (Kim et al. 2007; Kim et al. 2009). Leker and co-workers (2007) (Leker et al. 2007) demonstrated that long-term delivery of fibroblast growth factor-2 (bFGF) immediately after induction of cortical brain ischemia in SHRs resulted in the enhancement of NSC/NPC proliferation in the SVZ, white matter (the corpus callosum) and peri-infarct cortex throughout 90 days after ischemia and the improvement of motor function. These findings provide insights into developing new therapeutic approaches for brain protection through increasing endogenous NSCs/NPCs.

2.5 Hematopoietic Growth Factors Govern NSC/NPC Lineage Differentiation

Stem cell factor (SCF) and granulocyte-colony stimulating factor (G-CSF) were initially discovered as the essential hematopoietic growth factors that regulate the mobilization, proliferation and differentiation of hematopoietic stem cells/hematopoietic progenitor cells (HSCs/HPCs) (Ashman 1999) (Ripa and Kastrup 2008). Accumulating evidence has revealed that these hematopoietic growth factors may also play roles in the central nervous system (CNS). SCF (Zhao et al. 2007a) and G-CSF (Schneider et al. 2005; Zhao et al. 2007a) have been proven to pass through the blood-brain barrier in intact rodent brains. Systemic administration of SCF (Zhao et al. 2007b; Zhao et al. 2007a) and G-CSF (Schabitz et al. 2003; Six et al. 2003; Shyu et al. 2004; Schneider et al. 2005; Komine-Kobayashi et al. 2006; Zhao et al. 2007a) alone, or SCF in combination with G-CSF (Zhao et al. 2007b; Zhao et al. 2007a) has been shown to protect the brain against ischemic injury (Schabitz et al. 2003; Six et al. 2003; Shyu et al. 2004; Schneider et al. 2005; Komine-Kobayashi et al. 2006; Zhao et al. 2007a) and facilitate brain repair during chronic stroke (Zhao et al. 2007b). In addition, SCF and G-CSF appear to be involved in neuronal plasticity. Mice with mutations of SCF (Motro et al. 1996) or SCF receptor (Katafuchi et al. 2000) show impaired long-term potentiation (LTP) and spatial learning and memory. G-CSF deficient mice display cognitive impairment, LTP reduction, and poor neuronal networks in the hippocampus (Diederich et al. 2009). Interestingly, SCF receptor, ckit, and G-CSF receptor, GCSFR, have been found to be expressed in the NSCs/NPCs of the adult rodent brain (Schneider et al. 2005; Zhao et al. 2007a). However, the effects of SCF and G-CSF on the NSCs/NPCs remain largely unknown.

Using the approaches of cellular biology and molecular biology, we have determined the regulatory effects of SCF and G-CSF on lineage commitment of NSCs/NPCs. First of all, we observed that the receptors for SCF and G-CSF were expressed on the NSCs/NPCs in the ventricular zone of the developing rat brain at embryonic day 18 (Piao et al. 2012). The NSCs/NPCs were then isolated from the ventricular zone of the embryonic rat brain at E18, a time period in which both neurogenesis and gliogenesis take place. The NSCs/NPCs were then grown in cell culture dishes to form neurospheres. After forming the secondary neurospheres, NSCs/NPCs were disassociated into single cells and were then allowed to differentiate in a differentiation medium. SCF and G-CSF alone or in combination were added at the beginning of the differentiation stage of NSCs/NPCs. We found that SCF and G-CSF alone or in combination increased the number of TuJ1-positive neuronal cells and reduced GFAP-positive astrocytes (Piao et al. 2012), suggesting a regulative role of SCF and G-CSF on lineage switch of NSCs/NPCs. In line with our observation, other investigators have revealed that SCF (Jin et al. 2002) and G-CSF (Schneider et al. 2005), promote neurogenesis in vivo.

Very interestingly, SCF and G-CSF can promote neuronal fate determination of NSCs/NPCs at the proliferation stage. SCF and G-CSF alone or in combination were added to the dividing NSCs/NPCs in the secondary neurospheres for 3–4 days. Thereafter, the mitogens (bFGF and EGF) and the hematopoietic growth factors were then withdrawn, and the NSCs/NPCs were allowed to differentiate. We observed that only SCF in combination with G-CSF (SCF+G-CSF) treatment showed a significant increase in neurogenesis and reduction in gliogenesis (Piao et al. 2012). This observation led us to hypothesize that SCF+G-CSF promotes neuronal lineage commitment of NSCs/NPCs through the regulation of the cell cycle. To test this hypothesis, we examined the effects of SCF+G-CSF on the cell cycle and NSC/NPC propagation. The results of a cell cycle assay showed that SCF+G-CSF increased the population of NSCs/NPCs at the G1/G0 phase and decreased the population of NSCs/NPCs at the S phase (Piao et al. 2012), suggesting that

SCF+G-CSF causes the cell cycle arrest of NSCs/NPCs. To further confirm this result, BrdU was added into the NSCs/NPCs during the proliferating stage. We chose BrdU for labeling the dividing NSCs/NPCs at the S phase because BrdU is a thymidine analogue and can be incorporated into the newly synthesized DNA in S-phase cells. We found that SCF+G-CSF treatment during NSC/NPC proliferation resulted in a significant reduction of BrdU positive NSCs/NPCs, indicating that the NSC/NPC dividing is inhibited by SCF+G-CSF (Piao et al. 2012). Furthermore, when we examined NSC/NPC growth curve, we observed that NSC/NPC propagation was reduced by SCF+G-CSF (Piao et al. 2012). Together, these data suggest that SCF+G-CSF leads to cell cycle exit of NSCs/NPCs and inhibits NSC/NPC proliferation. Inducing cell cycle arrest of NSCs/NPCs is the critical and initial step to promote neuronal lineage differentiation of NSCs/NPCs. This notion is supported by the findings that pro-differentiation molecule-induced neuronal differentiation is preceded by cell cycle withdrawal (Henrique et al. 1997; Farah et al. 2000; Politis et al. 2008) and that promoting the cell cycle leads to inhibition of neuronal differentiation (Ohnuma et al. 2001).

How does SCF+G-CSF induce neuronal fate determination and commitment of NSCs/NPCs? To address this question, we examined the effects of SCF+G-CSF on the regulation of proneural bHLH transcription factors and Notch signaling pathway. Convincing evidence has shown that neurogenin 1 (Ngn1), one of the proneural bHLH transcription factors, promotes neuronal differentiation of NSCs/NPCs. Ngn1 has been found to be expressed in the ventricular zone, and Ngn1 only appears during cortical neurogenesis (Gradwohl et al. 1996; Sommer et al. 1996; Ma et al. 1997). In fact, Ngn1 acts as an activator of neuronal transcription factor that promotes neurogenesis and inhibits glial differentiation (Sun et al. 2001). Erythropoietin, one of the hematopoietic growth factors, has been shown to promote neuronal differentiation of NSCs/NPCs through Ngn1 mediation (Wang et al. 2006). By contrast, Notch1/Hes1 signaling contributes to maintaining NSCs/NPCs in an undifferentiated stage (Artavanis-Tsakonas et al. 1999; Mizutani and Saito 2005), enhances astroglial differentiation (Schmid et al. 2003; Anthony et al. 2005), and suppresses neuronal lineage commitment of NSCs/NPCs (Kageyama and Ohtsuka 1999). Based on these findings, we hypothesized that SCF+G-CSF promotes neuronal differentiation of NSCs/NPCs through acting the Ngn1 transcription and inhibiting the Notch1/Hes1 signaling. To test this hypothesis, we quantified gene expression of NSCs/NPCs with real-time RT-PCR after SCF+G-CSF treatment. The gene expression data showed that Ngn1 gene expression was significantly up-regulated by SCF+G-CSF. In addition, both Notch1 and Hes1 genes were significantly down-regulated by SCF+G-CSF (Piao et al. 2012). Moreover, when Ngn1 transcription was silenced by the siRNAs against Ngn1, SCF+G-CSF-induced enhancement of neuronal lineage commitment and inhibition of astroglial differentiation was significantly prevented (Piao et al. 2012). These findings suggest that Ngn1 is required for SCF+G-CSF-induced lineage determination and neuronal fate commitment of NSCs/NPCs. A putative model of the effects of SCF+G-CSF on cell cycle withdrawal and cell fate switch in NSCs/NPCs is presented in Fig. 2.2.

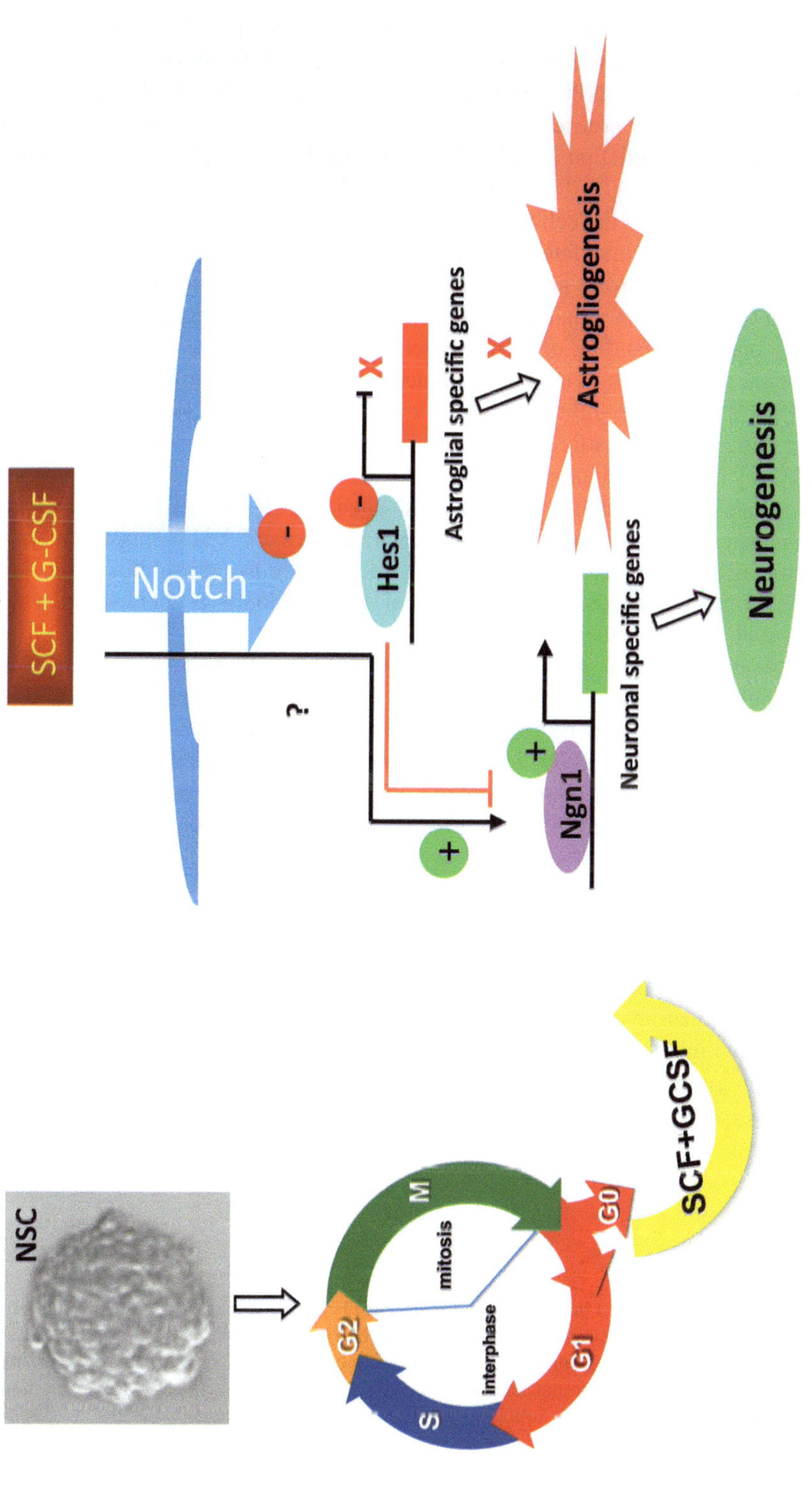

Fig. 2.2 A model for regulation of neuronal fate commitment of NSCs/NPCs by SCF+G-CSF during NSC/NPC dividing. SCF+G-CSF treatment during NSC/NPC proliferation (mitotic phase) coordinately arrests NSC/NPC cycle, suppresses astrocyte lineage commitment, and enhances neuronal fate differentiation through inhibiting Notch1/Hes 1 signaling and activating Ngn1. The dual effects of SCF+G-CSF on regulation of both NSC/NPC cycle and NSC/NPC lineage

Taken together, our research data support that hematopoietic growth factors, SCF+G-CSF, can govern NSC/NPC proliferation and differentiation and control lineage switch of NSCs/NPCs. SCF+G-CSF appears to display the dual effects on promoting the exit of proliferative NSCs/NPCs from the cell cycle and directing neuronal fate differentiation of NSCs/NPCs. It remains, however, to be further determined why and how SCF in combination with G-CSF but not SCF or G-CSF alone effectively stops the cell cycle for NSCs/NPCs and directs NSC/NPC differentiation into the neuronal fate, whether SCF+G-CSF facilitates survival and maturation of newborn neurons, and whether SCF+G-CSF-induced neurogenesis play a functional role in brain repair after brain injury. These mechanistic studies would advance our understanding of hematopoietic growth factors on neurogenesis and would help in developing a unique pharmaceutical therapy to reinforce endogenous NSCs/NPCs for repairing the brain in the condition of brain injury.

Other interesting questions to be addressed in future include whether SCF and G-CSF play a role in maintaining neurons in a differentiated stage and assuring neurons in a healthy and functioning status, and whether SCF and G-CSF are involved in pathogenesis of age-related neurodegenerative diseases. Addressing these open questions is important because a large body of evidence has shown that cell cycle re-entry of mature neurons leads to neuronal degeneration in Alzheimer's disease (Raina et al. 2004; McShea et al. 2007).

It is worth noting that the biological function of SCF and G-CSF is not restricted on regulation of the lineage commitment of NSCs/NPCs. Receptors for SCF and G-CSF have been found to be expressed not only on the NSCs/NPCs but also on the neurons (Schneider et al. 2005; Zhao et al. 2007a). In addition to the effects of SCF and G-CSF on neuroprotection as stated earlier in this chapter, our research group has recently demonstrated that SCF+G-CSF synergistically promotes neurite outgrowth (Su et al. 2013). Moreover, systemic administration of SCF+G-CSF in the phase of chronic stroke increases neuronal network formation in the cortex adjacent to the infarct cavities (Cui et al. 2013). These findings support a novel view concerning the role of SCF and G-CSF in the CNS that SCF+G-CSF has the capability to cease NSC/NPC dividing, guide NSCs/NPCs for neuronal lineage differentiation, promote neurite extension, enhance neuronal network formation, and protect neurons from injury. These research data suggest that SCF+G-CSF may have much broader biological function involved in brain development, brain plasticity, and brain repair after injury.

---

switch may be accomplished through the following processes: **a** Repress NSC/NPC proliferation and drive NSCs/NPCs out of cycle by inhibiting Notch1/Hes1 signaling and enhancing Ngn1; **b** Inhibit Notch1/Hes1 signaling and release the repressive effect of Hes1 on Ngn1 to repress astrogliogenesis and to trigger neuronal fate commitment; and **c** Directly activate Ngn1 to promote the differentiation of NSCs/NPCs into neurons. This schematic diagram summarizes the data of SCF+G-CSF on NSC/NPC differentiation published elsewhere. (Piao et al 2012)

Concluding Remarks

NSCs/NPCs do not reside in an "isolated" microenvironment in the brain, the factors or molecules produced by other organs or systems may also regulate the proliferation and differentiation of NSCs/NPCs. Hematopoietic growth factors, SCF and G-CSF, are produced by bone marrow stromal cells and fibroblasts (Heinrich et al. 1993; Watari et al. 1994) to govern bone marrow stem cell survival, proliferation and differentiation. The findings provided in this chapter supports that SCF and G-CSF are also involved in cell fate determination and commitment of NSCs/NPCs. This observation leads to the insight that the physiological and pathological changes in other systems may also affect NSC/NPC proliferation and differentiation in the CNS. In fact, the precise role of NSCs/NPCs in the setting of brain injury remains poorly understood. Our research data suggest that the early reaction of NSCs/NPCs in the condition of cortical brain ischemia is self-protective: escaping away from the area suffering from ischemic injury. Cell-cell interaction and communication play a key role in directing the migrating NSCs/NPCs to flee. In addition, focal brain ischemia-induced the increase in proliferation of NSCs/NPCs during the 1st week after brain ischemia appears to be critical for neuroprotection. Further clarification of the contribution of brain ischemia-induced NSC/NPC amplification in brain protection is critically important because it will provide crucial evidence needed to assist in developing new therapeutic strategies for the treatment of stroke.

Acknowledgements This work was partially supported by The National Institutes of Health, National Institute of Neurological Disorders and Stroke (NINDS), R01 NS060911.

References

Almeida RD, Manadas BJ, Melo CV, Gomes JR, Mendes CS, Graos MM, Carvalho RF, Carvalho AP, Duarte CB (2005) Neuroprotection by BDNF against glutamate-induced apoptotic cell death is mediated by ERK and PI3-kinase pathways. Cell Death Differ 12:1329–1343

Anthony TE, Mason HA, Gridley T, Fishell G, Heintz N (2005) Brain lipid-binding protein is a direct target of Notch signaling in radial glial cells. Genes Dev 19:1028–1033

Aoki J, Ohashi K, Mitsuhashi M, Murakami T, Oakes M, Kobayashi T, Doki N, Kakihana K, Sakamaki H (2014) Posttransplantation bone marrow assessment by quantifying hematopoietic cell-derived mRNAs in plasma exosomes/microvesicles. Clin Chem 60:675–682

Artavanis-Tsakonas S, Rand MD, Lake RJ (1999) Notch signaling: cell fate control and signal integration in development. Science 284:770–776

Arvidsson A, Kokaia Z, Lindvall O (2001) N-methyl-D-aspartate receptor-mediated increase of neurogenesis in adult rat dentate gyrus following stroke. Eur J Neurosci 14:10–18

Arvidsson A, Collin T, Kirik D, Kokaia Z, Lindvall O (2002) Neuronal replacement from endogenous precursors in the adult brain after stroke. Nat Med 8:963–970

Ashman LK (1999) The biology of stem cell factor and its receptor C-kit. Int J Biochem Cell Biol 31:1037–1051

Camussi G, Deregibus MC, Bruno S, Cantaluppi V, Biancone L (2010) Exosomes/microvesicles as a mechanism of cell-to-cell communication. Kidney Int 78:838–848

Chazal G, Durbec P, Jankovski A, Rougon G, Cremer H (2000) Consequences of neural cell adhesion molecule deficiency on cell migration in the rostral migratory stream of the mouse. J Neurosci 20:1446–1457

Cui L, Murikinati SR, Wang D, Zhang X, Duan WM, Zhao LR (2013) Reestablishing neuronal networks in the aged brain by stem cell factor and granulocyte-colony stimulating factor in a mouse model of chronic stroke. PLoS ONE 8:e64684

Curtis MA, Kam M, Nannmark U, Anderson MF, Axell MZ, Wikkelso C, Holtas S, van Roon-Mom WM, Bjork-Eriksson T, Nordborg C, Frisen J, Dragunow M, Faull RL, Eriksson PS (2007) Human neuroblasts migrate to the olfactory bulb via a lateral ventricular extension. Science 315:1243–1249

Diederich K, Sevimli S, Dorr H, Kosters E, Hoppen M, Lewejohann L, Klocke R, Minnerup J, Knecht S, Nikol S, Sachser N, Schneider A, Gorji A, Sommer C, Schabitz WR (2009) The role of granulocyte-colony stimulating factor (G-CSF) in the healthy brain: a characterization of G-CSF-deficient mice. J Neurosci 29:11572–11581

Egger B, Gold KS, Brand AH (2011) Regulating the balance between symmetric and asymmetric stem cell division in the developing brain. Fly (Austin) 5:237–241

Elmariah SB, Oh EJ, Hughes EG, Balice-Gordon RJ (2005) Astrocytes regulate inhibitory synapse formation via Trk-mediated modulation of postsynaptic GABAA receptors. J Neurosci 25:3638–3650

Farah MH, Olson JM, Sucic HB, Hume RI, Tapscott SJ, Turner DL (2000) Generation of neurons by transient expression of neural bHLH proteins in mammalian cells. Development 127:693–702

Farkas LM, Huttner WB (2008) The cell biology of neural stem and progenitor cells and its significance for their proliferation versus differentiation during mammalian brain development. Curr Opin Cell Biol 20:707–715

Filippov V, Kronenberg G, Pivneva T, Reuter K, Steiner B, Wang LP, Yamaguchi M, Kettenmann H, Kempermann G (2003) Subpopulation of nestin-expressing progenitor cells in the adult murine hippocampus shows electrophysiological and morphological characteristics of astrocytes. Mol Cell Neurosci 23:373–382

Gage FH (2000) Mammalian neural stem cells. Science 287:1433–1438

Gradwohl G, Fode C, Guillemot F (1996) Restricted expression of a novel murine atonal-related bHLH protein in undifferentiated neural precursors. Dev Biol 180:227–241

Hara H, Harada K, Sukamoto T (1993) Chronological atrophy after transient middle cerebral artery occlusion in rats. Brain Res 618:251–260

Heinrich MC, Dooley DC, Freed AC, B and L, Hoatlin ME, Keeble WW, Peters ST, Silvey KV, Ey FS, Kabat D et al (1993) Constitutive expression of steel factor gene by human stromal cells. Blood 82:771–783

Henrique D, Hirsinger E, Adam J, Le Roux I, Pourquie O, Ish-Horowicz D, Lewis J (1997) Maintenance of neuroepithelial progenitor cells by Delta-Notch signalling in the embryonic chick retina. Curr Biol 7:661–670

Jin K, Minami M, Lan JQ, Mao XO, Batteur S, Simon RP, Greenberg DA (2001) Neurogenesis in dentate subgranular zone and rostral subventricular zone after focal cerebral ischemia in the rat. Proc Natl Acad Sci U S A 98:4710–4715

Jin K, Mao XO, Sun Y, Xie L, Greenberg DA (2002) Stem cell factor stimulates neurogenesis in vitro and in vivo. J Clin Invest 110:311–319

Jin K, Sun Y, Xie L, Peel A, Mao XO, Batteur S, Greenberg DA (2003) Directed migration of neuronal precursors into the ischemic cerebral cortex and striatum. Mol Cell Neurosci 24:171–189

Kageyama R, Ohtsuka T (1999) The Notch-Hes pathway in mammalian neural development. Cell Res 9:179–188

Katafuchi T, Li AJ, Hirota S, Kitamura Y, Hori T (2000) Impairment of spatial learning and hippocampal synaptic potentiation in c-kit mutant rats. Learn Mem 7:383–392

Kempermann G, Wiskott L, Gage FH (2004a) Functional significance of adult neurogenesis. Curr Opin Neurobiol 14:186–191

Kempermann G, Jessberger S, Steiner B, Kronenberg G (2004b) Milestones of neuronal development in the adult hippocampus. Trends Neurosci 27:447–452

Kim HJ, Rowe M, Ren M, Hong JS, Chen PS, Chuang DM (2007) Histone deacetylase inhibitors exhibit anti-inflammatory and neuroprotective effects in a rat permanent ischemic model of stroke: multiple mechanisms of action. J Pharmacol Exp Ther 321:892–901

Kim HJ, Leeds P, Chuang DM (2009) The HDAC inhibitor, sodium butyrate, stimulates neurogenesis in the ischemic brain. J Neurochem 110:1226–1240
Komine-Kobayashi M, Zhang N, Liu M, Tanaka R, Hara H, Osaka A, Mochizuki H, Mizuno Y, Urabe T (2006) Neuroprotective effect of recombinant human granulocyte colony-stimulating factor in transient focal ischemia of mice. J Cereb Blood Flow Metab 26:402–413
Leker RR, Soldner F, Velasco I, Gavin DK, Androutsellis-Theotokis A, McKay RD (2007) Long-lasting regeneration after ischemia in the cerebral cortex. Stroke 38:153–161
Li B, Piao CS, Liu XY, Guo WP, Xue YQ, Duan WM, Gonzalez-Toledo ME, Zhao LR (2010) Brain self-protection: the role of endogenous neural progenitor cells in adult brain after cerebral cortical ischemia. Brain Res 1327:91–102
Lie DC, Song H, Colamarino SA, Ming GL, Gage FH (2004) Neurogenesis in the adult brain: new strategies for central nervous system diseases. Annu Rev Pharmacol Toxicol 44:399–421
Lois C, Alvarez-Buylla A (1994) Long-distance neuronal migration in the adult mammalian brain. Science 264:1145–1148
Lois C, Garcia-Verdugo JM, Alvarez-Buylla A (1996) Chain migration of neuronal precursors. Science 271:978–981
Luskin MB (1993) Restricted proliferation and migration of postnatally generated neurons derived from the forebrain subventricular zone. Neuron 11:173–189
Ma Q, Sommer L, Cserjesi P, Anderson DJ (1997) Mash1 and neurogenin1 expression patterns define complementary domains of neuroepithelium in the developing CNS and are correlated with regions expressing notch ligands. J Neurosci 17:3644–3652
McShea A, Lee HG, Petersen RB, Casadesus G, Vincent I, Linford NJ, Funk JO, Shapiro RA, Smith MA (2007) Neuronal cell cycle re-entry mediates Alzheimer disease-type changes. Biochim Biophys Acta 1772:467–472
Mizutani K, Saito T (2005) Progenitors resume generating neurons after temporary inhibition of neurogenesis by Notch activation in the mammalian cerebral cortex. Development 132:1295–1304
Mochizuki N, Takagi N, Onozato C, Moriyama Y, Takeo S, Tanonaka K (2008) Delayed injection of neural progenitor cells improved spatial learning dysfunction after cerebral ischemia. Biochem Biophys Res Commun 368:151–156
Motro B, Wojtowicz JM, Bernstein A, van der Kooy D (1996) Steel mutant mice are deficient in hippocampal learning but not long-term potentiation. Proc Natl Acad Sci U S A 93:1808–1813
Newman EA (2003) New roles for astrocytes: regulation of synaptic transmission. Trends Neurosci 26:536–542
Ohab JJ, Fleming S, Blesch A, Carmichael ST (2006) A neurovascular niche for neurogenesis after stroke. J Neurosci 26:13007–13016
Ohnuma S, Philpott A, Harris WA (2001) Cell cycle and cell fate in the nervous system. Curr Opin Neurobiol 11:66–73
Piao CS, Li B, Zhang LJ, Zhao LR (2012) Stem cell factor and granulocyte colony-stimulating factor promote neuronal lineage commitment of neural stem cells. Differentiation 83:17–25
Politis PK, Thomaidou D, Matsas R (2008) Coordination of cell cycle exit and differentiation of neuronal progenitors. Cell Cycle 7:691–697
Qu M, Buchkremer-Ratzmann I, Schiene K, Schroeter M, Witte OW, Zilles K (1998) Bihemispheric reduction of GABAA receptor binding following focal cortical photothrombotic lesions in the rat brain. Brain Res 813:374–380
Raina AK, Zhu X, Smith MA (2004) Alzheimer's disease and the cell cycle. Acta Neurobiol Exp (Wars) 64:107–112
Redecker C, Wang W, Fritschy JM, Witte OW (2002) Widespread and long-lasting alterations in GABA(A)-receptor subtypes after focal cortical infarcts in rats: mediation by NMDA-dependent processes. J Cereb Blood Flow Metab 22:1463–1475
Ren JM, Finklestein SP (2005) Growth factor treatment of stroke. Curr Drug Targets CNS Neurol Disord 4:121–125
Ripa RS, Kastrup J (2008) G-CSF therapy with mobilization of bone marrow stem cells for myocardial recovery after acute myocardial infarction—a relevant treatment? Exp Hematol 36:681–686

Schabitz WR, Schwab S, Spranger M, Hacke W (1997) Intraventricular brain-derived neurotrophic factor reduces infarct size after focal cerebral ischemia in rats. J Cereb Blood Flow Metab 17:500–506

Schabitz WR, Kollmar R, Schwaninger M, Juettler E, Bardutzky J, Scholzke MN, Sommer C, Schwab S (2003) Neuroprotective effect of granulocyte colony-stimulating factor after focal cerebral ischemia. Stroke 34:745–751

Schiene K, Bruehl C, Zilles K, Qu M, Hagemann G, Kraemer M, Witte OW (1996) Neuronal hyperexcitability and reduction of GABAA-receptor expression in the surround of cerebral photothrombosis. J Cereb Blood Flow Metab 16:906–914

Schmid RS, McGrath B, Berechid BE, Boyles B, Marchionni M, Sestan N, Anton ES (2003) Neuregulin 1-erbB2 signaling is required for the establishment of radial glia and their transformation into astrocytes in cerebral cortex. Proc Natl Acad Sci U S A 100:4251–4256

Schneider A, Kruger C, Steigleder T, Weber D, Pitzer C, Laage R, Aronowski J, Maurer MH, Gassler N, Mier W, Hasselblatt M, Kollmar R, Schwab S, Sommer C, Bach A, Kuhn HG, Schabitz WR (2005) The hematopoietic factor G-CSF is a neuronal ligand that counteracts programmed cell death and drives neurogenesis. J Clin Invest 115:2083–2098

Shyu WC, Lin SZ, Yang HI, Tzeng YS, Pang CY, Yen PS, Li H (2004) Functional recovery of stroke rats induced by granulocyte colony-stimulating factor-stimulated stem cells. Circulation 110:1847–1854

Six I, Gasan G, Mura E, Bordet R (2003) Beneficial effect of pharmacological mobilization of bone marrow in experimental cerebral ischemia. Eur J Pharmacol 458:327–328

Sommer L, Ma Q, Anderson DJ (1996) Beurogenins, a novel family of atonal-related bHLH transcription factors, are putative mammalian neuronal determination genes that reveal progenitor cell heterogeneity in the developing CNS and PNS. Mol Cell Neurosci 8:221–241

Su Y, Cui L, Piao C, Li B, Zhao LR (2013) The effects of hematopoietic growth factors on neurite outgrowth. PLoS ONE 8:e75562

Sun Y, Nadal-Vicens M, Misono S, Lin MZ, Zubiaga A, Hua X, Fan G, Greenberg ME (2001) Neurogenin promotes neurogenesis and inhibits glial differentiation by independent mechanisms. Cell 104:365–376

Suzuki SO, Goldman JE (2003) Multiple cell populations in the early postnatal subventricular zone take distinct migratory pathways: a dynamic study of glial and neuronal progenitor migration. J Neurosci 23:4240–4250

Touzani O, Roussel S, MacKenzie ET (2001) The ischaemic penumbra. Curr Opin Neurol 14:83–88

Wang L, Zhang ZG, Zhang RL, Jiao ZX, Wang Y, Pourabdollah-Nejad DS, LeTourneau Y, Gregg SR, Chopp M (2006) Neurogenin 1 mediates erythropoietin enhanced differentiation of adult neural progenitor cells. J Cereb Blood Flow Metab 26:556–564

Watari K, Ozawa K, Tajika K, Tojo A, Tani K, Kamachi S, Harigaya K, Takahashi T, Sekiguchi S, Nagata S et al (1994) Production of human granulocyte colony stimulating factor by various kinds of stromal cells in vitro detected by enzyme immunoassay and in situ hybridization. Stem Cells 12:416–423

Wichterle H, Garcia-Verdugo JM, Alvarez-Buylla A (1997) Direct evidence for homotypic, glia-independent neuronal migration. Neuron 18:779–791

Yan Q, Radeke MJ, Matheson CR, Talvenheimo J, Welcher AA, Feinstein SC (1997) Immunocytochemical localization of TrkB in the central nervous system of the adult rat. J Comp Neurol 378:135–157

Zhao LR, Singhal S, Duan WM, Mehta J, Kessler JA (2007a) Brain repair by hematopoietic growth factors in a rat model of stroke. Stroke 38:2584–2591

Zhao LR, Berra HH, Duan WM, Singhal S, Mehta J, Apkarian AV, Kessler JA (2007b) Beneficial effects of hematopoietic growth factor therapy in chronic ischemic stroke in rats. Stroke 38:2804–2811

Chapter 3
The Role of Endogenous Neural Stem Cells in Stroke

Qichuan ZhuGe, Linhui Ruan and Kunlin Jin

Contents

Abstract Stroke is the 4th leading cause of death and the leading cause of severe long-term disability worldwide, with no effective treatment for most cases. The development of new effective therapies is needed to improve functional neurological recovery in stroke patients. Researches in experimental stroke in animal models over the past decade demonstrate that ischemic stroke enhances endogenous neural stem cells proliferation in SVZ and SGZ and promotes SVZ NSCs migration to the ischemic infarct site, differentiation into functional mature neurons. Ischemic injury

K. Jin (✉) · L. Ruan · Q. ZhuGe
Zhejiang Provincial Key Laboratory of Aging and Neurological Disorder Research, First Affiliated Hospital, Wenzhou Medical University, Wenzhou, China
e-mail: kunlin.jin@unthsc.edu

K. Jin · L. Ruan
Department of Pharmacology and Neuroscience, University of North Texas Health Science Center, 3500 Camp Bowie Blvd., Fort Worth, TX 76107, USA

Q. ZhuGe
Department of Neurosurgery, The First Affiliated Hospital, Wenzhou Medical University, Wenzhou, China
e-mail: drzhuge@yahoo.com

L.-R. Zhao, J. H. Zhang (eds.), *Cellular Therapy for Stroke and CNS Injuries*, Springer Series in Translational Stroke Research, DOI 10.1007/978-3-319-11481-1_3

triggers endogenous neural stem cell proliferation by a variety of growth factors, morphogens and neurotransmitters. Neuroblast migration occurs through SDF-1, MCP-1, MMP production and through association with vasculature and endothelial cells. These promising findings in stroke have brought hope to the development of neurorestorative therapy which aims to enhance endogenous neurogenesis after ischemic stroke and thereby contribute to the functional recovery.

Abbreviations

BDNF	Brain-derived neurotrophic factor
DG	Dentate gyrus
EGF	Epidermal growth factor
EPO	Erythropoietin
FGF-2	Fibroblast growth factor-2
GABA	Gamma-aminobutyric acid
GFAP	Glial fibrillary acidic protein
IGF-1	Insulin-like growth factor-1
MCP-1	Monocyte chemoattractant protein-1
MMP	Matrix metalloproteinase
NMDA	N-Methyl-D-aspartate
NPC	Neural progenitor cells
NSCs	Neural stem cells
rt-PA	Recombinant tissue plasminogen activator
SVZ	Subventricular zone
SGZ	Subgranular zone
SDF	Stromal cell-derived factor
VEGF	Vascular endothelial growth factor

3.1 Introduction

Stroke is the 4th leading cause of death in the United States and the leading cause of severe long-term disability worldwide and the number of patients suffered from stroke is on the increase at present, with no effective treatment for most cases. Currently, thrombolysis with recombinant tissue plasminogen activator (rt-PA) remains the only FDA approved treatment for acute ischemic stroke, but the short therapeutic time window makes this treatment applicable to only a minority of stroke patients (Brott and Bogousslavsky 2000). Therefore, the development of new effective therapies is needed to improve functional neurological recovery in stroke patients.

Endogenous Neural stem cells (NSCs) resident in the brain neurogenic regions raise the possibility of developing neural repair strategies for stroke. Researches in experimental stroke in animal models over the past decade have demonstrated that ischemic stroke enhances endogenous neural stem cells proliferation in SVZ and SGZ of dentate gyrus and promotes SVZ NSC migration to the ischemic infarct regions (Jin et al. 2001; Thored et al. 2006). Moreover, stroke-induced neurogenesis has also been reported in the adult human brain (Jin et al. 2006). These promising findings in stroke have brought hope to develop neurorestorative therapy which aims to enhance endogenous neurogenesis after stroke and thereby contribute to the functional recovery. This chapter aims to provide an understanding the role of endogenous neural stem cells in stroke and to describe current knowledge of how stroke induces endogenous neural stem cells proliferation, migration and differentiation.

3.2 Proliferation of Endogenous Neural Stem Cells After Stroke

It is well accepted that NSCs are primarily present in the adult SVZ and SGZ of dentate gyrus. Based on the morphology, protein expression, proliferation kinetics, and differentiation potential, three types of SVZ cells have been identified: type A (neuroblasts), type B (GFAP-positive progenitors) and type C (transit amplifying cells) (Zhao et al. 2008). GFAP, Nestin and Sox2 -positive type B cells are considered as the primary NSCs in the SVZ. This population resides in the wall of lateral ventricle and gives rise to actively dividing type C cells. Type C cells, which are negative for GFAP but express Mash1 and EGF receptor, generate type A cells that migrate into the olfactory bulb via the rostral migratory stream (RMS). In the SGZ of dentate gyrus, two types of NSCs, type 1(GFAP+SOX2+) and type 2(GFAP-) NSCs are present, maintaining a reciprocal relationship in the DG of hippocampus (Suh et al. 2007).

Proliferation is the first step of endogenous NSCs' response to ischemic stroke injury. It is reported that ischemic injury alone is sufficient to promote the proliferation of endogenous NSCs and thereby expands the NSCs pool (Zhang et al. 2007). The expansion of the NSCs pool is an essential factor in the development of endogenous neurorestorative strategies after stroke. Stroke induces early expansion of the NSCs pool by increasing the proportion of proliferating cells and shortening the length of cell cycle (Zhang et al. 2008b). In the adult rat brain, the proportion of actively dividing SVZ NSCs is about 15–21 % and the length of cell cycle of this population is approximately 18–21 h (Zhang et al. 2006). Stroke increases the proportion of proliferating SVZ NSCs, starting 2 days (24 %) and reaches a maximum at 7 days (31 %) after stroke. Two weeks after stroke, the level of proliferation returns to the baseline. Concurrently, stroke changes the cell cycle length of NSCs. The length of cell cycle shortens to 11 h at 2 days after stroke, which is significantly shorter than cell cycle length of 18–21 h in normal brain. Alteration of the G1 phase of SVZ NSCs may contribute to stroke-induced changes of the cell cycle length. Furthermore, studies by Zhang and his colleagues also revealed that stroke transiently switches NSCs division from asymmetric to symmetric (Zhang et al. 2004b). Symmetric division gives rise to two identical daughter cells that stay in the SVZ to maintain the NSCs pool, whereas asymmetric division generates two different daughter cells and the basal daughter becomes a young migratory cell that migrates away. The switching NSCs division from asymmetric to symmetric amplifies the pool of NSCs.

Mediators of Stroke-Induced NSCs Proliferation
Although little is known about the exact molecular mechanisms underlying the regulation of endogenous NSC proliferation after stroke, several potential mediators are beginning to be identified. Here, we review current data on growth factors, morphogens and neurotransmitters that are involved in the regulation of endogenous NSC proliferation after stroke.

3.2.1 Growth Factors and Neurotrophic Factors

3.2.1.1 FGF-2

FGF-2 is a well-known growth factor that plays a role in neurogenesis in the adult brain. Studies have reported that FGF-2 expression in the brain increased significantly after ischemic stroke (Naylor et al. 2005). Overexpression of FGF-2 significantly increased the proliferation of progenitor cells after ischemic stroke in both FGF-2 deficient mice and wild-type mice (Yoshimura et al. 2001). Conversely, FGF-2 knockout mice showed a reduction of ischemia-induced progenitor proliferation when compared with wild type mice. Moreover, administration of bone marrow stromal cells engineered to produce FGF-2 (Ikeda et al. 2005) has been shown to decrease infarct size. Taken together, these findings suggest that FGF-2 plays a role in NPC proliferation and neuroprotection after ischemic stroke. Several potential mechanisms underlying the neurogenic effect of FGF-2 in the ischemic injured brain have been presented. These include upregulation of BDNF, induction of GDNF, and downregulation of the NMDA receptor (Mattson et al. 1993; Lenhard et al. 2002; Kiprianova et al. 2004).

3.2.1.2 EGF

EGF, a known mitogen involved in the proliferation of adult NPCs, has an effect similar to FGF-2 in regulating endogenous NSC proliferation after stroke. Importantly, the expression of EGF-receptor on type C cells or TAPs was found to be increased after ischemic stroke (Ninomiya et al. 2006). Ischemia also causes an upregulation of Heparin binding EGF (HB-EGF) (Tanaka et al. 2004), which is known to act through the EGF receptor to promote neurogenesis. Previous studies have demonstrated that exogenous EGF administration in ischemic animals rescued 20% of the interneurons that would have died after ischemia, suggesting the neurogenic role of EGF in the adult brain after ischemic injury (Teramoto et al. 2003). Furthermore, infusion of EGF together with FGF-2 into the brain of adult rats was found to promote DG and SVZ NPC proliferation after focal ischemic stroke (Nakatomi et al. 2002; Tureyen et al. 2005). Administration of HB-EGF enhanced postischemic neurogenesis and contributed to the improvement of functional recovery (Jin et al. 2004). Overexpression of HB-EGF by viral delivery also led to a significant improvement in neurological function after ischemic stroke, which was attributed to increased neurogenesis by HB-EGF (Sugiura et al. 2005).

3.2.1.3 IGF-1

IGF-1, primarily produced in the liver, plays a major role in brain development. Several studies have demonstrated that focal ischemia significantly increases IGF-1 expression, its receptor and binding proteins (Yan et al. 2006). Blockage of IGF-1 by intracerebroventricular administration of IGF-1 antibodies resulted in a significant inhibition of neural progenitor proliferation induced by ischemic stroke, suggesting that IGF-1 regulates neurogenesis after ischemia. Conversely, administering IGF-1 after ischemic stroke promoted neurogenesis (Dempsey et al. 2003; Zhang et al. 2004a) and reduced neuronal loss (Brywe et al. 2005). In vitro studies demonstrated that IGF-1 stimulated the proliferation of cultured NPCs via phosphorylation of the PI-3-kinase/Akt signaling pathway (Kalluri et al. 2007). Furthermore, IGF-1 also enhanced glycogen synthase kinase phosphorylation, suggesting its involvement in NPC survival (Kalluri et al. 2007).

3.2.1.4 VEGF

Vascular endothelial growth factor (VEGF) is the major angiogenetic growth factor, which can induce angiogenesis and vasculogenesis through interaction with the VEGF receptor on endothelial cells. Our previous study revealed that VEGF promotes the proliferation of NSCs both in vitro and in the adult rat brain (Jin et al. 2002). Wang et al. showed that VEGF overexpression in transgenic mice greatly promoted ischemia-induced neurogenesis (Wang et al. 2007). Furthermore, intravenous administration of VEGF after stroke promoted angiogenesis in the ischemic penumbra and improved neurological performance. In addition, we previously showed that VEGF ICV administration to rats after focal ischemic stroke reduced infarct volume and resulted in enhanced neurological recovery suggesting that VEGF induces neurogenesis and neuroprotection after ischemic stroke (Sun et al. 2003).

3.2.1.5 BDNF

BDNF, one of the most extensively studied neurotrophic factors, is necessary for NSC proliferation and differentiation in the adult brain. Two research groups have revealed that the expression of BDNF and its receptor increased after ischemic stroke(Arai et al. 1996; Kokaia et al. 1998). Intrastriatal infusion of BDNF before ischemia in adult rats increased the survival of neurons in the dorsolateral side of the striatum and resulted in improved functional recovery (Andsberg et al. 2002). Furthermore, infusion of human mesenchymal stem cells expressing the BDNF gene after ischemic stroke greatly reduced the infarct volume. Consistent with these observations, knockout of BDNF in mice resulted in larger infarct volumes after MCAO as the inhibition of endogenous BDNF after ischemic injury may decrease the survival of neurons (Endres et al. 2000; Larsson et al. 2002).

3.2.2 Notch Signaling

Notch signaling is involved in neurogenesis in the neurogenic regions of the intact adult brain. Recent studies have documented that expression of Notch and Hes1 in SVZ NPCs significantly is increased after stroke (Zhang et al. 2008a). Inhibition of the Notch signaling pathway with siRNA or γ-secretase inhibitor blocked stroke-induced neurogenesis. Moreover, an in vivo study demonstrated that administration of Notch ligand delta-like 4 (Dll4) together with FGF-2 after stroke significantly increased the rate of proliferation of SVZ neural progenitor cells (Androutsellis-Theotokis et al. 2006). The same study also revealed that the Notch pathway interacts with the Shh pathway in regulating NSCs (Balordi and Fishell 2007; Angot et al. 2008).

3.2.3 Shh and Wnt Signalings

The Shh signaling pathway is involved in the proliferation and maintenance of NSCs in the adult brain. It is also associated with EPO in mediating adult neurogenesis. EPO is known to regulate neurogenesis in both the adult normal and ischemic brains through its receptor EPOR in the adult SVZ. Infusion of EPO significantly increased ischemia-induced neurogenesis whereas blockage of the Shh pathway with cyclopamine or siRNA significantly suppressed EPO-increased neurogenesis (Liu et al. 2007). Wnt signaling promotes proliferation and neuronal differentiation in adult hippocampal progenitor cells in the DG (Lie et al. 2005). Expression of Wnt and BMP family genes in SVZ NSCs of adult rodents were altered after stroke. However, how these genes regulate the proliferation of endogenous NSCs after stroke remains to be determined (Morris et al. 2007).

3.2.4 Neurotransmitters

Neurotransmitters released from nerve terminals have been demonstrated to promote NPC proliferation in the adult brains. Studies have also shown the involvement of the glutamate signaling pathway in post-ischemic NSC proliferation. Abnormal glutamatergic neurotransmission, in particular disrupted glutamatergic receptor expression, has been reported to play a significant role in neuronal death after ischemic stroke. Several studies have revealed that activation of the NMDA receptor, an ionotropic glutamate receptor, blocks proliferation of NPCs while inhibition of NMDA receptors via antagonists promotes NPC proliferation (Cameron et al. 1995; Nacher et al. 2003). A subsequent study conducted by Kluska et al. also reported that administration of an NMDA antagonist during brain ischemic stroke induced by photothrombosis resulted in enhanced neurogenesis in the hippocampus of rats (Kluska et al. 2005). However, Bernabeu and Sharp reported that administration of NMDA and AMPA receptor antagonists prevented stroke-induced neurogenesis in the SGZ after transient global ischemia (Bernabeu and Sharp 2000). These conflicting results may be attributed to species-specific differences and the different models of ischemic stroke. The mechanism by which glutamatergic signaling pathway mediate stroke-induced neurogenesis is unclear. Several other neurotransmitters such as GABA and dopamine have also been reported to play significant roles in the intact adult brain. However, their roles in mediating ischemic stroke-induced neurogenesis still need further evaluation.

3.3 Migration of Endogenous Neural Stem Cells After Stroke

In order for proliferating NSCs to contribute to the functional recovery, it is necessary for these NSCs to migrate from birthplace to the ischemic region. In the normal adult brain, the SVZ neuroblasts are destined to follow rostral migratory stream to migrate to the olfactory bulb. However, many of these SVZ neuroblasts migrate from the SVZ through the brain parenchyma into the ischemic injury region after ischemic stroke, as revealed by several studies (Arvidsson et al. 2002; Jin et al. 2003; Thored et al. 2006). This redirected migration is associated with cellular interactions between immature migrating neuroblasts, astrocytic processes and blood vessels (Yamashita et al. 2006). However, ischemic stroke also up-regulate inhibitory molecules such as chondroitin sulfate proteoglycans (CSPGs), which blocks neuroblast migration (Carmichael 2005). Stroke not only up-regulates attractive factors for neuroblast migration but also forms peri-infarct scar and produces barrier molecule to inhibit neuroblast migration. The net neuroblast migration is dependent on the balance of inhibitory molecules and attractive factors.

The mechanism underlying stroke-induced redirected migration is unclear. Several receptor-ligand signaling pathways and molecular factors involved in the stroke-induced endogenous NSCs migration have been identified. These include: stromal cell-derived factor 1 (SDF-1) and CXC chemokine receptor 4 (CXCR4), monocyte chemoattractant protein-1 (MCP-1) and CC chemokine receptor 2 (CCR2), and matrix metalloproteases (MMP). Further, the neurovascular niche within SVZ and SDG is also reported to be closely associated with post-stroke neuroblasts migration.

3.3.1 SDF

SDF-1 (CXCL12), a member of the alpha chemokine family, has been demonstrated to play an essential role in the mobilization and homing of stem cells to bone marrow with its receptor CXCR4 (Hattori et al. 2003). A role of SDF-1 and CXCR4 in the directing migration of neuroblasts in brain has also been reported as well. Study by Robin et al. showed that CXCR 4 is expressed in the NPCs and migrating neuroblasts in stroke brain, while blocking SDF-1alpha by a neutralizing antibody against

CXCR 4 significantly attenuated stroke-enhanced NPC migration, suggesting that SDF-1alpha generated in the stroke hemisphere may guide NPC migration towards the ischemic boundary via binding to its receptor CXCR 4 in the NPC (Robin et al. 2006). Moreover, Ohab et al. also demonstrated that intraventricular administration of SDF-1 in ischemic mice promote neuroblast migration after stroke and contribute to behavioral recovery (Ohab et al. 2006).

3.3.2 MCP-1

Monocyte chemoattractant protein-1 (MCP-1), a chemokine of the CC family, was previously shown to increase the migration of neural progenitors in vitro (Widera et al. 2004). It is also reported to plays a critical role in neuroblast migration after focal cerebral ischemia. Study by Yan et al. demonstrated that ischemic stroke in adult rats induces increase of MCP-1 expression in the activated microglia and astrocytes, which lasted for more than 3 days of reperfusion (Yan et al. 2007). This study also found that the migrating neuroblasts in the ischemic brain express the MCP-1 receptor CCR2, and there was a significant decrease in the number of migrating neuroblasts in MCP-1 and CCR2 knockout mice.

3.3.3 MMP

Matrix metalloproteinases (MMP), a family of proteinases, are known to play a role in extracellular matrix remodeling and cell migration. Recently, MMP have also been implicated in neuroblast migration from the SVZ. MMP-3 and −9 have been shown to express in NPCs, and inhibition of MMPs resulted in reduction in post-stroke neuroblast migration (Barkho et al. 2008). Furthermore, the expression of MMP on the vascular also contributes to the localization of neuroblasts after stroke (Wang et al. 2006). These findings suggest that neuroblasts may "digest" their way through the extracellular matrix as they migrate via secreting MMP. However, the detailed cellular and molecular mechanism underlying is unknown.

3.3.4 Neurovascular Niche

Neurogenesis is associated with angiogenesis in an environment termed the neurovascular niche within the SGZ and SVZ in the adult brain. In addition, the vasculature also appears to be closely tied to the migration of neuroblasts to ischemic injury regions. The vasculature promotes localization of migrating neuroblasts in the peri-infarct cortex after stroke. Many migrating neuroblasts are found to localize specifically to blood vessels in areas of active vascular sprouting and remodeling in the peri-infarct cortex region (Ohab et al. 2006; Thored et al. 2007). By blocking angiogenesis with endostatin, a direct inhibitor of post-ischemic angiogenesis, Ohab et al. demonstrated that angiogenesis and neurogenesis are causally linked within the post-stroke neurovascular niche (Ohab et al. 2006). One week of endostatin treatment resulted in a significant decrease in the number of new born endothelial cells and overall vascular density, leading to a 10-fold reduction in neuroblasts.

The mechanism by which angiogenic vascular mediate the migration and localization of neuroblasts in the peri-infarct region after stroke remain to be identified. Vasculature may provide a scaffold on which neuroblasts can migrate (Kojima et al. 2010), as the close association of migrating neuroblasts with vascular endothelial cells is observed. Vasculature not only provides a scaffold for the migration of neuroblasts, but also promotes neuroblsts migration via secreting various growth and chemotactic factors, including BDNF, MMPs, angiopoietins, and SDF-1. Another clue demonstrated the role of neurovascular niche in mediating post-stroke neuroblasts migration is that the processes of angiogenesis and neurogenesis share several similar molecular signaling in the peri-infarct neurovascular niche, such as ephrin/EphB signaling and Semaphorin 3α and its receptor neuropilin 1 (Suchting et al. 2006). Taken together, these suggest that neurovascular niche plays an essential role in mediating neuroblasts migration after ischemic stroke.

3.4 Differentiation and Neuronal Function

What is the fate of endogenous neuroblasts that migrate to peri-infarct region after ischemic stroke? Although a great quantity of neuroblasts reached the ischemic injury region, only few of them survived and differentiated into mature neurons. It is reported that most of neuroblasts which migrate to the peri-infarct region appear to die, perhaps from a failure to integrate or due to the inflammatory milieu (Arvidsson et al. 2002). In the DG of hippocampus, the stroke-induced neuroblasts migrate into GCL. Most of the surviving neuroblasts differentiate into calbindin or NeuN positive mature neurons by 3–4 weeks after ischemic stroke (Jin et al. 2003). Only a small number of the newly generated cells differentiate into GFAP positive astrocytes in the GCL of hippocampus (Zhu et al. 2003). In the SVZ, many of the surviving cells differentiate into neurons, but the precise nature of the neurons in the striatum is controversial. Some research groups reported that most of surviving neuroblasts differentiate into mature striatal neurons in adult rats after stroke (Arvidsson et al. 2002; Parent et al. 2002). However, study by Liu et al. found that the stroke-induced adult-born neurons exclusively differentiated into calretinin-expressing interneurons (Liu et al. 2009). The reasons for these disparate findings are not entirely clear. Further studies using transgenic methods to label adult-born neurons are therefore needed to address this question.

A number of literature supports the concept that increased neurogenesis is related to functional recovery (Kondziolka et al. 2000; Zhang et al. 2003). Whether these adult-born new neurons could replace lost cells by integrating into the surviving brain circuitry and contribute to functional recovery is unclear. Some studies support the functional integration of a small portion of newly generated neurons that migrate to the injured striatum after cerebral ischemia in adult brain (Hou et al. 2008). But whether this limited neurogenesis and functional integration contribute to functional improvement after stroke is uncertain. Arvidsson et al. (2002) doubted on the functional significance of post-stroke neurogenesis, as <0.2% of neurons destroyed after stroke were replaced in the infarcted striatum and an even smaller percentage in the infarcted cortex. Although interventions that are aimed at increasing neurogenesis have been shown to improve functional outcome, but these treatments are not specific to neurogenesis (Ohab et al. 2006; Leker et al. 2007). The causal link between neurogenesisand behavioral recovery after stroke remains to be demonstrated. And the approaches to resolve this issue are need in the near future. Using of transgenic technique such as inducible Cre to specifically knockout stroke-induce migrating neuroblasts after stroke could provide insight into the direct effect of neurogenesis on functional improvement.

Conclusions

Ischemic stroke induces endogenous NSC proliferation within the SVZ, migration from the SVZ, and localization in the peri-infarct cortex and striatum. This whole process include three distinct spatiotemporal zones, each of them is associated with distinct molecular and cellular interactions. The mechanisms that trigger augmented endogenous NSC proliferation, migration and their differentiation into specific neural types after stroke remains to be identified. Ischemic injury triggers endogenous NSC proliferation by a variety of growth factors, morphogens and neurotransmitters. Neuroblast migration occurs through SDF-1, MCP-1, MMP production and through association with vasculature and endothelial cells. However, whether these neuroblasts could integrate into the surviving brain circuitry and contribute to functional recovery is controversial. Understanding the fundamental mechanisms underlying stroke-induced adult neurogenesis will thus provide the basis for endogenous NSC therapy for ischemic stroke.

References

Androutsellis-Theotokis A, Leker RR, Soldner F, Hoeppner DJ, Ravin R, Poser SW, Rueger MA, Bae SK, Kittappa R, McKay RD (2006) Notch signalling regulates stem cell numbers in vitro and in vivo. Nature 442:823–826

Andsberg G, Kokaia Z, Klein RL, Muzyczka N, Lindvall O, Mandel RJ (2002) Neuropathological and behavioral consequences of adeno-associated viral vector-mediated continuous intrastriatal neurotrophin delivery in a focal ischemia model in rats. Neurobiol Dis 9:187–204

Angot E, Loulier K, Nguyen-Ba-Charvet KT, Gadeau AP, Ruat M, Traiffort E (2008) Chemoattractive activity of sonic hedgehog in the adult subventricular zone modulates the number of neural precursors reaching the olfactory bulb. Stem Cells 26:2311–2320

Arai S, Kinouchi H, Akabane A, Owada Y, Kamii H, Kawase M, Yoshimoto T (1996) Induction of brain-derived neurotrophic factor (BDNF) and the receptor trk B mRNA following middle cerebral artery occlusion in rat. Neurosci Lett 211:57–60

Arvidsson A, Collin T, Kirik D, Kokaia Z, Lindvall O (2002) Neuronal replacement from endogenous precursors in the adult brain after stroke. Nat Med 8:963–970

Balordi F, Fishell G (2007) Hedgehog signaling in the subventricular zone is required for both the maintenance of stem cells and the migration of newborn neurons. J Neurosci 27:5936–5947

Barkho BZ, Munoz AE, Li X, Li L, Cunningham LA, Zhao X (2008) Endogenous matrix metalloproteinase (MMP)-3 and MMP-9 promote the differentiation and migration of adult neural progenitor cells in response to chemokines. Stem Cells 26:3139–3149

Bernabeu R, Sharp FR (2000) NMDA and AMPA/kainate glutamate receptors modulate dentate neurogenesis and CA3 synapsin-I in normal and ischemic hippocampus. J Cereb Blood Flow Metab 20:1669–1680

Brott T, Bogousslavsky J (2000) Treatment of acute ischemic stroke. N Engl J Med 343:710–722

Brywe KG, Mallard C, Gustavsson M, Hedtjarn M, Leverin AL, Wang X, Blomgren K, Isgaard J, Hagberg H (2005) IGF-1 neuroprotection in the immature brain after hypoxia-ischemia, involvement of Akt and GSK3beta? Eur J Neurosci 21:1489–1502

Cameron HA, McEwen BS, Gould E (1995) Regulation of adult neurogenesis by excitatory input and NMDA receptor activation in the dentate gyrus. J Neurosci 15:4687–4692

Carmichael ST (2005) Rodent models of focal stroke: size, mechanism, and purpose. NeuroRx 2:396–409

Dempsey RJ, Sailor KA, Bowen KK, Tureyen K, Vemuganti R (2003) Stroke-induced progenitor cell proliferation in adult spontaneously hypertensive rat brain: effect of exogenous IGF-1 and GDNF. J Neurochem 87:586–597

Endres M, Fan G, Hirt L, Fujii M, Matsushita K, Liu X, Jaenisch R, Moskowitz MA (2000) Ischemic brain damage in mice after selectively modifying BDNF or NT4 gene expression. J Cereb Blood flow Metab 20:139–144

Hattori K, Heissig B, Rafii S (2003) The regulation of hematopoietic stem cell and progenitor mobilization by chemokine SDF-1. Leuk Lymphoma 44:575–582

Hou SW, Wang YQ, Xu M, Shen DH, Wang JJ, Huang F, Yu Z, Sun FY (2008) Functional integration of newly generated neurons into striatum after cerebral ischemia in the adult rat brain. Stroke 39:2837–2844

Ikeda N, Nonoguchi N, Zhao MZ, Watanabe T, Kajimoto Y, Furutama D, Kimura F, Dezawa M, Coffin RS, Otsuki Y, Kuroiwa T, Miyatake S (2005) Bone marrow stromal cells that enhanced fibroblast growth factor-2 secretion by herpes simplex virus vector improve neurological outcome after transient focal cerebral ischemia in rats. Stroke 36:2725–2730

Jin K, Minami M, Lan JQ, Mao XO, Batteur S, Simon RP, Greenberg DA (2001) Neurogenesis in dentate subgranular zone and rostral subventricular zone after focal cerebral ischemia in the rat. Proc Natl Acad Sci U S A 98:4710–4715

Jin K, Zhu Y, Sun Y, Mao XO, Xie L, Greenberg DA (2002) Vascular endothelial growth factor (VEGF) stimulates neurogenesis in vitro and in vivo. Proc Natl Acad Sci U S A 99:11946–11950

Jin K, Sun Y, Xie L, Peel A, Mao XO, Batteur S, Greenberg DA (2003) Directed migration of neuronal precursors into the ischemic cerebral cortex and striatum. Mol Cell Neurosci 24:171–189

Jin K, Sun Y, Xie L, Childs J, Mao XO, Greenberg DA (2004) Post-ischemic administration of heparin-binding epidermal growth factor-like growth factor (HB-EGF) reduces infarct size and modifies neurogenesis after focal cerebral ischemia in the rat. J Cereb Blood Flow Metab 24:399–408

Jin K, Wang X, Xie L, Mao XO, Zhu W, Wang Y, Shen J, Mao Y, Banwait S, Greenberg DA (2006) Evidence for stroke-induced neurogenesis in the human brain. Proc Natl Acad Sci U S A 103:13198–13202

Kalluri HS, Vemuganti R, Dempsey RJ (2007) Mechanism of insulin-like growth factor I-mediated proliferation of adult neural progenitor cells: role of Akt. Eur J Neurosci 25:1041–1048

Kiprianova I, Schindowski K, von Bohlen O, Halbach O, Krause S, Dono R, Schwaninger M, Unsicker K (2004) Enlarged infarct volume and loss of BDNF mRNA induction following brain ischemia in mice lacking FGF-2. Exp Neurol 189:252–260

Kluska MM, Witte OW, Bolz J, Redecker C (2005) Neurogenesis in the adult dentate gyrus after cortical infarcts: effects of infarct location, N-methyl-D-aspartate receptor blockade and anti-inflammatory treatment. Neuroscience 135:723–735

Kojima T, Hirota Y, Ema M, Takahashi S, Miyoshi I, Okano H, Sawamoto K (2010) Subventricular zone-derived neural progenitor cells migrate along a blood vessel scaffold toward the post-stroke striatum. Stem Cells 28:545–554

Kokaia Z, Andsberg G, Yan Q, Lindvall O (1998) Rapid alterations of BDNF protein levels in the rat brain after focal ischemia: evidence for increased synthesis and anterograde axonal transport. Exp Neurol 154:289–301

Kondziolka D, Wechsler L, Goldstein S, Meltzer C, Thulborn KR, Gebel J, Jannetta P, DeCesare S, Elder EM, McGrogan M, Reitman MA, Bynum L (2000) Transplantation of cultured human neuronal cells for patients with stroke. Neurology 55:565–569

Larsson E, Mandel RJ, Klein RL, Muzyczka N, Lindvall O, Kokaia Z (2002) Suppression of insult-induced neurogenesis in adult rat brain by brain-derived neurotrophic factor. Exp Neurol 177:1–8

Leker RR, Soldner F, Velasco I, Gavin DK, Androutsellis-Theotokis A, McKay RD (2007) Long-lasting regeneration after ischemia in the cerebral cortex. Stroke 38:153–161

Lenhard T, Schober A, Suter-Crazzolara C, Unsicker K (2002) Fibroblast growth factor-2 requires glial-cell-line-derived neurotrophic factor for exerting its neuroprotective actions on glutamate-lesioned hippocampal neurons. Mol Cell Neurosci 20:181–197

Lie DC, Colamarino SA, Song HJ, Desire L, Mira H, Consiglio A, Lein ES, Jessberger S, Lansford H, Dearie AR, Gage FH (2005) Wnt signalling regulates adult hippocampal neurogenesis. Nature 437:1370–1375

Liu XS, Zhang ZG, Zhang RL, Gregg S, Morris DC, Wang Y, Chopp M (2007) Stroke induces gene profile changes associated with neurogenesis and angiogenesis in adult subventricular zone progenitor cells. J Cereb Blood Flow Metab 27:564–574

Liu F, You Y, Li X, Ma T, Nie Y, Wei B, Li T, Lin H, Yang Z (2009) Brain injury does not alter the intrinsic differentiation potential of adult neuroblasts. J Neurosci 29:5075–5087

Mattson MP, Kumar KN, Wang H, Cheng B, Michaelis EK (1993) Basic FGF regulates the expression of a functional 71 kDa NMDA receptor protein that mediates calcium influx and neurotoxicity in hippocampal neurons. J Neurosci 13:4575–4588

Morris DC, Zhang ZG, Wang Y, Zhang RL, Gregg S, Liu XS, Chopp M (2007) Wnt expression in the adult rat subventricular zone after stroke. Neurosci Lett 418:170–174

Nacher J, Alonso-Llosa G, Rosell DR, McEwen BS (2003) NMDA receptor antagonist treatment increases the production of new neurons in the aged rat hippocampus. Neurobiol Aging 24:273–284

Nakatomi H, Kuriu T, Okabe S, Yamamoto S, Hatano O, Kawahara N, Tamura A, Kirino T, Nakafuku M (2002) Regeneration of hippocampal pyramidal neurons after ischemic brain injury by recruitment of endogenous neural progenitors. Cell 110:429–441

Naylor M, Bowen KK, Sailor KA, Dempsey RJ, Vemuganti R (2005) Preconditioning-induced ischemic tolerance stimulates growth factor expression and neurogenesis in adult rat hippocampus. Neurochem Int 47:565–572

Ninomiya M, Yamashita T, Araki N, Okano H, Sawamoto K (2006) Enhanced neurogenesis in the ischemic striatum following EGF-induced expansion of transit-amplifying cells in the subventricular zone. Neurosci Lett 403:63–67

Ohab JJ, Fleming S, Blesch A, Carmichael ST (2006) A neurovascular niche for neurogenesis after stroke. J Neurosci 26:13007–13016

Parent JM, Vexler ZS, Gong C, Derugin N, Ferriero DM (2002) Rat forebrain neurogenesis and striatal neuron replacement after focal stroke. Ann Neurol 52:802–813

Robin AM, Zhang ZG, Wang L, Zhang RL, Katakowski M, Zhang L, Wang Y, Zhang C, Chopp M (2006) Stromal cell-derived factor 1alpha mediates neural progenitor cell motility after focal cerebral ischemia. J Cereb Blood Flow Metab 26:125–134

Suchting S, Bicknell R, Eichmann A (2006) Neuronal clues to vascular guidance. Exp Cell Res 312:668–675

Sugiura S, Kitagawa K, Tanaka S, Todo K, Omura-Matsuoka E, Sasaki T, Mabuchi T, Matsushita K, Yagita Y, Hori M (2005) Adenovirus-mediated gene transfer of heparin-binding epidermal growth factor-like growth factor enhances neurogenesis and angiogenesis after focal cerebral ischemia in rats. Stroke 36:859–864

Suh H, Consiglio A, Ray J, Sawai T, D'Amour KA, Gage FH (2007) In vivo fate analysis reveals the multipotent and self-renewal capacities of Sox2+ neural stem cells in the adult hippocampus. Cell Stem Cell 1:515–528

Sun Y, Jin K, Xie L, Childs J, Mao XO, Logvinova A, Greenberg DA (2003) VEGF-induced neuroprotection, neurogenesis, and angiogenesis after focal cerebral ischemia. J Clin Invest 111:1843–1851

Tanaka R, Yamashiro K, Mochizuki H, Cho N, Onodera M, Mizuno Y, Urabe T (2004) Neurogenesis after transient global ischemia in the adult hippocampus visualized by improved retroviral vector. Stroke 35:1454–1459

Teramoto T, Qiu J, Plumier JC, Moskowitz MA (2003) EGF amplifies the replacement of parvalbumin-expressing striatal interneurons after ischemia. J Clin Invest 111:1125–1132

Thored P, Arvidsson A, Cacci E, Ahlenius H, Kallur T, Darsalia V, Ekdahl CT, Kokaia Z, Lindvall O (2006) Persistent production of neurons from adult brain stem cells during recovery after stroke. Stem Cells 24:739–747

Thored P, Wood J, Arvidsson A, Cammenga J, Kokaia Z, Lindvall O (2007) Long-term neuroblast migration along blood vessels in an area with transient angiogenesis and increased vascularization after stroke. Stroke 38:3032–3039

Tureyen K, Vemuganti R, Bowen KK, Sailor KA, Dempsey RJ (2005) EGF and FGF-2 infusion increases post-ischemic neural progenitor cell proliferation in the adult rat brain. Neurosurgery 57:1254–1263; discussion 1254–1263

Wang L, Zhang ZG, Zhang RL, Gregg SR, Hozeska-Solgot A, LeTourneau Y, Wang Y, Chopp M (2006) Matrix metalloproteinase 2 (MMP2) and MMP9 secreted by erythropoietin-activated endothelial cells promote neural progenitor cell migration. J Neurosci 26:5996–6003

Wang Y, Jin K, Mao XO, Xie L, Banwait S, Marti HH, Greenberg DA (2007) VEGF-overexpressing transgenic mice show enhanced post-ischemic neurogenesis and neuromigration. J Neurosci Res 85:740–747

Widera D, Holtkamp W, Entschladen F, Niggemann B, Zanker K, Kaltschmidt B, Kaltschmidt C (2004) MCP-1 induces migration of adult neural stem cells. Eur J Cell Biol 83:381–387

Yamashita T, Ninomiya M, Hernandez Acosta P, Garcia-Verdugo JM, Sunabori T, Sakaguchi M, Adachi K, Kojima T, Hirota Y, Kawase T, Araki N, Abe K, Okano H, Sawamoto K (2006) Subventricular zone-derived neuroblasts migrate and differentiate into mature neurons in the post-stroke adult striatum. J Neurosci 26:6627–6636

Yan YP, Sailor KA, Vemuganti R, Dempsey RJ (2006) Insulin-like growth factor-1 is an endogenous mediator of focal ischemia-induced neural progenitor proliferation. Eur J Neurosci 24:45–54

Yan YP, Sailor KA, Lang BT, Park SW, Vemuganti R, Dempsey RJ (2007) Monocyte chemoattractant protein-1 plays a critical role in neuroblast migration after focal cerebral ischemia. J Cereb Blood Flow Metab 27:1213–1224

Yoshimura S, Takagi Y, Harada J, Teramoto T, Thomas SS, Waeber C, Bakowska JC, Breakefield XO, Moskowitz MA (2001) FGF-2 regulation of neurogenesis in adult hippocampus after brain injury. Proc Natl Acad Sci U S A 98:5874–5879

Zhang ZG, Jiang Q, Zhang R, Zhang L, Wang L, Arniego P, Ho KL, Chopp M (2003) Magnetic resonance imaging and neurosphere therapy of stroke in rat. Ann Neurol 53:259–263

Zhang J, Li Y, Chen J, Yang M, Katakowski M, Lu M, Chopp M (2004a) Expression of insulin-like growth factor 1 and receptor in ischemic rats treated with human marrow stromal cells. Brain Res 1030:19–27

Zhang R, Zhang Z, Zhang C, Zhang L, Robin A, Wang Y, Lu M, Chopp M (2004b) Stroke transiently increases subventricular zone cell division from asymmetric to symmetric and increases neuronal differentiation in the adult rat. J Neurosci 24:5810–5815

Zhang RL, Zhang ZG, Lu M, Wang Y, Yang JJ, Chopp M (2006) Reduction of the cell cycle length by decreasing G1 phase and cell cycle reentry expand neuronal progenitor cells in the subventricular zone of adult rat after stroke. J Cereb Blood Flow Metab 26:857–863

Zhang RL, LeTourneau Y, Gregg SR, Wang Y, Toh Y, Robin AM, Zhang ZG, Chopp M (2007) Neuroblast division during migration toward the ischemic striatum: a study of dynamic migratory and proliferative characteristics of neuroblasts from the subventricular zone. J Neurosci 27:3157–3162

Zhang RL, Zhang ZG, Chopp M (2008a) Ischemic stroke and neurogenesis in the subventricular zone. Neuropharmacology 55:345–352

Zhang RL, Zhang ZG, Roberts C, LeTourneau Y, Lu M, Zhang L, Wang Y, Chopp M (2008b) Lengthening the G(1) phase of neural progenitor cells is concurrent with an increase of symmetric neuron generating division after stroke. J Cereb Blood Flow Metab 28:602–611

Zhao C, Deng W, Gage FH (2008) Mechanisms and functional implications of adult neurogenesis. Cell 132:645–660

Zhu DY, Liu SH, Sun HS, Lu YM (2003) Expression of inducible nitric oxide synthase after focal cerebral ischemia stimulates neurogenesis in the adult rodent dentate gyrus. J Neurosci 23:223–229

Chapter 4
Bone Marrow Mesenchymal Stromal Cell Transplantation: A Neurorestorative Therapy for Stroke

Jieli Chen, Poornima Venkat and Michael Chopp

Contents

J. Chen (✉) · P. Venkat · M. Chopp
Department of Neurology, Henry Ford Hospital, E&R Building,
3091, Detroit, MI 48202, USA
e-mail: jieli@neuro.hfh.edu

P. Venkat · M. Chopp
Department of Physics, Oakland University, Rochester, MI, USA

L.-R. Zhao, J. H. Zhang (eds.), *Cellular Therapy for Stroke and CNS Injuries*,
Springer Series in Translational Stroke Research, DOI 10.1007/978-3-319-11481-1_4

Abstract A decade long focus on neuroprotection for stroke and neural injury, and its failure to translate into the clinical setting has led to a major shift of focus from neuroprotection to neurorestoration. Neurorestoration involves the remodeling and rekindling of neurovascular plasticity within the central nervous system which drive neurological recovery. Bone marrow-derived mesenchymal stem cell (BMSC) therapy is a promising cell-based neurorestorative therapy for stroke. This chapter provides an update on the use of BMSCs to promote neurorestorative effects in the sub-acute and chronic phases after stroke. The biological processes involved in promoting neurorestorative effects post ischemia are outlined, molecular mechanisms that promote neurogenesis, synaptogenesis, vascular and white matter remodeling, and neurovascular interactions and plasticity are discussed, the involvement of microRNA's in regulating neurorestorative mechanisms is introduced, and an update on clinical trials for BMSC treatment of stroke is presented.

Abbreviations

BMSCs	Bone marrow-derived mesenchymal stem cells
miRNA	microRNA
tPA	Tissue plasminogen activator
MSCs	Marrow stromal cells
BBB	Brain-blood-barrier
NeuN	Neuron-specific nuclear protein
GFP	Green fluorescent protein
IBZ	ischemic border zone
GFAP	Glial fibrillary acidic protein
VEGF	vascular endothelial growth factor
FGF2	Basic fibroblast growth factor
PGF	Placental growth factor
IGF	Insulin-like growth factor
BDNF	Brain derived neurotrophic factor
HGF	hepatocyte growth factor
Ang1	Angiopoietin-1
SVZ	sub-ventricular zone
GDNF	Glial cell-derived neurotrophic factor
NGF	Nerve growth factor
SDF	Stromal cell-derived factor 1
CBF	Cerebral blood flow
PDGF	Platelet derived growth factor
MMP	Matrix metalloproteinase
OPC's	Oligodendrocyte progenitor cells
OLs	Oligodendrocytes
T1DM	Type-one diabetes mellitus
MCAo	Middle cerebral artery occlusion
HBMSCs	Human BMSCs

4.1 Introduction

Ischemic stroke is a world wide concern as the third leading cause of death and long-term disability. Developing innovative approaches to effectively promote neuroprotection and neurorestoration after stroke are of prime research interest. Neuroprotective approaches aim to prevent or slow down ischemic cell damage by introducing therapeutic agents that can uphold the integrity of the neuronal circuit. Encouraging results using neuroprotective agents were reported in animal stroke models; however, these were not successful in the Phase III clinical trials, except for the only FDA approved thrombolytic agent, tissue plasminogen activator (tPA). Decades long focus on neuroprotection and its failure to translate into the clinical setting has led to a shift of interest to neurorestoration. In 2006 and 2011, the National Institutes of Neurological Disease and Stroke (NINDS) Stroke Progress Review Group identified post stroke neurorestoration as a major priority for stroke research. Today, we have available to us a vast amount of preclinical research data indicating that sub-acutely administered cell-based therapies can promote neurorestoration after stroke (Chen et al. 2001a, b; Liu et al. 2014).

This chapter provides an update on the use of bone marrow-derived mesenchymal stem cells (BMSCs) to promote recovery processes in the sub-acute and chronic phases after stroke. The biological process involved in promoting neurorestorative effects post ischemia will be outlined, molecular mechanisms that promote vascular and white matter remodeling discussed, the involvement of microRNA's in regulating neurorestorative mechanisms will be introduced briefly and an update on clinical trials presented.

4.2 Bone Marrow-Derived Mesenchymal Stem Cells (BMSCs) Biology

Mesenchymal stem cells, or marrow stromal cells (MSCs), are multipotent stromal cells that contain cells with a stem-cell-like character that allows them to differentiate into a variety of cell types, including bone, cartilage, adipocytes, and hematopoietic supporting tissues (Krebsbach et al. 1999). MSCs express CD73, CD90 and CD105 surface markers, and lack the expression of CD11b, CD14, CD19, CD34, CD45, CD79a and HLA-DR surface markers (Dominici et al. 2006). MSCs derived from bone marrow are referred to as BMSCs. In culture, the BMSCs can be separated from hematopoietic cells by their differential adhesion to the cultured plastic and their prolonged proliferative potential (Krebsbach et al. 1999). BMSCs are an excellent source of cells for treating a variety of central nervous system diseases.

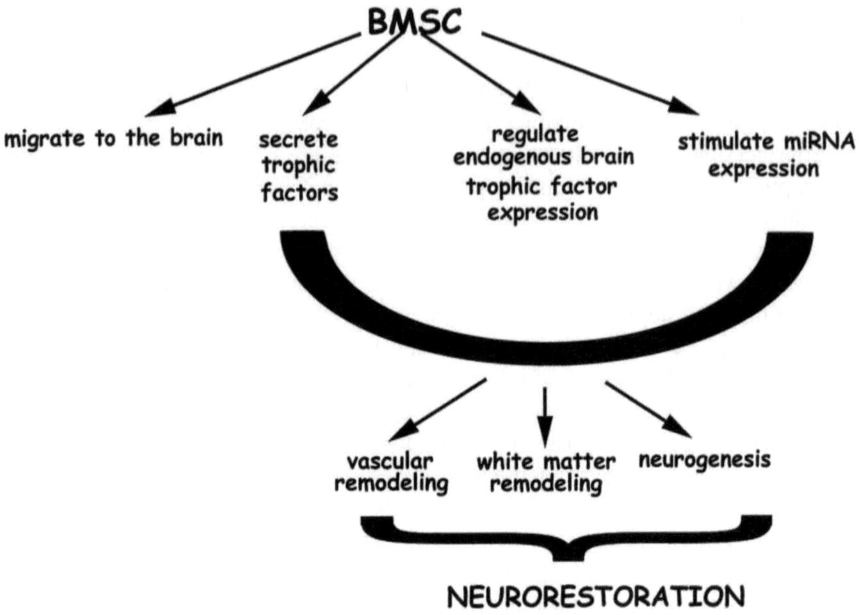

Fig. 4.1 BMSC therapy induces neurorestorative effects post stroke

4.3 BMSCs Therapy of Stroke

Figure 4.1 shows how BMSC therapy for stroke induces neurorestorative effects. The various effects of BMSC therapy contributing towards neurological recovery has been discussed in the following sections. Figure 4.2 summarizes the various contributors towards neurorestoration that are typically targets of cell therapy.

4.3.1 BMSC Migration into the Ischemic Brain

Many studies have confirmed the ability of BMSCs to migrate to the injury site (Lee et al. 2003; Wu et al. 2008). BMSC therapy initiated intravenously at 1 or 7 days after stroke in an animal stroke model promoted neurovascular remodeling resulting in improved neurological function post stroke (Chen et al. 2001a, 2003a; Shen et al. 2007). Administered BMSCs were found to survive and migrate to the ischemic brain. The disruption of the brain-blood-barrier (BBB) may contribute to the entry of BMSCs into ischemic brain compared to normal cerebral tissue thereby aiding BMSC migration to the ipsilateral hemisphere. BMSCs harvested from mouse were labeled using a nuclear fluorescence dye, bis-benzimide, for 24 h. The labeling of BMSCs enabled visualization of migration and differentiation of the implanted BMSCs. The labeled BMSCs were stereotactically transplanted into the ischemic striatum in mouse post MCAo (middle cerebral occlusion) model. Migration of the

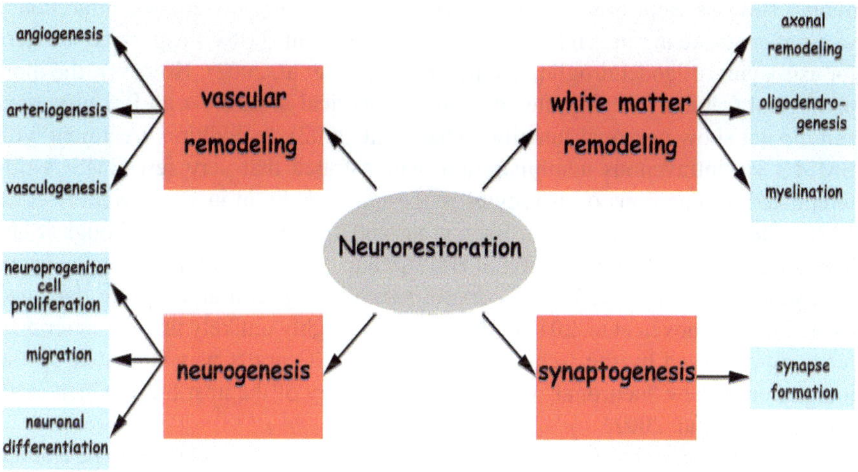

Fig. 4.2 The various aspects of neurorestoration that are targeted by cell therapy to improve neurological recovery post stroke

implanted BMSCs to the injured cortex and corpus callosum was observed. Four weeks post transplantation, some BMSCs were found to express neuronal markers like NeuN (neuron-specific nuclear protein) (Zhao et al. 2002). After 7 days of BMSC transplantation in animals that underwent transient MCAo and were treated at 24 h post stroke, migration of BMSCs to tissues supplied by the middle cerebral artery was observed (Wu et al. 2008). Another interesting study included transplantation of BMSCs derived from green fluorescent protein (GFP) expressing transgenic mice, 7 days post permanent MCAo in mice, followed by in vivo tracking of transplanted cells using fluorescence optical imaging (Shichinohe et al. 2004). The emitted green fluorescence was observed by exposing the skull and using in vivo fluorescence optical microscopy. Observations over the following 4–12 weeks revealed that the BMSCs migrated towards the ischemic injury site and immunohistochemistry analysis indicated accumulation in the ischemic border zone (IBZ) and some expressed neural phenotypes (Shen et al. 2007).

4.3.2 BMSC Differentiation

BMSCs have the ability to differentiate into neural and mesodermal cell lines (Zhao et al. 2002). BMSCs can stimulate nerve regeneration (Tohill et al. 2004). An in vitro study, in which BMSCs were co-cultured with fetal brain neural cells, suggests that differentiation of BMSCs to neurons can be promoted by close proximity to neural cells (Yan et al. 2005). When BMSCs were cultured in the presence of growth factors like BDNF or EGF, they expressed glial fibrillary acidic protein (GFAP), NeuN, and nestin (a marker for neural precursors), suggesting the capability of BMSCs to differentiate into neuronal cells (Sanchez-Ramos et al. 2000). Purified human BMSCs were grafted into the cortex surrounding the area of infarction 1 week after stroke in rats, and some transplanted human BMSCs expressed markers for astrocytes, oligodendroglia, and neurons (Zhao et al. 2002). However, the morphological features of the grafted cells were spherical in nature with few processes and did not show neuronal function (Zhao et al. 2002). In addition, treatment with BMSCs via intravenous administration demonstrated that very few BMSCs differentiated into neurons or astrocytes in the ischemic brain in vivo (Weaver et al. 1991; Chen et al. 2001a, 2003a; Shen et al. 2007; Li et al. 2008). Although it has been suggested that differentiation of BMSCs into cells of neural lineage may occur both in vitro and in vivo, BMSCs do not survive long term and replace damaged tissue (van Velthoven et al. 2012). Therefore, it is highly unlikely that the functional recovery observed by the ischemic rats with BMSC grafts was mediated by the integration of new "neuronal" cells into the circuitry of the host brain (Chen et al. 2001a; Zhao et al. 2002).

Only a small fraction of the injected BMSCs expressed neural phenotype and the mechanism of action of the therapeutic effects of BMSCs differ from that originally proposed for stem cells therapy. Stem cells being totipotent may replace injured/dead brain cells. The stem-cell like population in the BMSCs function in a similar manner and may differentiate into neural cell lines (Zhang et al. 2009; Shichinohe et al. 2010; Ding et al. 2011). However, these cells are a minor subpopulation of the BMSCs and do not contribute to the restoration of function, and only a very small percentage of the BMSCs assume parenchymal cell phenotype (Li et al. 2002). The neurological benefit induced by BMSC treatment appears to be derived primarily from the release of growth and trophic factors (van Velthoven et al. 2012).

4.3.3 BMSCs Secrete Trophic Factors

Exogenous BMSCs have been shown to produce, and more importantly, to stimulate the production of neuroprotective and neurorestorative factors in parenchymal cells. A broad range of mRNAs for angiogenic and arteriogenic cytokines are expressed by the BMSCs. Some of these are vascular endothelial growth factor (VEGF), basic fibroblast growth factor (FGF2), placental growth factor, insulin-like growth factor and Angiopoietin-1 (Ang1) (Chen et al. 2003b; Zacharek et al. 2007). Human BMSCs when cultured in the supernatant derived from ischemic rat brain extracts increase their production of several growth factors like brain derived neurotrophic factor (BDNF), VEGF and hepatocyte growth factor (HGF) (Chen et al. 2002a, b). The release of these growth factors contributes to neurological recovery by initiating cell repair mechanisms, increasing cell proliferation in the sub-ventricular zone (SVZ) and reducing apoptosis in the ischemic region (Li et al. 2002) and thereby improve functional benefit (Chen et al. 2002a, b, 2003b; Zacharek et al. 2007; Wakabayashi et al. 2010).

bFGF protects brain cells against toxins and promotes cell survival. Following focal cerebral ischemia in animal models, intravenous bFGF administration can reduce infarct volume and stop the down-regulation of anti-apoptotic protein Bcl-2 (Ay et al. 2001). bFGF therapy has neuroprotective effects when administered within a few hours after ischemic onset and reduces infarct volume; and neuro-restorative effects when administered a day after the ischemic attack by enhancing synaptogenesis and neuronal sprouting (Ay et al. 1999). BDNF is associated with neuronal survival, learning and memory, neuroplasticity and synaptic plasticity. Increase in BDNF has been correlated with improved behavioral recovery following brain injury. In an animal stroke model, BDNF showed therapeutic promise for functional recovery post stroke (Ploughman et al. 2009). Glial cell-derived neurotrophic factor (GDNF) promotes endogenous cell repair mechanisms, increases neurogenesis, neuroblast proliferation and migration from the SVZ and decreases apoptosis (Kobayashi et al. 2006; Yuan et al. 2013). Nerve growth factor (NGF) promotes cortical neurons against ischemic insults. The NGF/receptor system has been suggested to play an important role between astrocyte/neuron interaction under focal cerebral ischemia (Lee et al. 1998). Decrease in delayed death of hippocampal neurons post ischemia has also been reported upon treatment with NGF (Lee et al. 1998). VEGF promotes angiogenesis and is also capable of exerting neuroprotective effects (Sondell et al. 1999). VEGF decreases post ischemic lesion volume, apoptosis, and overall neuronal damage (Hayashi et al. 1998). Hence, enhancing trophic factor and growth factor production may be the mechanism of action in BMSC therapy. BMSC treatment increases the release of neurotrophins and angiogenic growth factors like bFGF, BDNF, GDNF, HGF, NGF, VEGF, etc. from parenchymal cells and thereby promotes endogenous brain repair mechanisms (Chen et al. 2001a; Hardy et al. 2008).

The intravenously administered BMSCs migrate to the brain and stimulate the endogenous cells (mainly astrocytes and endothelial cells) to produce growth factors, which in turn promote angiogenesis and vascular stabilization, partially mediated by VEGF/Flk1 and Ang1/Tie2 pathways (Zacharek et al. 2007). Insulin-like growth factor 1 (IGF-1) is a natural activator of the AKT signaling pathway. It is also known to stimulate cellular growth and proliferation and inhibit apoptosis. When cells were subjected to ischemic insult, elevated IGF-1 mRNA expression levels were observed (Wakabayashi et al. 2010). Upon BMSC therapy, enhanced expression levels of IGF-1 mRNA, IGF-1 and IGF-1R immuno-reactive cells was observed, suggesting that the neurological functional benefit may also be derived from an IGF-1 mediated self repair mechanism (Wakabayashi et al. 2010).

4.3.4 BMSC Treatment of Stroke Induces Neurorestorative Effects and Increases Interaction/Coupling Among Restorative Events

The conceptual model of the neurovascular unit focuses on illustrating the dynamic interactions among the brain neurons, micro-vascular endothelium, extracellular matrix, astrocytes and pericytes. While the neurovascular unit has gained importance as a key component of stroke damage, another model called the vascular neural network, encompassing the neurovascular unit and the vasculature supplying and interacting with the neural structures of the brain, is gaining attention. Stroke can inflict extensive damage to the vasculature of the brain along with damage to the neural circuit. Hence, the vascular neural network is a potential target for neurorestorative therapies (Zhang et al. 2012). The vascular neural network is a collection of cell types and structures inclusive of capillary endothelial cells, astrocytes, pericytes and neurons that constitute the neurovascular unit and structures connecting with arteries and arterioles upstream of the cerebral microcirculation, like vascular smooth muscle cells, arterial endothelial cells, perivascular nerves and venules downstream (Zhang et al. 2012). The vascular neural network can potentially target all five therapeutic aspects of stroke recovery; retardation of injury, reperfusion of ischemic penumbra, reduction of tissue injury, regeneration of new cells and reorganization of vasculature and neural networks (Zhang et al. 2012). Following brain injury, similar to how cortical gray matter recovery is mediated by a neurovascular niche, white matter recovery may be mediated by an oligovascular niche that promotes oligodendrogenesis and angiogenesis (Pham et al. 2012). These concepts are evolving with accumulating literature and influencing the neurorestorative approach towards stroke recovery.

Angiogenesis and neurogenesis in the neurovascular niches of the CNS play a key role in recovery. Neurogenesis and angiogenesis are linked together by growth factors and chemokines and trophic effects of Stromal cell-derived factor 1 (SDF1) and Ang1 which are up-regulated by blood vessels within the niche, and improve behavioral recovery (Ohab et al. 2006). Results from an animal stroke model indicated improved post stroke neurological recovery that was partly attributed to the migration of newly born neurons in the SVZ to the cortex, neurogenesis derived from GFAP-expressing progenitor cells in the SVZ, and migration of neuroblasts to a neurovascular niche in peri-infarct cortex (Ohab et al. 2006).

BMSC treatment of stroke induces expression of many angiogenic and trophic factors, and thereby regulates interaction/coupling between restorative events. The neurorestorative events include: vascular remodeling, white matter remodeling, neurogenesis and synaptic plasticity which may collectively contribute towards BMSC treatment induced neurorestoration after stroke.

4.3.4.1 BMSC Treatment of Stroke Increases Vascular Remodeling

Vascular remodeling includes angiogenesis and arteriogenesis and plays a major role in regulating cerebral blood flow (CBF). Cerebral blood flow regulation and the re-establishment of functional microvasculature via angiogenesis and arteriogenesis in the IBZ helps maintain neural function and creates a hospitable microenvironment for neuronal plasticity leading to functional recovery (Plate 1999; Chen et al. 2003b; Pratt et al. 2004). Enhanced vascular remodeling improves blood supply to ischemic cells and mediates the generation of restorative trophic factors.

Vascular remodeling leads to the formation of mature and functional blood vessels and hence has gained much importance as a therapeutic target.

Angiogenesis, defined as the process of growth of new blood vessels from pre-existing vessels, has been associated with improved long-term recovery of stroke patients (Arenillas et al. 2007; Navarro-Sobrino et al. 2011). Tissue ischemia and hypoxia stimulate angiogenesis hours after the ischemic attack and angiogenesis may continue for weeks (Dor and Keshet 1997). Several growth factors like VEGF, endothelial nitric oxide synthase (eNOS), bFGF, platelet derived growth factor (PDGF) and Ang-1 mediate and tightly regulate angiogenesis (Greenberg 1998). The physiological sequence of angiogenesis that occurs in the penumbra of the ischemic brain is as follows. Angiogenic growth factors bind to specific receptors located on the brain endothelial cells and stimulate endothelial cell proliferation, migration and sprouting from the existing vessel towards ischemic brain area (Greenberg 1998). The first step in this process is vasodilation initiated by NO (Carmeliet 2000) which in combination with the increase in VEGF expression increases vascular permeability allowing extravasation of plasma proteins that lay down a provisional scaffold for the migration of endothelial cells for vascular sprouting. Next, the sprouting endothelial cells form blood vessel tube-like structures. The initial vascular plexus forms mature vessels by sprouting, branching, pruning and promoting differential growth of endothelial cells, and recruitment of supporting cells, such as pericytes and smooth muscle cells (Folkman and D'Amore 1996; Risau 1997). This step involves the dissociation of smooth muscle cells and loosening the extracellular matrix which enwraps the mature vessel. Once the path of sprouting has been established, proliferating endothelial cells migrate to distant sites. Angiopoietin (Ang) -2, an inhibitor of Ang1/Tie2 signaling, may be involved in facilitating the detachment of pericyte cells from endothelial cells, while the matrix metalloproteinase (MMP) family of proteinases degrade matrix molecules and further weaken the vessel integrity (Feng et al. 2009). Once new blood vessel networks are formed, Ang1 which activates Tie2 receptors helps to stabilize networks initiated by VEGF. The Ang1/Tie2 interaction promotes pericyte recursion and mediates maturation of neovessels into more complex and functional vasculature.

Arteriogenesis is another process of restoring blood supply to an ischemic region when the main supply artery is blocked. Collateral vessels develop from pre-existing arterioles that adapt to supply the same end tissue as the blocked artery. While angiogenesis and arteriogenesis share some common features, they have quite distinctive pathways. Vessel collateralization is initiated by fluid shear stress due to pressure differences between perfusion territories (Erdo and Buschmann 2007). The circulating monocytes attract, adhere to the endothelium and invade to pave way for the collateral vessel development (Schaper and Buschmann 1999). These monocytes also produce growth factors like VEGF and proteolytic enzymes that enable migration and division of smooth muscle cells (Scholz et al. 2001). The smooth muscle cells coat the pre-existing capillaries transforming them into vessels with larger diameters (Buschmann and Schaper 2000; van Royen et al. 2001). Arteriogenesis and collateral vessel growth have also gained interest as therapeutic targets.

Studies indicate that BMSCs can participate in angiogenesis and increase arteriogenesis (Kinnaird et al. 2004; Zhu et al. 2011). Having a perivascular/vascular location, BMSCs contribute to vascular formation and function and might be harnessed to regenerate a blood supply to injured tissues (Watt et al. 2013). In vitro, BMSCs have also been reported to play an important role in angiogenesis in a time and dose dependent manner affecting all stages of angiogenesis from formation to maturation of vessels (Duffy et al. 2009). BMSCs implanted intravenously at 24 h post transient MCAo, expressed increased angiogenic factors, such as VEGF/FLK1 and Angiopoietin-1/Tie2 in the IBZ associated with enhanced functional of angiogenesis, vessel density and arteriogenesis as well as improved neurological functional recovery (Chen et al. 2003b; Zacharek et al. 2007; Wu et al. 2008; Cui et al. 2009). In a chronic limb ischemia model, treatment with autologous BMSCs significantly increased VEGF expression and increased vessel density (Hoffmann et al. 2010), improved blood flow and enhanced arteriogenesis (Zhu et al. 2011). BMSCs aid in brain repair by inducing a neovascular response by selectively targeting the injured tissue and improving blood flow to the ischemic zone via arteriogenesis and angiogenesis (Cui et al. 2009). These angiogenic and arteriogenic vessels are stimulated to produce trophic and growth factors that in turn contribute to brain plasticity and recovery of post stroke neurological function (Kinnaird et al. 2004). These results suggest that BMSCs induce and promote angiogenesis and arteriogenesis in the ischemic brain tissue to improve blood supply and alleviate neurological deficits.

4.3.4.2 BMSC Treatment of Stroke Increases White Matter Remodeling

White matter remodeling includes oligodendrogenesis, myelination and axonal regeneration. Oligodendrocyte progenitor cells (OPC's) produced in the SVZ differentiate to form oligodendrocytes (OLs). While OPC's do not have myelin forming ability, the mature oligodendrocytes do. In the ischemic brain, these mature oligodendrocytes may restore the myelin sheath lost during the ischemic attack and also myelinate the newly formed/sprouted axons. Hence, oligodendrogenesis is critical to restore and improve axonal function and neuronal communication (Zhang et al. 2013).

Oligodendrogenesis plays a key role in remyelination of axons and neural communication. Oligodendrocytes are at a higher risk of damage from ischemia since white matter has limited blood supply when compared to gray matter, and there is very little collateral blood flow in deep white matter (Back et al. 2002). Loss of myelin is of primary concern caused by oligodendrocyte damage, as injured oligodendrocytes can no longer produce myelin. Detrimental inflammatory responses, like increased MMP-9 and MMP-2 expression have been reported to further worsen white matter damage (Chen et al. 2011a). Following an ischemic stimulus, a greater fraction of the neural stem cells in the SVZ form OPC's which in turn differentiate into mature myelin producing oligodendrocytes in the corpus callosum or striatum in the brain. This is central to the repair damaged axons as well as to coating the newly sprouting axons with myelin. Myelin is a dielectric by nature with electrical insulating properties. The myelin sheath on the axons is not continuous and exposes the axon only at the nodes of Ranvier. This facilitates the much faster saltatory conduction over continuous wave conduction as the electrical impulses jump from one node to the next. There is mounting evidence associating increased OPCs and oligodendrocytes expression with improved neurological function post stroke in animal models.

Ischemia subjects axons to extensive damage impairing neuronal communication, and poor axonal regeneration and neuroplasticity have been associated with poor functional recovery (Walmsley and Mir 2007). Axonal remodeling begins 2–3 weeks post stroke and has been detected even at 28 days post the ischemic attack (Liu et al. 2009, 2010). Promoting axonal regeneration improves brain plasticity and can restore previously lost functions (Papadopoulos et al. 2006). Successful axonal remodeling is central to promoting brain repair and nerve regeneration.

White matter remodeling is enhanced by intra-carotid BMSC transplantation which is facilitated by axonal sprouting and re-myelination in the cortical IBZ and corpus callosum (Shen et al. 2006). BMSC implantation into the ischemic brain significantly reduced the transient MCAo-induced cortical loss and thinning of the white matter and enhanced cortical beta-III-tubulin immunoreactivity (Chen et al. 2008). Significantly increased thickness of individual axons and myelin, and areas of the corpus callosum and the numbers of white matter bundles in the striatum also were detected in the IBZ of BMSCs-treated rats compared to stroked rats (Li et al. 2006). BMSC therapy also improved structural neuroplasticity and increased axonal outgrowth from healthy brain tissue (Andrews et al. 2008). BMSCs injected into the lateral ventricle of rats at 1 day post permanent MCAo improved neurological functional recovery by increasing synaptogenesis indicated by an increase in Synaptophysin expression in the IBZ (Weng et al. 2008). Intravenous administration of combination of BMSC with Niaspan in type-one diabetes mellitus (T1DM) stroke rats significantly increases myelin and axonal density in the striatum of white matter of the IBZ, as well as increases Synaptophysin and SMI-31 (a pan-axonal neurofilament marker) expression in the ischemic brain compared to T1DM-MCAo control (Ye et al. 2013). BMSC treatment of stroke also increases the number of oligodendrocyte precursor cells in the corpus callosum and enhances the areas of the corpus callosum in both hemispheres (Shen et al. 2006). These results suggest that BMSCs facilitate axonal sprouting and remyelination in the cortical IBZ and corpus callosum, which may underlie neurological functional improvement caused by BMSC treatment.

Corticospinal tract fibers originating from the contralesional motor cortex sprout into the denervated spinal cord after stroke and play an important role in motor functional outcome after stroke. BMSCs treatment significantly increases biotinylated dextran amine-labeled contralesional axons sprouted into the denervated spinal cord (Liu et al. 2008). The interaction of axonal remodeling between the bilateral-hemisphere also regulates functional outcome after stroke. Stroke induces spontaneous inter- and intra-cortical axonal plasticity originating from the ipsilesional motor cortex in adult rats. BMSC treatment of stroke enhances axonal plasticity and bilateral-hemisphere axonal remodeling (Liu et al. 2010). The axonal remodeling is significantly correlated with spontaneous functional recovery after stroke and recovery is significantly enhanced by BMSCs (Liu et al. 2010).

4.3.4.3 BMSC Treatment of Stroke Increases Neurogenesis

Neurogenesis is the generation of neurons from neural stem and progenitor cells primarily occurring in the sub ventricular zone (SVZ) and sub granular zone (SGZ) of the hippocampus. Post ischemic neurogenesis replaces damaged neurons and restores neuronal communication. Cell apoptosis and cell proliferation are tightly regulated and maintain the progenitor cell population (Shehadah et al. 2010). However, in a rat stroke model it was observed that following an ischemic attack, neurogenesis is increased and the greatly expanded neural progenitor cells migrate to the ischemic boundary regions and some differentiate into mature neurons (Parent et al. 2002). Based on an animal model of stroke, it has been reported that 48 h after stroke, the neural progenitor cell expansion in the SVZ is increased by the following modifications in the cell cycle (Zhang et al. 2006b). The cell cycle length is shortened from 19 h in non stroke cells to 15.3 h in stroke cells by reducing the duration of the G1 phase. A larger number of SVZ cells from the stroke group thus, re-enter the cell cycle after mitosis. Hence, the neurorestorative sequence is tightly controlled, starting with neuronal progenitor cell proliferation leading up to their differentiation into mature neurons. Thus, BMSC treatment of stroke not only regulates vascular and white matter remodeling, but also increases neurogenesis (Chen et al. 2003a; Zhang et al. 2004; Li et al. 2008; Yoo et al. 2008; Bao et al. 2011). Increased neurogenesis was observed associated with increased cellular proliferation in the SGZ and SVZ, and migration of NPCs to the IBZ and their differentiation into mature neurons. BMSC therapy increases neurogenesis, and migration and differentiation of neural cells, and also decreases apoptosis, as well as promotes neurological recovery (Yoo et al. 2008; Bao et al. 2011).

4.3.4.4 BMSC Treatment of Stroke Increases Synaptogenesis

The process of formation of new synapses is called synaptogenesis. Synaptic plasticity involves strengthening of synapses, via which, neurons communicate, and remodeling of axons, dendrites, etc. which in concert, improve brain functionality. Neurons communicate with other neurons and cells via the synapse by passing neurotransmitters across the synaptic cleft or communicating through electrical signals across gap junctions. Significant increase in synaptogenesis and improved functional outcome have been reported using several cell-based therapies, including intravenous or intracarotid artery BMSC treatment of stroke (Shen et al. 2006; Zhang et al. 2006a; Hayase et al. 2009; Ding et al. 2011, 2013; Gutierrez-Fernandez et al. 2013). This enhanced synaptogenesis was correlated with improvement in functional recovery (Gutierrez-Fernandez et al. 2013). Enhancing synaptogenesis improves neuronal communication and is an important target for neurorestorative therapy. The mechanism of BMSC treatment induced synaptogenesis may be related with BMSC secretion or stimulation of endogenous sonic hedgehog (Shh), endogenous tPA, NGF and Noggin expression (Shen et al. 2006; Ding et al. 2011; Shen et al. 2011; Ding et al. 2013).

4.3.5 Preconditioning of BMSCs and Treatment of Stroke

Hypoxia preconditioning is a strategy that enhances cell survival and apoptotic tolerance of neural progenitors and BMSCs in the ischemic brain (Theus et al. 2008; Francis and Wei 2010; Wei et al. 2012). In adult rats, transplantation of BMSCs pre-treated with sub-lethal hypoxia (0.5 % oxygen) at 24 h after MCAo, had several benefits including increased angiogenesis and neurogenesis, decreased inflammation by inhibition of pro-inflammatory cytokines, decreased microglia activity in the brain and enhanced trophic factor production, all of which reinforced the benefits of BMSC therapy (Wei et al. 2012). Rapid and efficient functional recovery derived from similar beneficial effects and neuronal differentiation of neural progenitors from embryonic stem cells, resulted when hypoxia pre-conditioned embryonic stem cells were administered to ischemic rats (Theus et al. 2008). Treatment of stroke with BMSCs derived from stroke rats (Isch-BMSCs) also exhibited better functional outcome compared with BMSCs derived from normal rats (nor-BMSCs) (Zacharek et al. 2010). The enhanced beneficial effects may be related with the Isch-BMSC increased expression of Ang1/Tie2, VEGF/Flk1, bFGF, and GDNF expression and enhances angiogenesis, arterial density, and axonal regeneration compared with Nor-BMSC (Zacharek et al. 2010).

4.3.6 Adverse Effect of BMSC Treatment in Diabetes Population

The benefits of BMSC therapy in stroke have been well established, but when BMSC therapy was extended to diabetic subjects, adverse effects were reported (Chen et al. 2011b). When type 1 diabetic rats were subjected to stroke using the 2 h transient middle cerebral artery occlusion model and treated with BMSCs at 24 h post MCAo, no functional benefit was observed On the contrary, adverse reactions, including increased mortality, elevated BBB leakage, hemorrhage and vascular damage were observed. Elevated macrophage expression (indicated by ED1 marker) and angiogenin expression were associated with these detrimental effects. Clearly a treatment that works well in normal subjects did not translate to diabetic stroke subjects. In addition, BMSC treatment in T1DM stroke rats also increases arteriosclerosis-like vascular changes (Chen et al. 2011b), which may increase the likelihood of secondary stroke. Therefore, caution should be exercised while employing BMSCs treatment in diabetic stroke patients and the long term vascular changes should be investigated.

4.3.7 Combination BMSCs with Pharmacological Therapy

Another promising cell therapy strategy is the use of combination therapy. The strength of combination therapy lies in its ability to additively combine the positive effects of multiple treatments and at the same time overcome any adverse effects of one treatment. Combination therapy of Niaspan, an extended release formulation of Niacin, with BMSC not only significantly decreased adverse effects like BBB leakage, vascular damage, Angiogenin, matrix metalloproteinase 9 (MMP9) and ED1 expression, but also improved white matter remodeling and synaptic plasticity that are beneficial to recovery (Yan et al. 2013; Ye et al. 2013). Following 2 h MCAo, the animals were treated with BMSCs at 24 h and daily for 14 days orally with Niaspan. Niaspan is effective in reducing neurological deficits post stroke by promoting axonal remodeling, angiogenesis, and arteriogenesis (Chen et al. 2007, 2009; Cui et al. 2010; Yan et al. 2012). Another pharmacological drug that has been used in combination with BMSC is Sodium Ferulate (Zhao et al. 2013). This combination therapy significantly decreased infarction size and promoted exogenous BMSCs differentiation into neural-and astrocytic-like cells, as well as enhanced repair capacity of brain parenchymal cells by promoting glucose metabolism and endogenous neurogenesis after stroke. Combination treatment of stroke using BMSCs with statins also enhances BMSC migration into the ischemic brain, amplifies arteriogenesis and angiogenesis, and improves functional outcome after stroke (Cui et al. 2009; Pirzad Jahromi et al. 2012).

4.3.8 Role of microRNA (miRNA) in BMSC Therapy

microRNA's (miRNAs) are small non-coding RNA molecules that play a key role gene expression regulation at the post-transcriptional level. They act as molecular master switches and can regulate the translation of hundreds of genes. Acting either alone or together with other miRNAs, they regulate expression of many genes, biological pathways, and complex biological networks within cells. They are gaining importance as "molecular rheostats" capable of facilitating tissue repair because of their ability to fine-tune and switch regulatory circuits mediating tissue repair via hypoxia-response, inflammation and angiogenesis (Sen 2011). To therapeutically target miRNAs, challenges like mode of delivery and potential off-target effects have to be overcome. Cell-based therapy appears to be an effective strategy of manipulating miRNA expression (Juranek et al. 2013).

BMSCs administered intravenously secrete microvesicles that contain and transport enriched miRNA/ mRNA/ proteins/ etc. depending on parent phenotype, from origin to target cells (Collino et al. 2011; Katsuda et al. 2013). These miRNAs are in turn capable of stimulating endogenous brain cells to express and release miRNA, resulting in enhanced neurorestorative effects post stroke (Juranek et al. 2013). In post stroke BMSC therapy, miRNAs can regulate neurite outgrowth (Juranek et al. 2013) and control neuronal progenitor cell proliferation and differentiation (Lim et al. 2010; Wang et al. 2013). miR-34a that is down regulated during BMSC neuronal differentiation, mediates neuronal precursor motility, that is vital for homing of stem cells to target tissue (Chang et al. 2011). Other miRNAs important for BMSC differentiation include miR-96, miR-124 and miR-199a, that are differentially expressed during osteogenic, adipogenic, and chondrogenic induction of hBMSCs (Laine et al. 2012). The involvement of miRNAs in regulating neurogenesis is important. Apart from mediating the trans-differentiation of BMSCs into functional neurons, they also help to restore lost and damaged neurons in the injured brain (Lim et al. 2010).

Microvesicles enable two way communications between BMSCs and injured cells via miRNA exchange; aiding neuronal differentiation and induction of regenerative pathways in injured cells. Manipulation of these extracellular microvesicles to mediate intracellular communication can be used to yield therapeutically beneficial effects. These microvesicles secreted by BMSCs while in resting or active state, can be manipulated to deliver miRNAs specifically to injured cells to enhance their recovery (Wang et al. 2013). For example, a novel miRNA treatment strategy for malignant glioma uses miR146b (known to have anti tumor properties). BMSCs transfected with miR -146b secrete exosomes that are employed as a delivery vehicle to decrease tumor volume (Katakowski et al. 2013). When BMSCs were exposed to ischemic rat brain tissue extracts derived from post MCAo animals, increased miR-133b expression was observed in the secreted exosomes (Juranek et al. 2013; Xin et al. 2013b). These exosomes can transfer miR133b from BMSCs to neurons and astrocytes which resulted in a significant increase in neurite branch number and total neurite length (Xin et al. 2012; Juranek et al. 2013).

Therapeutic efficiency of BMSC therapy may also be enhanced by manipulating the miRNAs produced by these BMSCs to mimic or antagonize desired microRNA actions. miR126 plays an important role in angiogenesis and is expressed in endothelial cells in blood vessels and capillaries (Nikolic et al. 2010). Since increasing angiogenesis is a desired effect in cell therapy, transplantation of BMSCs over-expressing miR126 was shown to improve therapeutic efficiency and cardiac function after myocardial ischemia by increasing the secretion of angiogenic factors, improving hypoxia resistance and activating the Notch signaling pathway ligand Delta-like-4 (Huang et al 2013). In diabetic subjects, chronic inflammatory responses interfering with wound healing has been associated with abnormal downregulation of miR 146a (Xu et al. 2012). Post-myocardial infarction treatment using BMSC showed an increase in miR 146a expression and decrease in inflammatory factor expression.

Involvement of miRNAs in a broad range of repair mechanisms, like stem cell differentiation, neovascularization, control of apoptosis, cardiac remodeling, etc. has been reported in myocardial infarction therapy using BMSCs (Wen et al. 2012). Functional recovery has been associated with the direct transfer of pro-survival miR-210 from BMSCs to host cardiomyocytes in the ischemic heart. In spite of improved cardiac function and structural remodeling, BMSC therapy in myocardial infarction is curtailed by the low survival rates of transplanted BMSCs in the ischemic myocardium (Wen et al. 2012). Therefore, it has been suggested to use miRNAs as novel regulators and therapeutic modulators of individual cardiovascular miRNAs of MSCs to improve therapeutic efficiency (Wen et al. 2012). In stroke treatment, exosomes derived from BMSCs improve functional recovery, enhance angiogenesis, neurogenesis and neurite remodeling, and hold promise as a novel treatment strategy for stroke (Xin et al. 2013a).

4.4 An Update on Clinical Trials of BMSC Treatment of Stroke

BMSCs therapy has been widely tested in clinical trials for cardiovascular, immune related and neurological disorders. With approximately 79 registered clinical trial sites for the evaluation of BMSC therapy world wide (Malgieri et al. 2010), cell therapy using BMSC is considered safe with encouraging therapeutic results (Bang et al. 2005; Sykova et al. 2006).

The results of clinical trials for stroke treatment using autologous BMSCs administered intravenously indicated promising results with respect to safety, feasibility and functional recovery (Bang et al. 2005; Suarez-Monteagudo et al. 2009; Lee et al. 2010). In South Korea, a group of stroke patients with cerebral infarcts in the middle cerebral arterial region were treated intravenously with autologous BMSCs (Bang et al. 2005). The study included serial evaluations and comparisons with a control group of patients who did not receive BMSC therapy over a 1 year period. The results assure safety of the treatment and suggest improvement in functional recovery. The follow-up long-term evaluation report presented higher survival rates in the BMSC treated group of patients when compared to the non-BMSC treated control group. No significant side effects were reported further assuring the safety of autologous BMSCs therapy delivered using the intravenous route (Lee et al. 2010). In the United States, a Phase I/II clinical trial to test the safety and efficacy of allogenic BMSCs to treat ischemic stroke has been initiated by researchers at University of California, San Diego with other collaborators (NCT01297413). In Spain, a Phase I/II clinical trial tested the feasibility, safety and improvement of neurological outcome in middle cerebral arterial stroke patients treated intra-arterially with autologous bone marrow mononuclear cells on day 5 and 9 post stroke (Moniche et al. 2012). Over the 6 month follow up time frame, no adverse effects like deaths/ tumor formation/ stroke recurrence were reported, apart from two isolated partial seizures at 3 months.

Conclusions and Acknowledgements

The major physiological mechanisms for BMSC treatment induced neurorestoration post stroke, include angiogenesis, arteriogenesis, neurogenesis and white matter remodeling. BMSC therapy also induces the interaction and coupling between these various neurorestorative events. The role of microRNAs in modulating biological pathways is also of prime interest to understand mechanisms of neurorestorative effects and to improve BMSC therapies for stroke. An update on clinical trials for BMSC therapy has also been presented.

Acknowledgements Research reported in this publication was supported by National Institute on Aging under award number RO1AG031811 (JC), RO1AG 037506 (MC) and R41NS080329 (JC), and National Institute of Neurological Disorders and Stroke (NINDS) under award number RO1NS083078 (JC)

References

Andrews EM, Tsai SY, Johnson SC, Farrer JR, Wagner JP, Kopen GC, Kartje GL (2008) Human adult bone marrow-derived somatic cell therapy results in functional recovery and axonal plasticity following stroke in the rat. Exp Neurol 211:588–592

Arenillas JF, Sobrino T, Castillo J, Davalos A (2007) The role of angiogenesis in damage and recovery from ischemic stroke. Curr Treat Options Cardiovasc Med 9:205–212

Ay H, Ay I, Koroshetz WJ, Finklestein SP (1999) Potential usefulness of basic fibroblast growth factor as a treatment for stroke. Cerebrovasc Dis 9:131–135

Ay I, Sugimori H, Finklestein SP (2001) Intravenous basic fibroblast growth factor (bFGF) decreases DNA fragmentation and prevents downregulation of Bcl-2 expression in the ischemic brain following middle cerebral artery occlusion in rats. Brain Res Mol Brain Res 87:71–80

Back SA, Han BH, Luo NL, Chricton CA, Xanthoudakis S, Tam J, Arvin KL, Holtzman DM (2002) Selective vulnerability of late oligodendrocyte progenitors to hypoxia-ischemia. J Neurosci 22:455–463

Bang OY, Lee JS, Lee PH, Lee G (2005) Autologous mesenchymal stem cell transplantation in stroke patients. Ann Neurol 57:874–882

Bao X, Wei J, Feng M, Lu S, Li G, Dou W, Ma W, Ma S, An Y, Qin C, Zhao RC, Wang R (2011) Transplantation of human bone marrow-derived mesenchymal stem cells promotes behavioral recovery and endogenous neurogenesis after cerebral ischemia in rats. Brain Res 7:103–113

Buschmann I, Schaper W (2000) The pathophysiology of the collateral circulation (arteriogenesis). J Pathol 190:338–342

Carmeliet P (2000) Mechanisms of angiogenesis and arteriogenesis. Nat Med 6:389–395

Chang SJ, Weng SL, Hsieh JY, Wang TY, Chang MD, Wang HW (2011) MicroRNA-34a modulates genes involved in cellular motility and oxidative phosphorylation in neural precursors derived from human umbilical cord mesenchymal stem cells. BMC Med Genomics 4:65

Chen J, Li Y, Wang L, Zhang Z, Lu D, Lu M, Chopp M (2001a) Therapeutic benefit of intravenous administration of bone marrow stromal cells after cerebral ischemia in rats. Stroke 32:1005–1011

Chen J, Sanberg PR, Li Y, Wang L, Lu M, Willing AE, Sanchez-Ramos J, Chopp M (2001b) Intravenous administration of human umbilical cord blood reduces behavioral deficits after stroke in rats. Stroke 32:2682–2688

Chen X, Katakowski M, Li Y, Lu D, Wang L, Zhang L, Chen J, Xu Y, Gautam S, Mahmood A, Chopp M (2002a) Human bone marrow stromal cell cultures conditioned by traumatic brain tissue extracts: growth factor production. J Neurosci Res 69:687–691

Chen X, Li Y, Wang L, Katakowski M, Zhang L, Chen J, Xu Y, Gautam SC, Chopp M (2002b) Ischemic rat brain extracts induce human marrow stromal cell growth factor production. Neuropathology 22:275–279

Chen J, Li Y, Katakowski M, Chen X, Wang L, Lu D, Lu M, Gautam SC, Chopp M (2003a) Intravenous bone marrow stromal cell therapy reduces apoptosis and promotes endogenous cell proliferation after stroke in female rat. J Neurosci Res 73:778–786

Chen J, Zhang ZG, Li Y, Wang L, Xu YX, Gautam SC, Lu M, Zhu Z, Chopp M (2003b) Intravenous administration of human bone marrow stromal cells induces angiogenesis in the ischemic boundary zone after stroke in rats. Circ Res 92:692–699

Chen J, Cui X, Zacharek A, Jiang H, Roberts C, Zhang C, Lu M, Kapke A, Feldkamp CS, Chopp M (2007) Niaspan increases angiogenesis and improves functional recovery after stroke. Ann Neurol 62:49–58

Chen JR, Cheng GY, Sheu CC, Tseng GF, Wang TJ, Huang YS (2008) Transplanted bone marrow stromal cells migrate, differentiate and improve motor function in rats with experimentally induced cerebral stroke. J Anat 213:249–258

Chen J, Cui X, Zacharek A, Ding GL, Shehadah A, Jiang Q, Lu M, Chopp M (2009) Niaspan treatment increases tumor necrosis factor-alpha-converting enzyme and promotes arteriogenesis after stroke. J Cereb Blood Flow Metab 29:911–920

Chen J, Cui X, Zacharek A, Cui Y, Roberts C, Chopp M (2011a) White matter damage and the effect of matrix metalloproteinases in type 2 diabetic mice after stroke. Stroke 42:445–452

Chen J, Ye X, Yan T, Zhang C, Yang XP, Cui X, Cui Y, Zacharek A, Roberts C, Liu X, Dai X, Lu M, Chopp M (2011b) Adverse effects of bone marrow stromal cell treatment of stroke in diabetic rats. Stroke 42:3551–3558

Collino F, Bruno S, Deregibus MC, Tetta C, Camussi G (2011) Chapter fourteen–microRNAs and mesenchymal stem cells. In: Gerald L (ed) Vitamins & hormones, vol 87. Academic, Massachusetts, pp 291–320

Cui X, Chopp M, Zacharek A, Roberts C, Lu M, Savant-Bhonsale S, Chen J (2009) Chemokine, vascular and therapeutic effects of combination Simvastatin and BMSC treatment of stroke. Neurobiol Dis 36:35–41

Cui X, Chopp M, Zacharek A, Roberts C, Buller B, Ion M, Chen J (2010) Niacin treatment of stroke increases synaptic plasticity and axon growth in rats. Stroke 41:2044–2049

Ding J, Cheng Y, Gao S, Chen J (2011) Effects of nerve growth factor and Noggin-modified bone marrow stromal cells on stroke in rats. J Neurosci Res 89:222–230

Ding X, Li Y, Liu Z, Zhang J, Cui Y, Chen X, Chopp M (2013) The sonic hedgehog pathway mediates brain plasticity and subsequent functional recovery after bone marrow stromal cell treatment of stroke in mice. J Cereb Blood Flow Metab 33:1015–1024

Dominici M, Le Blanc K, Mueller I, Slaper-Cortenbach I, Marini F, Krause D, Deans R, Keating A, Prockop D, Horwitz E (2006) Minimal criteria for defining multipotent mesenchymal stromal cells. The International Society for Cellular Therapy position statement. Cytotherapy 8:315–317

Dor Y, Keshet E (1997) Ischemia-driven angiogenesis. Trends Cardiovasc Med 7:289–294

Duffy GP, Ahsan T, O'Brien T, Barry F, Nerem RM (2009) Bone marrow-derived mesenchymal stem cells promote angiogenic processes in a time- and dose-dependent manner in vitro. Tissue Eng Part A 15:2459–2470

Erdo F, Buschmann IR (2007) Arteriogenesis: a new strategy of therapeutic intervention in chronic arterial disorders. Cellular mechanism and experimental models. Orv Hetil 148:633–642

Feng Y, Wang Y, Pfister F, Hillebrands JL, Deutsch U, Hammes HP (2009) Decreased hypoxia-induced neovascularization in angiopoietin-2 heterozygous knockout mouse through reduced MMP activity. Cell Physiol Biochem 23:277–284

Folkman J, D'Amore PA (1996) Blood vessel formation: what is its molecular basis? Cell 87:1153–1155

Francis KR, Wei L (2010) Human embryonic stem cell neural differentiation and enhanced cell survival promoted by hypoxic preconditioning. Cell Death Dis 1:22

Greenberg DA (1998) Angiogenesis and stroke. Drug News Perspect 11:265–270

Gutierrez-Fernandez M, Rodriguez-Frutos B, Ramos-Cejudo J, Teresa Vallejo-Cremades M, Fuentes B, Cerdan S, Diez-Tejedor E (2013) Effects of intravenous administration of allogenic bone marrow- and adipose tissue-derived mesenchymal stem cells on functional recovery and brain repair markers in experimental ischemic stroke. Stem Cell Res Ther 4:11

Hardy SA, Maltman DJ, Przyborski SA (2008) Mesenchymal stem cells as mediators of neural differentiation. Curr Stem Cell Res Ther 3:43–52

Hayase M, Kitada M, Wakao S, Itokazu Y, Nozaki K, Hashimoto N, Takagi Y, Dezawa M (2009) Committed neural progenitor cells derived from genetically modified bone marrow stromal cells ameliorate deficits in a rat model of stroke. J Cereb Blood Flow Metab 29:1409–1420

Hayashi T, Abe K, Itoyama Y (1998) Reduction of ischemic damage by application of vascular endothelial growth factor in rat brain after transient ischemia. J Cereb Blood Flow Metab 18:887–895

Hoffmann J, Glassford AJ, Doyle TC, Robbins RC, Schrepfer S, Pelletier MP (2010) Angiogenic effects despite limited cell survival of bone marrow-derived mesenchymal stem cells under ischemia. Thorac Cardiovasc Surg 58:136–142

Huang F, Zhu X, Hu XQ, Fang ZF, Tang L, Lu XL, Zhou SH (2013) Mesenchymal stem cells modified with miR-126 release angiogenic factors and activate Notch ligand Delta-like-4, enhancing ischemic angiogenesis and cell survival. Int J Mol Med 31:484–492

Juranek JK, Geddis MS, Song F, Zhang J, Garcia J, Rosario R, Yan SF, Brannagan TH, Schmidt AM (2013) RAGE deficiency improves postinjury sciatic nerve regeneration in type 1 diabetic mice. Diabetes 62:931–943

Katakowski M, Buller B, Zheng X, Lu Y, Rogers T, Osobamiro O, Shu W, Jiang F, Chopp M (2013) Exosomes from marrow stromal cells expressing miR-146b inhibit glioma growth. Cancer Lett 335:201–204

Katsuda T, Kosaka N, Takeshita F, Ochiya T (2013) The therapeutic potential of mesenchymal stem cell-derived extracellular vesicles. Proteomics 13:1637–1653

Kinnaird T, Stabile E, Burnett MS, Lee CW, Barr S, Fuchs S, Epstein SE (2004) Marrow-derived stromal cells express genes encoding a broad spectrum of arteriogenic cytokines and promote in vitro and in vivo arteriogenesis through paracrine mechanisms. Circ Res 94:678–685

Kobayashi T, Ahlenius H, Thored P, Kobayashi R, Kokaia Z, Lindvall O (2006) Intracerebral infusion of glial cell line-derived neurotrophic factor promotes striatal neurogenesis after stroke in adult rats. Stroke 37:2361–2367

Krebsbach PH, Kuznetsov SA, Bianco P, Robey PG (1999) Bone marrow stromal cells: characterization and clinical application. Crit Rev Oral Biol Med 10:165–181

Laine SK, Alm JJ, Virtanen SP, Aro HT, Laitala-Leinonen TK (2012) MicroRNAs miR-96, miR-124, and miR-199a regulate gene expression in human bone marrow-derived mesenchymal stem cells. J Cell Biochem 113:2687–2695

Lee TH, Kato H, Chen ST, Kogure K, Itoyama Y (1998) Expression of nerve growth factor and trkA after transient focal cerebral ischemia in rats. Stroke 29:1687–1696

Lee J, Kuroda S, Shichinohe H, Ikeda J, Seki T, Hida K, Tada M, Sawada K, Iwasaki Y (2003) Migration and differentiation of nuclear fluorescence-labeled bone marrow stromal cells after transplantation into cerebral infarct and spinal cord injury in mice. Neuropathology 23:169–180

Lee JS, Hong JM, Moon GJ, Lee PH, Ahn YH, Bang OY (2010) A long-term follow-up study of intravenous autologous mesenchymal stem cell transplantation in patients with ischemic stroke. Stem Cells 28:1099–1106

Li Y, Chen J, Chen XG, Wang L, Gautam SC, Xu YX, Katakowski M, Zhang LJ, Lu M, Janakiraman N, Chopp M (2002) Human marrow stromal cell therapy for stroke in rat: neurotrophins and functional recovery. Neurology 59:514–523

Li Y, McIntosh K, Chen J, Zhang C, Gao Q, Borneman J, Raginski K, Mitchell J, Shen L, Zhang J, Lu D, Chopp M (2006) Allogeneic bone marrow stromal cells promote glial-axonal remodeling without immunologic sensitization after stroke in rats. Exp Neurol 198:313–325

Li WY, Choi YJ, Lee PH, Huh K, Kang YM, Kim HS, Ahn YH, Lee G, Bang OY (2008) Mesenchymal stem cells for ischemic stroke: changes in effects after ex vivo culturing. Cell Transplant 17:1045–1059

Lim PK, Patel SA, Gregory LA, Rameshwar P (2010) Neurogenesis: role for microRNAs and mesenchymal stem cells in pathological states. Curr Med Chem 17:2159–2167

Liu Z, Li Y, Zhang X, Savant-Bhonsale S, Chopp M (2008) Contralesional axonal remodeling of the corticospinal system in adult rats after stroke and bone marrow stromal cell treatment. Stroke 39:2571–2577

Liu Z, Zhang RL, Li Y, Cui Y, Chopp M (2009) Remodeling of the corticospinal innervation and spontaneous behavioral recovery after ischemic stroke in adult mice. Stroke 40:2546–2551

Liu Z, Li Y, Zhang ZG, Cui X, Cui Y, Lu M, Savant-Bhonsale S, Chopp M (2010) Bone marrow stromal cells enhance inter- and intracortical axonal connections after ischemic stroke in adult rats. J Cereb Blood Flow Metab 30:1288–1295

Liu X, Ye R, Yan T, Yu SP, Wei L, Xu G, Fan X, Jiang Y, Stetler RA, Liu G, Chen J (2014) Cell based therapies for ischemic stroke: from basic science to bedside. Prog Neurobiol 115C:92–115

Malgieri A, Kantzari E, Patrizi MP, Gambardella S (2010) Bone marrow and umbilical cord blood human mesenchymal stem cells: state of the art. Int J Clin Exp Med 3:248–269

Moniche F, Gonzalez A, Gonzalez-Marcos JR, Carmona M, Pinero P, Espigado I, Garcia-Solis D, Cayuela A, Montaner J, Boada C, Rosell A, Jimenez MD, Mayol A, Gil-Peralta A (2012) Intra-arterial bone marrow mononuclear cells in ischemic stroke: a pilot clinical trial. Stroke 43:2242–2244

Navarro-Sobrino M, Rosell A, Hernandez-Guillamon M, Penalba A, Boada C, Domingues-Montanari S, Ribo M, Alvarez-Sabin J, Montaner J (2011) A large screening of angiogenesis biomarkers and their association with neurological outcome after ischemic stroke. Atherosclerosis 216:205–211

Nikolic I, Plate KH, Schmidt MH (2010) EGFL7 meets miRNA-126: an angiogenesis alliance. J Angiogene Res 2:9

Ohab JJ, Fleming S, Blesch A, Carmichael ST (2006) A neurovascular niche for neurogenesis after stroke. J Neurosci 26:13007–13016

Papadopoulos CM, Tsai SY, Cheatwood JL, Bollnow MR, Kolb BE, Schwab ME, Kartje GL (2006) Dendritic plasticity in the adult rat following middle cerebral artery occlusion and Nogo-A neutralization. Cereb Cortex 16:529–536

Parent JM, Vexler ZS, Gong C, Derugin N, Ferriero DM (2002) Rat forebrain neurogenesis and striatal neuron replacement after focal stroke. Ann Neurol 52:802–813

Pham LD, Hayakawa K, Seo JH, Nguyen MN, Som AT, Lee BJ, Guo S, Kim KW, Lo EH, Arai K (2012) Crosstalk between oligodendrocytes and cerebral endothelium contributes to vascular remodeling after white matter injury. Glia 60:875–881

Pirzad Jahromi G, Seidi S, Sadr SS, Shabanzadeh AP, Keshavarz M, Kaka GR, Hosseini SK, Sohanaki H, Charish J (2012) Therapeutic effects of a combinatorial treatment of simvastatin and bone marrow stromal cells on experimental embolic stroke. Basic Clin Pharmacol Toxicol 110:487–493

Plate KH (1999) Mechanisms of angiogenesis in the brain. J Neuropathol Exp Neurol 58:313–320

Ploughman M, Windle V, MacLellan CL, White N, Dore JJ, Corbett D (2009) Brain-derived neurotrophic factor contributes to recovery of skilled reaching after focal ischemia in rats. Stroke 40:1490–1495

Pratt PF, Medhora M, Harder DR (2004) Mechanisms regulating cerebral blood flow as therapeutic targets. Curr Opin Investig Drugs 5:952–956

Risau W (1997) Mechanisms of angiogenesis. Nature 386:671–674

Sanchez-Ramos J, Song S, Cardozo-Pelaez F, Hazzi C, Stedeford T, Willing A, Freeman TB, Saporta S, Janssen W, Patel N, Cooper DR, Sanberg PR (2000) Adult bone marrow stromal cells differentiate into neural cells in vitro. Exp Neurol 164:247–256

Schaper W, Buschmann I (1999) Arteriogenesis, the good and bad of it. Eur Heart J 20:1297–1299

Scholz D, Cai WJ, Schaper W (2001) Arteriogenesis, a new concept of vascular adaptation in occlusive disease. Angiogenesis 4:247–257

Sen CK (2011) MicroRNAs as new maestro conducting the expanding symphony orchestra of regenerative and reparative medicine. Physiol Genomics 43:517–520

Shehadah A, Chen J, Zacharek A, Cui Y, Ion M, Roberts C, Kapke A, Chopp M (2010) Niaspan treatment induces neuroprotection after stroke. Neurobiol Dis 40:277–283

Shen LH, Li Y, Chen J, Zhang J, Vanguri P, Borneman J, Chopp M (2006) Intracarotid transplantation of bone marrow stromal cells increases axon-myelin remodeling after stroke. Neuroscience 137:393–399

Shen LH, Li Y, Chen J, Zacharek A, Gao Q, Kapke A, Lu M, Raginski K, Vanguri P, Smith A, Chopp M (2007) Therapeutic benefit of bone marrow stromal cells administered 1 month after stroke. J Cereb Blood Flow Metab 27:6–13

Shen LH, Xin H, Li Y, Zhang RL, Cui Y, Zhang L, Lu M, Zhang ZG, Chopp M (2011) Endogenous tissue plasminogen activator mediates bone marrow stromal cell-induced neurite remodeling after stroke in mice. Stroke 42:459–464

Shichinohe H, Kuroda S, Lee JB, Nishimura G, Yano S, Seki T, Ikeda J, Tamura M, Iwasaki Y (2004) In vivo tracking of bone marrow stromal cells transplanted into mice cerebral infarct by fluorescence optical imaging. Brain Res Brain Res Protoc 13:166–175

Shichinohe H, Kuroda S, Maruichi K, Osanai T, Sugiyama T, Chiba Y, Yamaguchi A, Iwasaki Y (2010) Bone marrow stromal cells and bone marrow-derived mononuclear cells: which are suitable as cell source of transplantation for mice infarct brain? Neuropathology 30:113–122

Sondell M, Lundborg G, Kanje M (1999) Vascular endothelial growth factor has neurotrophic activity and stimulates axonal outgrowth, enhancing cell survival and Schwann cell proliferation in the peripheral nervous system. J Neurosci 19:5731–5740

Suarez-Monteagudo C, Hernandez-Ramirez P, Alvarez-Gonzalez L, Garcia-Maeso I, de la Cuetara-Bernal K, Castillo-Diaz L, Bringas-Vega ML, Martinez-Aching G, Morales-Chacon LM, Baez-Martin MM, Sanchez-Catasus C, Carballo-Barreda M, Rodriguez-Rojas R, Gomez-Fernandez L, Alberti-Amador E, Macias-Abraham C, Balea ED, Rosales LC, Del Valle Perez L, Ferrer BB, Gonzalez RM, Bergado JA (2009) Autologous bone marrow stem cell neurotransplantation in stroke patients. An open study. Restor Neurol Neurosci 27:151–161

Sykova E, Jendelova P, Urdzikova L, Lesny P, Hejcl A (2006) Bone marrow stem cells and polymer hydrogels-two strategies for spinal cord injury repair. Cell Mol Neurobiol 26:1113–1129

Theus MH, Wei L, Cui L, Francis K, Hu X, Keogh C, Yu SP (2008) In vitro hypoxic preconditioning of embryonic stem cells as a strategy of promoting cell survival and functional benefits after transplantation into the ischemic rat brain. Exp Neurol 210:656–670

Tohill M, Mantovani C, Wiberg M, Terenghi G (2004) Rat bone marrow mesenchymal stem cells express glial markers and stimulate nerve regeneration. Neurosci Lett 362:200–203

van Royen N, Piek JJ, Buschmann I, Hoefer I, Voskuil M, Schaper W (2001) Stimulation of arteriogenesis; a new concept for the treatment of arterial occlusive disease. Cardiovasc Res 49:543–553

van Velthoven CT, Kavelaars A, Heijnen CJ (2012) Mesenchymal stem cells as a treatment for neonatal ischemic brain damage. Pediatr Res 71:474–481

Wakabayashi K, Nagai A, Sheikh AM, Shiota Y, Narantuya D, Watanabe T, Masuda J, Kobayashi S, Kim SU, Yamaguchi S (2010) Transplantation of human mesenchymal stem cells promotes functional improvement and increased expression of neurotrophic factors in a rat focal cerebral ischemia model. J Neurosci Res 88:1017–1025

Walmsley AR, Mir AK (2007) Targeting the Nogo-A signalling pathway to promote recovery following acute CNS injury. Curr Pharm Des 13:2470–2484

Wang XQ, Zhu XJ, Zou P (2013) Research progress of mesenchymal stem cell-derived microvesicle. Zhongguo Shi Yan Xue Ye Xue Za Zhi 21:227–230

Watt SM, Gullo F, van der Garde M, Markeson D, Camicia R, Khoo CP, Zwaginga JJ (2013) The angiogenic properties of mesenchymal stem/stromal cells and their therapeutic potential. Br Med Bull 108:25–53

Weaver JD, Stetten G, Littlefield JW (1991) Partial trisomies in two spontaneously arising long-lived human keratinocyte lines. In vitro Cell Dev Biol 27A:670–675

Wei L, Fraser JL, Lu ZY, Hu X, Yu SP (2012) Transplantation of hypoxia preconditioned bone marrow mesenchymal stem cells enhances angiogenesis and neurogenesis after cerebral ischemia in rats. Neurobiol Dis 46:635–645

Wen Z, Zheng S, Zhou C, Yuan W, Wang J, Wang T (2012) Bone marrow mesenchymal stem cells for post-myocardial infarction cardiac repair: microRNAs as novel regulators. J Cell Mol Med 16:657–671

Weng JS, Liu N, Du HW, Chen RH, Zhang YX, Wang JH, Huang HP (2008) Effects of bone marrow-derived mesenchymal stem cells transplantation on recovery of neurological functions and expression of synaptophysin in focal cerebral infarction in rats. Xi Bao Yu Fen Zi Mian Yi Xue Za Zhi 24:34–37

Wu J, Sun Z, Sun HS, Weisel RD, Keating A, Li ZH, Feng ZP, Li RK (2008) Intravenously administered bone marrow cells migrate to damaged brain tissue and improve neural function in ischemic rats. Cell Transplant 16:993–1005

Xin H, Li Y, Buller B, Katakowski M, Zhang Y, Wang X, Shang X, Zhang ZG, Chopp M (2012) Exosome-mediated transfer of miR-133b from multipotent mesenchymal stromal cells to neural cells contributes to neurite outgrowth. Stem Cells 30:1556–1564

Xin H, Li Y, Cui Y, Yang JJ, Zhang ZG, Chopp M (2013a) Systemic administration of exosomes released from mesenchymal stromal cells promote functional recovery and neurovascular plasticity after stroke in rats. J Cereb Blood Flow Metab 33:1711–1715

Xin H, Li Y, Liu Z, Wang X, Shang X, Cui Y, Zhang ZG, Chopp M (2013b) miR-133b promotes neural plasticity and functional recovery after treatment of stroke with multipotent mesenchymal stromal cells in rats via transfer of exosome-enriched extracellular particles. Stem Cells 31:2737–2746

Xu J, Wu W, Zhang L, Dorset-Martin W, Morris MW, Mitchell ME, Liechty KW (2012) The role of microRNA-146a in the pathogenesis of the diabetic wound-healing impairment: correction with mesenchymal stem cell treatment. Diabetes 61:2906–2912

Yan BB, Wang LL, Qiao JF, Peng HS, Cao JY, Li HL, Jin LH (2005) Study of mechanism of differentiation of bone stromal stem cells into neurons in vitro. Xi Bao Yu Fen Zi Mian Yi Xue Za Zhi 21:301–304

Yan T, Chopp M, Ye X, Liu Z, Zacharek A, Cui Y, Roberts C, Buller B, Chen J (2012) Niaspan increases axonal remodeling after stroke in type 1 diabetes rats. Neurobiol Dis 46:157–164

Yan T, Ye X, Chopp M, Zacharek A, Ning R, Venkat P, Roberts C, Lu M, Chen J (2013) Niaspan attenuates the adverse effects of bone marrow stromal cell treatment of stroke in type one diabetic rats. PloS One 8:e81199

Ye X, Yan T, Chopp M, Zacharek A, Ning R, Venkat P, Roberts C, Chen J (2013) Combination BMSC and Niaspan treatment of stroke enhances white matter remodeling and synaptic protein expression in diabetic rats. Int J Mol Sci 14:22221–22232

Yoo SW, Kim SS, Lee SY, Lee HS, Kim HS, Lee YD, Suh-Kim H (2008) Mesenchymal stem cells promote proliferation of endogenous neural stem cells and survival of newborn cells in a rat stroke model. Exp Mol Med 40:387–397

Yuan M, Wen SJ, Yang CX, Pang YG, Gao XQ, Liu XQ, Huang L, Yuan QL (2013) Transplantation of neural stem cells overexpressing glial cell line-derived neurotrophic factor enhances Akt and Erk1/2 signaling and neurogenesis in rats after stroke. Chin Med J 126:1302–1309

Zacharek A, Chen J, Cui X, Li A, Li Y, Roberts C, Feng Y, Gao Q, Chopp M (2007) Angiopoietin1/Tie2 and VEGF/Flk1 induced by MSC treatment amplifies angiogenesis and vascular stabilization after stroke. J Cereb Blood Flow Metab 27:1684–1691

Zacharek A, Shehadah A, Chen J, Cui X, Roberts C, Lu M, Chopp M (2010) Comparison of bone marrow stromal cells derived from stroke and normal rats for stroke treatment. Stroke 41:524–530

Zhang J, Li Y, Chen J, Yang M, Katakowski M, Lu M, Chopp M (2004) Expression of insulin-like growth factor 1 and receptor in ischemic rats treated with human marrow stromal cells. Brain Res 1030:19–27

Zhang C, Li Y, Chen J, Gao Q, Zacharek A, Kapke A, Chopp M (2006a) Bone marrow stromal cells upregulate expression of bone morphogenetic proteins 2 and 4, gap junction protein connexin-43 and synaptophysin after stroke in rats. Neuroscience 141:687–695

Zhang RL, Zhang ZG, Lu M, Wang Y, Yang JJ, Chopp M (2006b) Reduction of the cell cycle length by decreasing G1 phase and cell cycle reentry expand neuronal progenitor cells in the subventricular zone of adult rat after stroke. J Cereb Blood Flow Metab 26:857–863

Zhang J, Li Y, Zhang ZG, Lu M, Borneman J, Buller B, Savant-Bhonsale S, Elias SB, Chopp M (2009) Bone marrow stromal cells increase oligodendrogenesis after stroke. J Cereb Blood Flow Metab 29:1166–1174

Zhang JH, Badaut J, Tang J, Obenaus A, Hartman R, Pearce WJ (2012) The vascular neural network–a new paradigm in stroke pathophysiology. Nat Rev Neurol 8:711–716
Zhang R, Chopp M, Zhang ZG (2013) Oligodendrogenesis after cerebral ischemia. Front Cell Neurosci 29,7:201 (eCollection)
Zhao LR, Duan WM, Reyes M, Keene CD, Verfaillie CM, Low WC (2002) Human bone marrow stem cells exhibit neural phenotypes and ameliorate neurological deficits after grafting into the ischemic brain of rats. Exp Neurol 174:11–20
Zhao Y, Lai W, Xu Y, Li L, Chen Z, Wu W (2013) Exogenous and endogenous therapeutic effects of combination Sodium Ferulate and bone marrow stromal cells (BMSCs) treatment enhance neurogenesis after rat focal cerebral ischemia. Metab Brain Dis 28:655–666
Zhu CJ, Dong JX, Li J, Zhang MJ, Wang LP, Luo L (2011) Preliminary study on the mechanism of acupoint injection of bone marrow mesenchymal stem cells in improving blood flow in the rat of hind limb ischemia. J Tradit Chin Med 31:241–245

Chapter 5
Cord Blood as a Treatment for Stroke

Alison E. Willing and E. A. Foran

Contents

A. E. Willing (✉) · E. A. Foran
Center for Excellence in Aging and Brain Repair, Department of Neurosurgery and Brain Repair, Morsani College of Medicine, University of South Florida,
12901 Bruce B. Downs Blvd MDC 78, Tampa, FL 33612, USA
e-mail: awilling@health.usf.edu

L.-R. Zhao, J. H. Zhang (eds.), *Cellular Therapy for Stroke and CNS Injuries*,
Springer Series in Translational Stroke Research, DOI 10.1007/978-3-319-11481-1_5

Abstract Over the last few decades, there has been an explosion in stem cell research. The investigation of umbilical cord blood (UCB) cells as a treatment for stroke is even more recent. Ease of collection and the ability to maintain their stem cell properties post-cryopreservation made these cells very attractive candidates for treatment development initially. UCB cells have many advantages including a wide variety of cell types present, including hematopoietic stem cells, mesenchymal stem cells, endothelial progenitor cells, lymphocytes, and monocytes, which enhances their ability to modulate multiple targets impacted by neurodegenerative processes. Although the precise mechanisms of action are still being researched, UCB cells have been shown to benefit functional recovery and also reduce infarct size post-stroke. They have also demonstrated an ability to provide these benefits when administered peripherally and within 24–48 h post-stroke, which immensely expands the current treatment window of 3–4 h for tissue plasminogen activator. This chapter highlights the current research with UCB cells in the development of a novel treatment for stroke and demonstrates the great therapeutic potential of these cells.

Abbreviations

AAS	Antibiotic antimycotic solution
APB	Adult peripheral blood
BDNF	Brain derived neurotrophic factor
BFU-E	Erythroid burst-forming units
BM	Bone marrow
CDC	Center for disease control and prevention
CFU-G	Granulocyte-macrophage colony-forming units
CFU-GEM	Granulocyte/erythrocyte/macrophage/megakaryocyte colony-forming units
CLP	Common lymphoid progenitor
CNS	Central nervous system
CNTF	Ciliary neurotrophic factor
CSF	Colony stimulating factor
CX3CR1	Fractalkine receptor
ECFCs	Endothelial colony forming cells
EGF	Epidermal growth factor
EPCs	Endothelial progenitor cells
Epo	Erythropoietin
ESMSCs	Embryonic stem cell-derived mesenchymal stem cells
ext34-	Exterior CD34-negative
FDA	Food and drug administration
FGF	Fibroblast growth factor
G-CSF	Granulocyte colony stimulating factor
Gal-C	Galactocerebrocide
GAP43	Neural associated growth protein 43
GFAP	Glial fibrillary acidic protein
GM-CSF	Granulocyte and macrophage colony stimulating factor

GVHD	Graft-versus-host disease
hEGF	Human epidermal growth factor
HGF	Hepatic growth factor
HLA	Human leukocyte antigen
HSC	Hematopoeitic stem cells
hT cells	Jurkat T-cells
IA	Intrarterial
ICAM	Intercellular adhesion molecule
ICV	Intracerebroventricular
IFN	Interferon
Ig	Immunoglobulin
IGF	Insulin-like growth factor
IGFBP2	Insulin-like growth factor binding protein 2
IL	Interleukin
int34+	Interior CD34-positive
IS	Intrastriatal
IV	Intravenous(ly)
LDH	Lactic acid dehydrogenase
LIF	Leukemia inhibitory factor
LIR-8	Leukocyte immunoglobulin-like receptor-8
MAP	Microtubule associated protein
MARCO	Macrophage receptor with collagenous structure
MCAO	Middle cerebral artery occlusion
MHC	Major histocompatibility complex
MNC	Mononuclear cell
mNSS	Modified neurological severity score
MPC	Myeloid progenitor cells
MRI	Magnetic resonance imaging
MSCs	Mesenchymal stem cells
NeuN	Neuron-specific neural protein
NF-200	Neurofilament heavy
NGF	Nerve growth factor
NK	Natural killer
NO	Nitric oxide
NSC	Neural stem cell
NT4/5	Neurotrophin 4/5
OCT4	Octomer-binding transcription factor-4
OGD	Oxygen glucose deprivation
PBS	Phosphate buffered saline
PDGF	Platelet-derived growth factor
Prdx	Peroxiredoxin
qRT-PCR	Quantitative real time-polymerase chain reaction
RA	Retinoic acid
ROS	Reactive oxygen species
SCF	Stem cell factor

SDF	Stromal cell-derived factor
SOX2	Sex-determining region Y-box two
SSEA	Stage specific embryonic antigen
SVZ	Subventricular zone
TGF	Transforming growth factor
tPA	Tissue plasminogen activator
TPO	Thrombopoietin
TRA	Tumor rejection antigen
TuJ1	III β-tubulin
UCB	Umbilical cord blood
VCAM	Vascular adhesion molecule
VEGF	Vascular endothelial growth factor
vWF	von Willebrand factor

5.1 Introduction

According to the Center for Disease Control and Prevention (CDC), stroke is among the top five causes of death (Hoyert and Xu 2012). There are upwards of 700,000 strokes diagnosed each year (Go et al. 2013), in the United States alone. Ischemic stroke is the most common type of stroke, with 87% of strokes being ischemic and 13% either intracerebral and subarachnoid hemorrhagic. Currently there is only one U.S. Food and Drug Administration (FDA) approved treatment for ischemic stroke that addresses the underlying pathophysiology of the event, tissue plasminogen activator (tPA). While this drug was first used as a thrombolytic therapy for experimental coronary artery thrombosis, it was later shown that it could be used in animals to reduce deficits resulting from embolic stroke (Zivin 1985). Ten years later, the results of the clinical trial of recombinant tPA for the treatment of acute ischemic stroke were published in the New England Journal of Medicine (NINDS rtPA Stroke Study Group 1995). They found that 3 months after stroke, patients who received tPA within 3 h of stroke onset demonstrated increased recovery when compared to the patients from the control group (stroke + placebo). The tPA provided 30% improved probability of having little to no disability at 3 months post-stroke. In June of 1996, the FDA approved tPA for the "management of acute ischemic stroke in adults", under the brand name, Alteplase (Activase) (Walton et al. 1996). The major problem with tPA as a treatment for ischemic stroke is that it must be administered within 3 h of stroke onset for benefit and 95% of stroke victims miss this narrow therapeutic window. Further, it cannot be used for hemorrhagic stroke. This lack of sufficient treatment for stroke has led many researchers to seek alternative therapies, including the use of stem cells or other cell sources. Cells derived from umbilical cord blood (UCB) have shown promising pre-clinical results as an experimental therapy for stroke, demonstrating reduced infarct size and improved functional recovery.

5.1.1 History of Cord Blood Transplants

Research using UCB cells is a relatively recent development in the stem cell field. Ethical concerns over the use of embryonic and fetal stem cells have led to an increase in efforts to find suitable alternative sources with fewer ethical concerns. In the late 1950s and early 1960s, scientists began to examine UCB as a source for such cells. In the late 1980s, Broxmeyer and fellow researchers conducted one of the first detailed empirical examinations of UCB cells as a cell transplant therapy for hematologic disease (Broxmeyer et al. 1989). These researchers examined the cellular composition of UCB, looking for cells with stem cell characteristics. Without a direct assay for hematopoietic stem cells (HSCs) at that time, they relied on bioassays for proliferating cells. They identified granulocyte-macrophage colony-forming units (CFU-GM), erythroid burst-forming units (BFU-E), and granulocyte/erythrocyte/macrophage/megakaryocyte colony-forming units (CFU-GEMM) in the umbilical cord blood, indicating that HSCs were present. In 1988, the first UCB transplant was performed in France for the treatment of a pediatric patient with Fanconi anemia (Broxmeyer et al. 1989), a condition that causes the failure of hematocyte production in bone marrow (Auerbach 1995). Since this initial use in hematologic myeloid disease, UCB cells have been used to treat other hematopoietic diseases such as aplastic anemia, acute leukemia, haemoglobinopathies, lymphoma, thalassemia, and sickle cell disease (Rosenkranz and Meier 2011). Initially the cells were used for pediatric transplants because single units of UCB were thought to have insufficient cell quantities for adult hematopoietic reconstitution (Rosenkranz and Meier 2011). By increasing infused cell dosage, UCB transplants for hematological diseases in adults developed quickly (Ballen et al. 2013). Eventually the use of more than one UCB unit allowed for a wider patient selection, reduced intensity conditioning prior to transplant, and an increase in probability of disease free survival from 30 to 50 %.

Success with UCB transplantations for hematologic diseases, can be partially attributed to the lower immunogenicity and incidence of graft-versus-host disease (GVHD) compared to bone marrow sources, due to the naïveté of the cell populations (Gluckman et al. 1997). The low immunogenicity is an attractive quality for a cell therapeutic because it limits the likelihood of severe adverse effects subsequent the transplant.

5.1.2 Collection and Storage of Cord Blood Cells

The use of UCB for hematopoietic transplants led researchers to further examine the properties and mechanisms of action of the cells in experimental conditions. Much of what we know about UCB cells is from cell culture research, either simply exploring the characteristics of the cells or investigating their actions in in vitro models of disease. One of the first important issues to be addressed was the optimal collection and storage procedures, in order to obtain and maintain the maximum number of viable cells. Much of this research came from Broxmeyer's research group and while they were interested in developing UCB cells as a treatment for blood diseases, this research is applicable to all uses of UCB cells.

The umbilical cord is the circulatory connection between the maternal placenta and the fetus while in the womb. Developed from the allantois and fetal yolk sac, the umbilical cord allows for nourishment to pass from mother to fetus (Ali and Al-Mulla 2012). The umbilical cord contains three blood vessels, two arteries and one vein, and is approximately 6 cm around and 50–60 cm long. The blood is collected from the placental stump of the umbilical vein (Whedon and Wujcik 1997). Compared to bone marrow, UCB collection is a simple and painless procedure for the donor.

The goal of the first major study on UCB was to develop an alternative source of stem cells to bone marrow-derived cells for myoablative and nonmyoablative diseases (Broxmeyer et al. 1989). The UCB was collected from the maternal end of the transected cord, with the placenta still in situ because the uterine contractions increased blood flow. This type of collection is considered an open collection procedure, using gravity or in situ means such as blood pressure to collect UCB from the umbilical stump (Bertolini et al. 1995). The alternative procedure is to collect the blood by needle aspiration from an ex situ placenta (Broxmeyer et al. 1989). When the open and closed collection procedures were compared, open collection procedures had a bacterial infection rate of 12.5% as opposed to 3.3% in closed collection procedure samples (Bertolini et al. 1995). The ex situ, or closed, collection of UCB is generally considered less invasive and therefore preferable, especially since both methods produce a similar cell yield and lack of serious adverse events (Ballen 2005).

Other considerations that affect cell yield have also been examined. Vaginal deliveries have been shown to yield significantly larger volumes of UCB compared to cesarean sections (Bertolini et al. 1995). Further, cell yield is influenced by the length of time taken to clamp the cord; clamping the cord 30 s after delivery reduces blood volume by half, from 77 ± 23 ml to 31 ± 18 ml.

One of the critical issues that had to be determined was how long the UCB cells remained viable after collection. To address this issue, Broxmeyer and colleagues collected UCB and combined it with an anticoagulant before storing the blood at 4°C, room temperature (≈25°C), or in an incubator at 37°C (Broxmeyer et al. 1989). Cell survival was examined after 24, 48, and 72 h. The researchers found that BFU-E, CFU-GM, and CFU-GEMM progenitor cells, at all three temperatures tested, survived well after 1 day. At 48 h, significant cell loss was observed in units stored at 37°C. After 3 days, cord blood units stored at 4°C and room temperature still maintained progenitor cell levels similar to freshly harvested UCB, but UCB at 37°C was devoid of progenitor cells. This finding demonstrated a particularly long cell survival period, making the transport of UCB for research and treatment easier. However, cryopreservation extends the shelf-life of the UCB cells even more.

The viability of the UCB progenitor cells has now been examined after almost 25 years of cryopreservation (Broxmeyer et al. 2011). Even after 23.5 years of cryopreservation, UCB samples maintained efficient recovery of progenitor cells that proliferated and generated colonies to a similar extent as freshly collected UCB.

Further, CD34+ HSCs from these cryopreserved samples engrafted in NOD/SCID mice with long-term repopulation and self-renewal success.

Initial attempts at preserving the UCB progenitors long term were not particularly successful. Using various methods (ammonium chloride cell lysis, gravity sedimentation, centrifugation and methylcellulose) to eliminate the erythrocytes from the UCB prior to cryopreservation yielded 50–90 % progenitor cell loss and, as a result, UCB was preserved as whole blood to avoid major progenitor cell loss (Broxmeyer et al. 1989). Six years later, two procedures for separating out the mononuclear cell (MNC) fraction by density gradient centrifugation (Ficoll-hypaque and Hanks balanced salt solution with 3 % gelatin) proved to have very good cell recovery rates, 86–92 %, when the blood was fractioned within 12 h of collection (Bertolini et al. 1995). The density gradient method became the method most commonly used to isolate a UCB cell fraction rich in progenitors.

These studies paved the way for the establishment of cord blood banks globally in the early 1990s. These banks recruit consenting donors, process and preserve the collected cord blood, and also allocate its distribution for transplantation and research (Ballen 2005).

5.1.3 Expansion of Umbilical Cord Blood Stem Cells

Stem cells are highly proliferative cells, capable of significant expansion and differentiation. One of the main reasons UCB was first used in hematopoietic reconstitution was because of the presence of stem and progenitor cells (Broxmeyer et al. 1989). However, one of the main drawbacks to UCB transplantation is that the numbers of certain stem and progenitor cells seem to be lower than the amount necessary for a single unit transplant for an adult. For example, the embryonic-like stem cells found in UCB are not plentiful, but maintain high proliferation capacity. In theory, these cells could be expanded, differentiated, or cultured to meet the requirements of a study or treatment. Gilmore and associates demonstrated that UCB cells could be expanded in vitro for approximately 9 week and maintained as long as 19 weeks (Gilmore et al. 2000). Further, the CD34+ UCB cells demonstrated greater expansion when cultured in serum-containing media with *flt*-3 ligand and thrombopoietin (TPO) growth factors, stem cell factor (SCF), and interleukin (IL)-6 versus serum-containing medium with only *flt*-3 ligand and TPO or serum-free medium only. The CD34+ UCB HSCs were capable of expansion, 100 times the original number of cells, over approximately 9 weeks in culture. These investigators also demonstrated that cord blood plasma produced comparable results as fetal calf serum. Together, the use of cytokines, serum or plasma could allow for sufficient expansion of single UCB units for therapeutic use.

Broxmeyer et al. found similar results using UCB derived myeloid progenitor cells (MPCs) (Broxmeyer et al. 1992). They cultured the UCB MPCs with colony stimulating factors (CSFs), examined the in vitro cell expansion and compared UCB cell expansion to BM-derived MPC expansion. The UCB cells demonstrated greater expansion than BM-derived cells and gave rise to a wider variety of cell types, suggesting the presence of earlier stage progenitor cells within UCB. Early stage progenitor cells are advantageous for multiple research purposes because of their expansion and differentiation potential.

More recently, another research group examined the expansion of UCB-derived MSCs in the presence or absence of insulin-like growth factor binding protein 2 (IGFBP2) (Ong et al. 2012). The UCB MNCs were co-cultured with embryonic stem cell-derived mesenchymal stem cells (ESMSCs) with cytokines, SCF, *flt* ligand, and TPO (Ong et al. 2012). They observed that IGFBP2 significantly increased the expansion of hematopoietic stem cells (HSCs; CD34+/CD38-/CD90+) from the UCB MNC fraction, but did not significantly increase the total number of nucleated cells. The presence of IGFBP2 in culture with UCB MNCs also led to an observed reduction of CD3+ and CD19+ lymphocytes. They concluded that their method of co-culturing UCB MNCs with ESMSCs in the presence of IGFBP2 was successful in significant HSC expansion within the culture, but not beneficial to overall ex vivo expansion.

5.2 Cellular Composition of Umbilical Cord Blood Mononuclear Fraction

UCB is a heterogeneous collection of cells containing a variety of progenitor cells, such as the CFU-GMs, BFU-Es, and CFU-GEMMs discussed earlier as well as HSCs and mesenchymal stem cells (MSCs). UCB also contains red blood cells, plasma cells, and white blood cells. The UCB preparations that have been used as an experimental treatment for stroke vary. Understanding the cellular composition of UCB is imperative to predict outcomes and manipulate mechanisms of action, especially once employed in clinical therapeutics (see Table 5.1).

5.2.1 Hematopoieitic Stem Cells

The initial use of UCB for treatment of hematopoietic disorders was based on the ability of HSCs in the UCB to reconstitute the immune system and therefore be a viable substitute for bone marrow. The HSCs are characterized by their expression of CD34, CD45, and CD133, which are hematopoietic lineage-specific antigens (McGuckin et al. 2003; Ali and Al-Mulla 2012). HSCs are not only present in UCB, but also seemed to be in a very early stage of development, as they maintain a very high proliferation rate even after over 20 years of cryopreservation (Broxmeyer et al. 2011). When McGuckin et al. (2003) examined the characteristics of UCB immature cell populations using expression of the lineage-specific antigens CD34, CD38, CD117, CD133, CD164, and Thy-1, compared to bone marrow-derived HSCs, they found that UCB contained more cells expressing an interiorCD34 (int34+) marker while being negative for exterior CD34 (ext34⁻). They also determined that the

UCB Cells

T-lymphocyte[1]	B-lymphocyte[2]	Monocyte[3]	Hematopoeitic Stem Cell[4]	Mesenchymal Stem Cell[5]	Other UCB Stem Cells[6]	Endothelial Progenitor Cell[7]	Mononuclear Fraction[8]
Phenotype Naïve/Immature	**Phenotype** Naïve/Immature	**Phenotype** Naïve/Immature	**Phenotype** CD34+ Cells	**Phenotype** Non-hematopoetic lineage stem cells	**Phenotype** "Embryonic-like" stem cells Umbilical cord stem cells	**Phenotype** Naïve/Immature	**Phenotype** Mixture of stem/progenitor cells and agranular leukocytes
Surface Antigen Expression CD3+/CD38+ CD4+/CDRA+ CD8+/CDRA+ CD3+/TCR-α/β CD45RA+ CD3-/CD8+ CD3-/CD7+	**Surface Antigen Expression** CD19+/CD20+ CD19+/CD24+/CD5+ CD19+/CD20+/Ig- CD5+/CD19+	**Surface Antigen Expression** Mac1+ CD14+ CD11b+	**Surface Antigen Expression** CD34+ CD45+ CD133+ CD34+/CD38- intCD34+/extCD34-/CD133-/CD38low/- intCD34+/extCD34-/extCD117low	**Surface Antigen Expression** CD34-/CD45-/CD133- CD74+/CD184+ Increased CD44+ vs. APB-MSCs Decreases CD105 vs. APB-MSCs	**Surface Antigen Expression** OCT4+ SSEA-3+/SSEA-4+ TRA-1-60+/TRA-1-81+ NANOG+ Low HLA-ABC/HLA-DQ & Low CD40/CD80/CD86	**Surface Antigen Expression** *Fresh* CD14+/CD45+ CD34-/CD133- VEGFR-2+/V-cadherin+/CD105+ *Cryopreserved* CD31+/CD133+/VEGFR-2+/vWF+	**Surface Antigen Expression** *Mixed due to multiple cell types*
Quantity within UCB 55-60% MNC Composition CD3-/CD8+ only in UCB CD3-/CD7+ only in UCB CD3+/CD38+ greater than in APB	**Quantity within UCB** 25-30% MNC Composition Similar % of B-cells vs. APB, but greater absolute # of cells	**Quantity within UCB** <0.2% MNC Composition Similar % of Monocytes vs. APB, but greater absolute # of cells CD14+ Monocytes 4.05 ± 1.03 % of nucleated cells	**Quantity within UCB** Greater #s than in APB	**Quantity within UCB**	**Quantity within UCB** 0.1% of the MNC fraction	**Quantity within UCB** 15Xs more than in APB	
Cytokine expression *High expression:* IL-10 *low expression:* IL-2	**Cytokine expression**	**Cytokine expression** CD14+ Monocytes express: IL-1β IL-6 IL-12 TNF-α This expression is increased with the addition of LPS and IFN-γ *in vitro*.	**Cytokine expression**	**Cytokine expression** *High expression of:* IL6, IL8, TIMP-1, & TIMP-2 *Also express:* ENA-78, GM-CSF, GRO, IL-1b, IL-6, IL-8, MCP-1, OSM, VEGF, FGF-4, FGF-7, FGF-9, GCP-2, IGFBP-1, IGFBP-2, IGFBP-3, IGFBP-4, IP-10, LIF, MIF, MIP-3a, osteoprotegerin, PARC, PIGF, TGF-b2, TGF-b3,	**Cytokine expression** Unrestricted somatic stem cells (USSC) express: GM-CSF HGF IL-1β, IL-6, IL-8, IL-11, IL-12, IL-15 LIF M-CSF SCF SDF-1α TGF-1β VEGF	**Cytokine expression**	**Cytokine expression** MCP-1 IL-8 TIMP-1 NGF-β IL-3 RANTES *After 10 days in vitro:* ENA-78, MDC, OSM, & VEGF *After 28 days in vitro:* IL-6, TNF-α, IL-1α, IL-10, TNF-β, GM-CSF, IFN-γ, MCP-2, & MCP-3

Table 5.1 Cellular characteristics of cord blood cells. The cord blood has a variety of white blood cells including T cells, B cells, monocytes, HSCs, MSCs, other stem cells, EPCs. These are usually found within the MNC fraction. The information in this table were obtained frome following references: [1](Harris et al. 1992; Bacchetta et al. 1994; Wedgwood et al. 1997; D'Arena et al. 1998). [2](Harris et al. 1992; Wedgwood et al. 1997; D'Arena et al. 1998; McGuckin et al. 2003). [3](Harris et al. 1992; Wedgwood et al. 1997; D'Arena et al. 1998; Morgado et al. 2008; Jiang et al. 2010). [4](McGuckin et al. 2003; Sanberg et al. 2005). [5](Liu and Hwang 2005; Martins et al. 2009). [6](Kogler et al. 2005; McGuckin et al. 2005; Zhao et al. 2006; Ali and Al-Mulla 2012). [7](Eggermann et al. 2003; Ingram et al. 2004; Vanneaux et al. 2010). [8](Newman et al. 2006).

UCB cells lacked exteriorized CD38 while expressing CD133. This pattern of antigen expression (intCD34$^+$/extCD34$^-$/CD133$^+$/CD38$^{low/-}$) was more highly expressed in UCB than bone marrow and indicated a more naïve phenotype. In addition to this population of very immature progenitor cells, another more primitive cell subset was also observed. These cells were quiescent stem/progenitor cells and potentially capable of long-term engraftment and ectodermal differentiation; they expressed intCD34$^+$/extCD34$^-$/extCD117^{low} (McGuckin et al. 2003)

5.2.2 Mesenchymal Stem Cells

Mesenchymal stem cells (MSCs) are also present in UCB. These cells are capable of differentiating into mesodermal cell lineages, including adipogenic, chondrogenic and osteogeneic cells (Bieback et al. 2004; Ali and Al-Mulla 2012). These cells can be identified by their lack of typical hematopoietic antigens, CD34, CD45, and CD133. Although MSCs are found in low numbers in UCB, they are highly proliferative (Ali and Al-Mulla 2012). The MSC can be isolated from the MNC fraction, forming fibroblastoid cells in culture. MSC identity was confirmed through flow cytometry by expression of CD29, CD44, CD73, and HLA-class I (Bieback et al. 2004), although there was decreased fluorescent intensity of the antigen expression compared to bone marrow-derived MSCs. These MSCs also displayed a slightly different antigen profile, with a lower expression of SH2/CD105 and a higher expression of CD106. The isolated cells could be induced to differentiate into osteogenic, adipogenic and chondrogenic lineages, confirming their mesenchymal stem/progenitor status.

5.2.3 Endothelial Progenitors

Endothelial progenitor cells (EPCs) are also found in UCB (Ingram et al. 2004). These are highly proliferative cells involved in the vascular genesis, angiogenesis and arteriogenesis. There are no specific antigens that identify all EPCs but most express CD34, CD133 and vascular endothelial growth factor receptor (VEGFR)-2 (Khakoo and Finkel 2005). Unfortunately these cells also share many surface antigens with HSCs (Asahara and Kawamoto 2004), making bioassays the only way to infer their presence. In general, these cells express endothelial surface antigens including CD31, CD105, CD141, CD144, and CD146 as well as von Willebrand Factor (vWF) but do not express CD14 and CD45.

There have been a number of studies that have compared EPCs from UCB, adult peripheral blood (APB) and bone marrow. UCB contained fifteen times more EPCs than adult peripheral blood; these cells formed larger colonies and could be expanded more than 100 passages without significant senescence or decreased telomere length (Ingram et al. 2004). When the CD34+ cells isolated from the UCB MNC fraction were cultured, they initially expressed CD14 and CD45 (Eggermann et al. 2003) and exhibited a monocyte-like phenotype. This expression profile indicates neither progenitor nor endothelial cell lineage. However, after 6 days in vitro, cells began to demonstrate endothelial phenotype, VEGFR-2+, vascular endothelial (VE)-cadherin+, and CD105+, although they remained CD34- and CD133- and CD14+ and CD45+. After about 10 days in culture, the number of cells demonstrating a endothelial-spindle shaped morphology increased.

EPCs have also been found in cryopreserved UCB (Vanneaux et al. 2010). EPCs from fresh and cryopreserved UCB were compared. Daily assessments of potential endothelial colonies were made, looking for cells with an adherent cobblestone-like morphology. The EPCs from the cryopreserved UCB contained significantly fewer endothelial colony forming cells (ECFCs) than in fresh UCB, but they had a shorter average expansion time. Even so, all cells had high CD31, CD133, VEGFR-2, and vWF expression as well as arterial and venous endothelial cell genes. EPCs from both sources were found to be capable of forming tube-like vessels and participate in in vitro wound healing, forming a cell monolayer over scar tissue. Vanneaux et al. also examined the effects of cryopreservation on cell viability and telomere length and found that fresh and cryopreserved UCB cells had no significant differences in viability or in telomere length.

5.2.4 Other Stem and Progenitor Cells

Not only does UCB contain HSCs, MSCs and EPCs, there is evidence that the UCB has additional stem cell populations. Small non-hematopoietic cells have been found in UCB (McGuckin et al. 2005; Zhao et al. 2006); this population of cells expresses markers typically associated with embryonic stem cells, such as the homeobox protein NANOG, octomer-binding transcription factor-4 (OCT4), and sex determining region Y-box two (SOX2) (Ali and Al-Mulla 2012). McGuckin et al. found a population of highly proliferative embryoid-like cells from lineage-restricted immunomagnetic-separated UCB cells (McGuckin et al. 2005). These cells exhibited exponential growth for approximately 5 weeks, after an initial week-long lag phase. Even secondary non-adherent cells, from the original UCB cell selection, once replated were capable of similarly high growth levels; these cells express some embryonic-specific antigens: OCT4, stage specific embryonic antigens three and four (SSEA-3, SSEA-4 respectively), tumor rejection antigen (TRA)-1-60, and TRA-1-81). During this study, the researchers were able to differentiate this cell type into hepatogenic cells (McGuckin et al. 2005). Another research group also isolated stem-like cells that strongly expressed embryonic markers NANOG, OCT4, SSEA-3, SSEA-4 as well as TRA-1–60 and TRA-1-81 (Zhao et al. 2006); these cells did not express markers of HSCs or MSCs. The cells also weakly expressed human leukocyte antigen (HLA) -ABC, HLA-DQ, CD40, CD80, and CD86. This is a similar immunogenic profile as embryonic stem cells, which have low immunogenicity. These cells were able to differentiate into neuronal-like cells, endothelial cells and functional pancreatic-islet cells.

5.2.5 *Immune Cells*

UCB also has a full complement of immune cells in addition to the stem cell compartment. There are many naïve T-lymphocytes, B-lymphocytes, and monocytes in the UCB MNC. These important components of UCB likely play a large role in the low incidence of GVHD with UCB transplants (D'Arena et al. 1998). When UCB immune cells were compared to immune cells from adult peripheral blood using flow cytometry, the percentage of B cells and monocytes was similar between the two sources, but T cells were elevated in UCB.

5.2.5.1 Lymphocytes

T and B cells are cells of the adaptive immune system. Common lymphoid progenitor (CLP) cells are cells with the capacity to develop into B-lymphocytes, T-lymphocytes, and natural killer (NK) cells (Haddad et al. 2004). In UCB these CLPs demonstrate considerable potential for lymphocyte generation and express a particular antigen profile: CD45RA+/CD38-/CD7+. Haddad et al. demonstrated that these CLPs from UCB can be further identified by their antigen profile, in order to predict their differentiation lineage. They found that UCB CLPs expressing CD34+/CD45RA^{hi}/CD7+ were more likely to develop into T cells and CD/34+/CD45RA^{hi}/Lin-CD10+ were more likely to differentiate into B-cells when studied in vitro. They observed that to some degree these cells retained potential for alternative lymphocyte lineage differentiation, but tended towards T/NK cell or B cell lineage, respectively. These findings suggest that many of the lymphocytes found in UCB are in an immature state as precursor cells.

Approximately 55–60 % of the UCB MNC fraction is CD3+ T cells (Wedgwood et al. 1997). When the number and phenotype of UCB T cells was compared to T cells from adult blood using flow cytometry, there was both a greater percentage and a greater number of CD3+/CD38+ immature T lymphocytes in UCB (D'Arena et al. 1998). Further, there were more naïve CD4+ and CD8+ T cells, as identified by their co-expression of the CD45RA+ antigen. In adult peripheral blood samples there were more mature CD4+ cells that co-expressed CD45RO. These authors concluded that UCB contained more T cells and that they were an immature phenotype compared to adult peripheral blood.

Harris and colleagues also investigated the phenotype and immunologic profile of the T cells present in UCB (Harris et al. 1992). While they found fewer CD3+ T lymphocytes in UCB compared to adult peripheral blood samples, the majority of UCB-derived T cells expressed CD3/TCR-α/β, indicative of an immature phenotype, as well as the CD45RA naïve phenotype found in approximately 90 % of UCB T cells. They also found another population of CD3-/CD8+ and CD3-/CD7+ T cells in UCB, but not in adult peripheral blood. Both T cells and NK cells lacked the lytic activity of normal adult lymphocytes, but this activity did develop after about 1 week in culture. UCB T lymphocytes also did not respond to IL-2 or HLA stimulation the same way mature T cells did and their own HLA expression was significantly lower in UCB cells versus adult peripheral blood samples. This lack of mature phenotype may contribute to the therapeutic effects demonstrated by the UCB cells.

B lymphocytes are responsible for the recognition of antigens and production of plasma and memory cells assisting in the immune system's recognition of foreign molecules (Solomon et al. 1999). These plasma cells from activated B cells secrete antibodies, or immunoglobulins (Ig). Fetal B cells are produced from differentiated precursor cells found in the liver and depending on their stage of differentiation, their expression of various surface antigens changes (Abbas et al. 2011). B cell maturity can be attributed to surface antigen expression of CD19, CD20, and Ig expression (Edwards and Cambridge 2006). The first stage of B cell differentiation is marked by the expression of CD19+/CD20-Ig- (pro-B-cell phenotype), continuing differentiation to express CD19+/CD20+/Ig- indicating a pre-B-cell, and becoming an immature B lymphocyte positive for expression of Ig (CD19+/CD20+/Ig+).

Approximately 25–30 % of the UCB MNC is CD20+ B lymphocytes (Wedgwood et al. 1997). In studies that have compared B lymphocytes from UCB and adult peripheral blood, there appear to be similar numbers of B cells as determined by percentage of cells expressing CD19+ and CD20+ antigens (Harris et al. 1992; D'Arena et al. 1998). Interestingly, about half of the B lymphocytes expressed a CD5+ immature phenotype. When immunoglobulin expression was examined, UCB-derived B cells had greater expression of IgM than adult peripheral blood control samples (Wedgwood et al. 1997). Using flow cytometry and PCR analysis the UCB-derived B cells that expressed CD20+ were found not to express IgG. These findings would indicate that the B cells found in UCB are in the pre-B-cell stage. Further, the IgM antibodies bind oxidized polysaccharides, glycoconjugates and phospholipids (Binder and Silverman 2005) on damaged cells, targeting the cells for destruction. In this way, UCB-derived B cells could limit the overall inflammatory response to stroke.

5.2.5.2 Monocytes

Monocytes (CD14+) are cells of the innate immune system. Adult monocytes remove apoptotic cells, scavenge toxins, and differentiate into macrophage and dendritic cells. They produce both pro- and anti- inflammatory cytokines, growth factors such as VEGF, reactive oxygen species (ROS), nitric oxide, prostaglandins, complement and proteolytic enzymes (see references (Geissmann et al. 2008; Auffray et al. 2009) for recent reviews). These cells may express major histocompatibility (MHC) class II antigen and CD11c (Van Voorhis et al. 1983) and therefore can function as antigen presenting cells. During acute inflammation, such as occurs after stroke, the cells are mobilized from bone marrow by interaction of monocyte chemoattractant protein (MCP)-1 with the CCR2 receptor (van Furth and Cohn 1968), and from the spleen in a CCR2-independent manner (Swirski et al. 2009). There appear to be two types of monocytes in the adult – the CD14+ CD16 cells which express CCR2 and produce IL—10 (Ziegler-Heitbrock et al. 1992; Weber et al. 2000; Geissmann et al. 2008) and the CD14+ CD16+ monocytes, which express high levels of CX3CR1 (fractalkine receptor) and secrete tumor necrosis factor (TNF)-α (Belge et al. 2002). These latter cells have been implicated in the acute inflammatory response following myocardial infarction in the mouse while the former infiltrate the ischemic heart later and are associated with tissue remodeling and inhibition of inflammation (Nahrendorf et al. 2007).

Monocytes within UCB are indistinguishable from monocytes harvested from adult peripheral blood in terms of phagocytic activity, ROS production, ability to degrade bacteria, production of many cytokines and expression of CD14 and CD16 antigens (Gille et al. 2009; Filias et al. 2011). Unlike monocytes from adult blood, however, UCB monocytes are not stimulated by hepatocyte growth factor (HGF) (Jiang et al. 2001). Stimulation of UCB monocytes with IL-4 and granulocyte and macrophage colony stimulating factor (GM-CSF) produces fewer dendritic cells compared to stimulation of adult monocytes (Liu et al. 2001). UCB monocytes also produce less IL-1β, TNF-α, integrin α5 subunit, intercellular adhesion molecule (ICAM)-1, and IL-10 than adult monocytes (Le et al. 1997; Brichard et al. 2001; Jiang et al. 2001). While less IL-10 production would suggest that immunosuppressive function of the UCB monocytes could be blunted compared to adult monocytes, UCB MNCs administered 48 h post-MCAO increase protein expression of IL-10 in the spleen and the brain (Vendrame et al. 2005; Vendrame et al. 2006). It is not clear whether the origin of the IL-10 is monocytes within the UCB MNC, other UCB cell populations or endogenous immune cells within the transplanted animals.

5.2.6 Cord Blood-Derived Neural Cell Lines

While the early work with UCB cells focused on expansion of the cells to treat hematologic disease, the more recent focus on developing a treatment for neurologic disease or injury has required different tools. Sanchez-Ramos and associates were the first to report neural gene and protein expression from UCB-derived cells in vitro (Sanchez-Ramos et al. 2001). Using retinoic acid (RA) and nerve growth factor (NGF), they observed cellular development of UCB cells into neural-like cells that expressed glial fibrillary acidic protein (GFAP) and β-tubulin III, expressed by astrocytes and neurons, respectively. They also observed a significant increase in neural gene mRNA expression for pleiotrophin, glypican-4, neural associated growth protein 43 (GAP43), neuronal pentraxin II, and neuronal PAS1 and a significant decrease in hematopoietic genes HLA class II, attractin, macrophage receptor with collagenous structure (MARCO), leukocyte immunoglobulin-like receptor-8 (LIR-8), CD1c, erythropoietin (Epo) and its receptor. Some of the upregulated gene expression coincided with protein expression, as they also detected significant levels of Mushashi-1, β-tubulin III, pleiotrophin, and neuron-specific neural protein (NeuN) proteins. These results were confirmed and extended to describe cells with a neuronal-like morphology in addition to expression of NeuN, neurofilament, microtubule associated protein (MAP)-2 and GFAP (Ha et al. 2001; Ha et al. 2003).

Later work suggested that it was CD133+ primitive stem cells present in the UCB that could produce neural cells (Jang et al. 2004). In another study, CD45- cells exposed to fibroblast growth factor (FGF)2 and human epidermal growth factor (hEGF) for 7 days could express β-tubulin III, GFAP, and galactocerebroside (Gal-C) (Bicknese et al. 2002). Similarly, lineage negative cells from UCB exposed to FGF4, SCF and *flt3*-ligand expressed nestin, neurofilament, A2B5 and Sox2 (Chua et al. 2009). Brain-derived neurotrophic factor (BDNF) was shown to induce neural-like differentiation of UCB MSCs (Lim et al. 2008). Based on these studies, it was a reasonable supposition that the UCB stem cells could produce neural cells and studies to develop UCB derived neural cell lines began.

One of the first attempts to develop such a cell line was reported by researchers in Poland. Bużańska et al first expanded CD34-/CD45- UCB cells in vitro and after sufficient expansion, differentiated the cells with neurobasal media containing RA and BDNF (Buzanska et al. 2002). The resulting cloned cells were positive for nestin, a marker associated with central nervous system (CNS) stem and progenitor cells. From these nestin positive cells, Bużańska et al. were able to further differentiate the UCB-derived cells with RA and BDNF into three neural progenitor cell types, expressing some neuronal and glial markers: β-tubulin III (neuronal marker), GFAP (astrocytic marker), and Gal-C, (oligodendrocytic marker). Four years later, this research group reported that they had established a neural stem cell (NSC) line from the CD34-/CD45- MNCs in UCB (Bużańska et al. 2006; Habich et al. 2006). Culturing the non-adhesive cells with different combinations of epidermal growth factor (EGF), vitamin B27, FGF2, leukemia inhibitory factor (LIF), and antibiotic antimycotic solution (AAS) with high, low or serum free media, researchers were able to establish a NSC lineage. Cells cultured with serum-free media, EGF, LIF, and FGF2 remained in an undifferentiated state as clumps of small round cells without development of an adherent basal layer. Cells cultured in low-serum, EGF, LIF and FGF2 developed an adhesive basal layer and maintained some undifferentiated free-floating cells, which expressed genes related to self-renewal (Wnt, Lif, and TGF-β pathway genes), NSCs (including Notch-2, -3, and-4), and neural commitment (Sox2). Both serum-free and low serum culture conditions yielded slower but continuous rates of expansion up to 20 days in vitro. High serum cultures yielded mostly adherent cell morphologies, and exhibited high initial growth rate, but became senescent. Researchers observed their NSC-line to be clonogenic, at a rate of about 10%, and demonstrated neural marker expression of nestin, GFAP, and neurofilament heavy (NF-200), when cultured in low-serum media conditions. Once this cell line was developed this research group was able to induce neuronal differentiation; the best conditions for generation of neurons were low serum and co-culture with rat astrocytes or hippocampal slices (Jurga et al. 2006). They also demonstrated that these cells expressed neurotransmitters and their receptors and were electrically active (Sun et al. 2005); unfortunately, the cells did not have the necessary sodium channels to generate an action potential. It was not clear if these cells were just not mature enough to have developed this ion channel or were not truly functional neurons.

5.3 Effect of Cord Blood Cells After Stroke

In the first study to use UCB cells as a therapeutic for stroke, researchers intravenously (IV) administered UCB cells to rats at two time points, 24 h and 7 days post-Middle cerebral artery occlusion (MCAO) (Chen et al. 2001). They found that their treatment improved rat functional recovery and reduced neurological deficits when administered at 24 h post-MCAO. Although they did not see an infarct volume reduction post-treatment, they were able to identify UCB cells within the stroked brain mostly localized around the site of injury. These results indicated that UCB cells were capable of stimulating functional recovery post stroke, but the mechanisms by which this occurred were unclear.

5.3.1 Which Cord Blood Cells are Most Therapeutic?

The UCB cell preparations reported in the literature have varied across research groups, most likely as a function of the UCB's heterogeneous nature and the hypothesized mechanism of action. Many researchers choose to employ the MNC fraction or to select a specific cell type, such as MSCs, HSCs or the population of embryonic-like stem cells found in UCB.

Nystedt et al examined the efficacy of UCB CD34+ cells in both transient MCAO using the intraluminal filament method, which produces an infarct encompassing striatum and cortex, and a permanent distal MCAO, in which the infarct is limited to cortex (Nystedt et al. 2006). They transplanted UCB-derived CD34+ cells (5×10^5 cells and 2×10^6, for transient and permanent occlusions, respectively). They failed to observe any change in infarct volume, but the treatments did provide a degree of functional recovery in both MCAO models.

Recently, a study examined the effects of three different fractions of UCB cells administered IV 24 h after stroke (Boltze et al. 2012a). Boltze et al. specifically examined the differences between the use of MNCs, CD34+ cells, and CD34- cells in the treatment of experimental stroke, using both human Jurkat T-cells (hT cells) and non-cell-containing vehicle as controls. Animals administered any type of UCB cells exhibited increased motor recovery compared to those animals that received the hT cells or vehicle control group. CD34+ and CD34- cells improved motor recovery at 7 days post-transplant, whereas, the UCB MNCs took longer to improve motor performance (17 days post-transplant). Although the MNCs did not demonstrate this effect until later in recovery, they exhibited a more pronounced recovery than either the CD34+ and CD34- cell treated groups. MNC treated rats also exhibited decreased CNS tissue loss, significantly more than the CD34+ treatment group, and more than the CD34- treatment group, although not statistically significant. These investigators concluded that UCB MNCs were the best choice for neuroprotection post-stroke, indicating that the combination of the cell types within the MNC enhances the neuroprotective effects of UCB on post-ischemic injury.

We recently published a similar study, but comparing the ability of UCB MNCs to induce stroke recovery to that of the naïve immune cells present in the MNC fraction (Womble et al. 2014). The most profound observation in this study was that almost the entire effect of MNC transplantation was replicated by administering UCB-derived CD14+ monocytes. Similarly, removal of these cells from the MNC fraction reduced recovery and prevented the reduction in infarct size that we normally obtain. These data suggest that the stem cell populations present in the MNC have only a small role in neural repair.

5.3.2 *Dose of Cells*

An important consideration in the use of UCB cells for cell therapy is the number of cells required to elicit treatment effects. The first study of UCB cells to treat stroke administered 3×10^6 cells IV (Chen et al. 2001). The first study to systematically determine the best dose to use was not performed until 3 years later when we examined the cell dose necessary for IV UCB cell benefits when administered at 24-h post-MCAO (Vendrame et al. 2004). Functional recovery and infarct volume reduction was optimal when UCB cells were given at a dose of 10^7 cells, compared to lower (10^5, 10^6) and higher (between 3×10^7 and 5×10^7) doses. However, when the cells were administered at 48-h post-MCAO, IV cell transplants only required 10^6 cells (Newcomb et al. 2006). When UCB cell doses of 1×10^6 cells are administered at 48-h post-MCAO they consistently improve function and reduce neural injury (Rowe et al. 2012). Examining tissue for phosphorylated (p)Akt, a protein involved in cell survival pathways, within white matter of stroked brains, investigators found that UCB cell treatment increased ipsilateral pAkt expression at 72 and 96 h post-ischemia. The cells not only reduced infarct volume, especially in the white matter, but also led to an increase in expression of the antioxidant, peroxiredoxin (Prdx)4 and reduced apoptosis as indicated by activated caspase 3 expression post ischemia.

5.3.3 *Timing of Administration*

One of the major problems with the rtPA treatment for stroke is the very narrow time window it offers: less than 4 h (NINDS rtPA Stroke Study Group 1995). One of the hopes for cell therapy in stroke is that it will allow for a more realistic treatment window to reduce the functional deficits caused by stroke. Various time points have been examined, immediate cell therapy administration and administration within hours or days of stroke onset (Newcomb et al. 2006). Studies suggest that an immediate administration of UCB cells does not demonstrate the same benefits, seen when administered at later time points. Although immediate cell administration has been shown to reduce functional deficits (Borlongan et al. 2004), this immediate administration would provide a smaller therapeutic time window than currently available.

The timing of cell administration affects the efficacy of the treatment (Newcomb et al. 2006). Researchers administered UCB cells IV to rats at one of six time points:

3, 24, 48, and 72 h, 7 days, and at 4 weeks post-ischemia. Previous empirical evidence concluded that 10^7 UCB cells were necessary for infarct volume reduction when administered within 24 h of stroke onset (Vendrame et al. 2004). However, Newcomb et al found that 10^6 UCB cells were sufficient to limit the infarct size and prevent cell loss in the penumbra, when given at 48 h post-ischemia. Not only was infarct size smaller, but rats that received cells at 48 h post-MCAO significantly improved motor function, compared to rats with MCAO only or cells administered at other time points. When the cells were delivered at 48 h post stroke, there was a significant reduction in inflammation and apoptosis, which led to reduced infarct volume. This therapeutic time window of 48 h post-ischemic onset would greatly expand the ability to treat and potentially alleviate much of the functional and behavioral damage that currently occurs in stroke victims.

This study was replicated by other investigators. Boltze and colleagues administered UCB MNCs IV at five experimental time windows, post-MCAO: 4, 24, 72, 120 h, and 14 days (Boltze et al. 2012b); the control group was treated with phosphate buffered saline (PBS) vehicle at 24 h post-MCAO. Sensorimotor function and recovery were measured with the modified Neurological Severity Score (mNSS), RotaRod, and BeamWalk assays, every 4 days beginning 1 day post-MCAO and ending 27 days post-MCAO. They found that UCB cells administered within 72 h of ischemic injury reduced stroke-related functional deficits and reduced infarct size even though the IV administered cells were not found engrafted in the cortex.

5.3.4 Route of Administration

Just as there are multiple methods of drug administration for pharmacologic treatment of disease, research for cell therapy explores different methods for cell administration. The chosen route of administration for UCB cells has a few variations, due to an evolution of understanding about the UCB cell mechanisms of action in vivo. There are four main routes of cell administration: intrastriatal (IS), intracerebroventricular (ICV), intravenous (IV), and intrarterial (IA). In both IS and ICV, the cells are administered directly into the CNS, whereas IV and IA routes of administration are peripheral. Although direct, IS delivery is an invasive technique for cell administration as it disturbs the skull, meninges, and parenchyma. After ischemic injury, the necrotic core is a toxic environment for transplanted cells and their survival rate is usually poor (Bühnemann et al. 2006), hence placement near the infarct, but not in it, is critical. IS and IV are the most common methods for UCB cell administration in experimental stroke therapy. Although both methods demonstrate positive effects, such as infarct volume reduction and improved behavioral recovery, IV has become the preferred route of administration for UCB cells in stroke.

The use of IS cell administration has demonstrated success in reducing infarct volume and improving neurobehavioral recovery after an ischemic incident (Koh et al. 2008). The cultured UCB MSC cells were administered by stereotaxic-IS injection, 14 days post-MCAO. The cells were injected ipsilateral to the lesion, in a dose of 6×10^5 UCB MSCs per animal. Animals that received UCB cell treatment showed improved neurological function, based on Neurlogical Deficit Scores, compared to control subjects. The treated animals also displayed reduced infarct volume compared to controls, as measured by magnetic resonance imaging (MRI). Although researchers found some expression of neural markers (NeuN), suggesting neural differentiation of implanted cells, the UCB cells predominantly maintained an undifferentiated state.

We also investigated the optimal route of administration for UCB cell therapy for the treatment of stroke (Willing et al. 2003). Given at 24-h post-MCAO, UCB cells were administered either IV (1×10^6 cells in 10 μl) or IS (250,000 cells in 2.5 μl). Overall, IV delivery produced less cellular debris in the injured striatum and increased functional recovery that was stable over time compared to the IS or stroke control groups, even though there was no evidence of infiltration of the UCB cells into the CNS. We had hypothesized that the cells would infiltrate the infarcted brain since the earlier demonstration that the stroked brain produced chemotaxic signals that attracted UCB MNCs (Chen et al. 2001) that included production of growth regulated oncogene/cytokine-induced neutrophil chemoattractant (GRO/CINC)-1,MCP-1, macrophage inflammatory protein (MIP)-1 (Newman et al. 2005; Jiang et al. 2008). Further, blocking these chemokines reduced UCB homing toward infarcted brain tissue.

A later study addressed the issue of penetration of the transplanted cells into the CNS (Borlongan et al. 2004). The premise was that mannitol, an osmotic that would open the blood brain barrier (BBB), would allow more cell infiltration into the injured brain. While the combination of UCB therapy and mannitol significantly reduced behavioral dysfunction and infarct size as measured 3 days post-MCAO compared to even the group treated with UCB cells alone, they still found few transplanted cells in the brain. Even so, treatment was still able to induce trophic factor expression in the infarcted brain. The benefit of mannitol administration was likely by changing osmotic pressure and minimizing edema as opposed to increasing contact of the UCB cells with neural substrates. It also suggests that the effect of the cells was most likely indirect either through the release of circulating factors or stimulating endogenous cells to produce trophic factors.

5.4 Underlying Mechanism of Action

Initially it was thought that the cells would act as replacement cells for necrotic tissue, especially if administered directly into the brain; it was believed that it was necessary for the cells to localize or be administered directly to the site of injury in order to ameliorate that injury, replacing necrotic cells in the infarct core. However, this view has shifted. Scientists now believe that UCB cells act indirectly through immune modulation and soluble factors (Vendrame et al. 2005). Phenomena such as angiogenesis, neurogenesis, the reduction or cessation of apoptosis and inflammation, as well as increased trophic support have all been associated with UCB cell treatment of stroke.

5.4.1 Cell Replacement

Much of the support for this theoretical mechanism of action stems from the in vitro research of UCB cell transdifferentiation. Although UCB cells are capable of differentiating into cells from endo-, meso-, and ecto-dermal cell lineages under proper cell culture conditions, the evidence is lacking for effective in vivo transdifferentiation and cell replacement. Based on the evidence of UCB cell high proliferation rate and ability to differentiate into multiple cell lineages, we chose to directly implant the UCB cells in the subventricular zone (SVZ) of the rat neonate (Zigova et al. 2002) because it is one of the regions of the adult brain known to maintain neurogenesis throughout life and these new cells typically migrate to the olfactory bulb via the rostral migratory stream. The implanted UCB cells were found mostly in the SVZ but also observed in the corpus callosum and nearby parenchyma; they were not found in the contralateral hemisphere. Survival of the cells was approximately 20 %. Most of the transplanted cells were observed to express GFAP, exhibit astrocyte-like morphology, and were found in the corpus callosum and parenchyma. Expression of neuronal proteins was a rare occurrence, with only about 0.2 % of cells expressing a human antigen and neuron-specific class III β-tubulin (TuJ1). In the adult striatum of NOD/SCID mice, no UCB cells were observed 1 mo after grafting even though 5 % of the UCB cells were present in the striatum at 5 days post- transplant and expressed early neuronal antigens (TuJ1) (Walczak et al. 2007). After MCAO, results are not much different. In the Chen et al. paper, survival of UCB cells within the infarcted hemisphere was approximately 1 % of the cells administered (Chen et al. 2001); few of these cells expressed neural antigens. In subsequent studies, few UCB cells have been found to survive in the brain of stroked rats or mice. Even in studies that have used UCB derived cells that have undergone neural induction prior to transplantation into infarcted cortex, there was a paucity of cells (Kozłowska et al. 2007).

5.4.2 Cord Blood Cells Provide Trophic Support

One of the advantages of using a cell therapy to treat an ischemic stroke is that the living cells are protein factories that are regulated by the microenvironment surrounding them. This cannot be replicated by the administration of a trophic cocktail. As such, the UCB cells can provide potent trophic influences to the injured brain. Studies that have examined the secretory products of the cells report the expression of a myriad of growth factors, cytokines, and chemokines in these cells.

When PC12 cells were exposed to oxygen glucose deprivation (OGD) followed by reoxygenation in the presence or absence of UCB MNCs, mRNA for nerve growth factor (NGF), VEGF, and FGF2 increased in the UCB cells (Arien-Zakay et al. 2009). The CD34+ UCB cells have been shown to express both NGF and the trkA receptor (Bracci-Laudiero et al. 2003). In addition to NGF, mRNA for the other neurotrophins, brain derived neurotrophic factor (BDNF) and neurotropin 4/5 (NT4/5), has also been observed in these cells (Fan et al. 2005). Zwart et al demonstrated that UCB MSCs secreted transforming growth factor (TGF)-β, ciliary neurotrophic factor (CNTF) as well as NT3 and BDNF (Zwart et al. 2009). We confirmed these results and showed that platelet derived growth factor (PDGF)-BB, granulocyte colony stimulating factor (G-CSF) and granulocyte and macrophage colony stimulating factor (GM-CSF) were also secreted constitutively (Chen et al. 2010). In a mouse model of ataxia, UCB cells were shown to express insulin like growth factor (IGF)-1 and VEGF (Zhang et al. 2011). Similarly, UCB MSCs were shown to secrete interleukin (IL)-6, VEGF, hepatocyte growth factor (HGF), and TGF-β (Park et al. 2009). Even the growth factors, and not just the neurotrophins, have been shown to have beneficial effects in rodent models of stroke (Dhandapani and Brann 2003; Schabitz et al. 2003; Zhu et al. 2005; Kawada et al. 2006; Shang et al. 2011; Shen et al. 2012; Wang et al. 2013a).

In addition to trophic/growth factors, we have shown that the UCB cells produce many chemokines and cytokines. The two most prevalent chemokines produced by the UCB cells are IL-8 (CXCL8) and MCP-1 (CCL2) (Newman et al. 2006) and these are produced regardless of the culture conditions the cells were exposed to. The environment of the cells, however, determines whether stromal derived factor (SDF)-1b, also referred to as CXCL12) and leptin are secreted. Hau et al demonstrated that CCL5 (RANTES), CCL3 (MIP-1α), CCL4 (MIP-1β) and CXCL10 (interferon gamma-induced protein 10 or IP-10) all increased in cultures of hypoxic neurons to which UCB cells were added (Hau et al. 2008). Cytokines that the cells produce include IL-1α, IL-6, IL-11, and small amounts of interferon (IFN)-γ and TNF-α (Chen et al. 2010). The cells also produce a number of molecules that are essential for the interaction of the cells with other cells and extracellular matrix, including E-selectin, L-selectin, intercellular adhesion molecule (ICAM), and vascular adhesion molecule (VCAM). While these adhesion molecules are essential for the trafficking of the UCB immune cells from the blood stream into the tissue, they may also be responsible for the ability of the UCB cells to enhance neuritic outgrowth in culture (Chen et al. 2010) and stimulate dendritic growth in other animal models of CNS aging or disease (Shahaduzzaman et al. 2013; Willing et al. In Press).

Any of these trophic factors, chemokines or cytokines may modulate processes as diverse as inflammation, apoptosis, angiogenesis, neurogenesis or the redox system. This is not an exhaustive list, but studies have been performed to examine UCB effects on each of these.

5.4.3 *Cord Blood-Neural Cell Interactions*

There have also been a number of studies to determine how UCB cells interact with neural cells in the injured brain (see Fig. 5.1). While in vivo studies of UCB cell efficacy examine infarct size, and infer that UCB cells saved neurons when infarct size decreases, there have been fewer studies that have actually measured neuronal survival. In an in vitro model of hypoxia, UCB cells added to the neuronal culture after hypoxia were able to decrease apoptotic cell death to the level observed in normoxic cultures (Hau et al. 2008). In another study, glutamate was applied to either

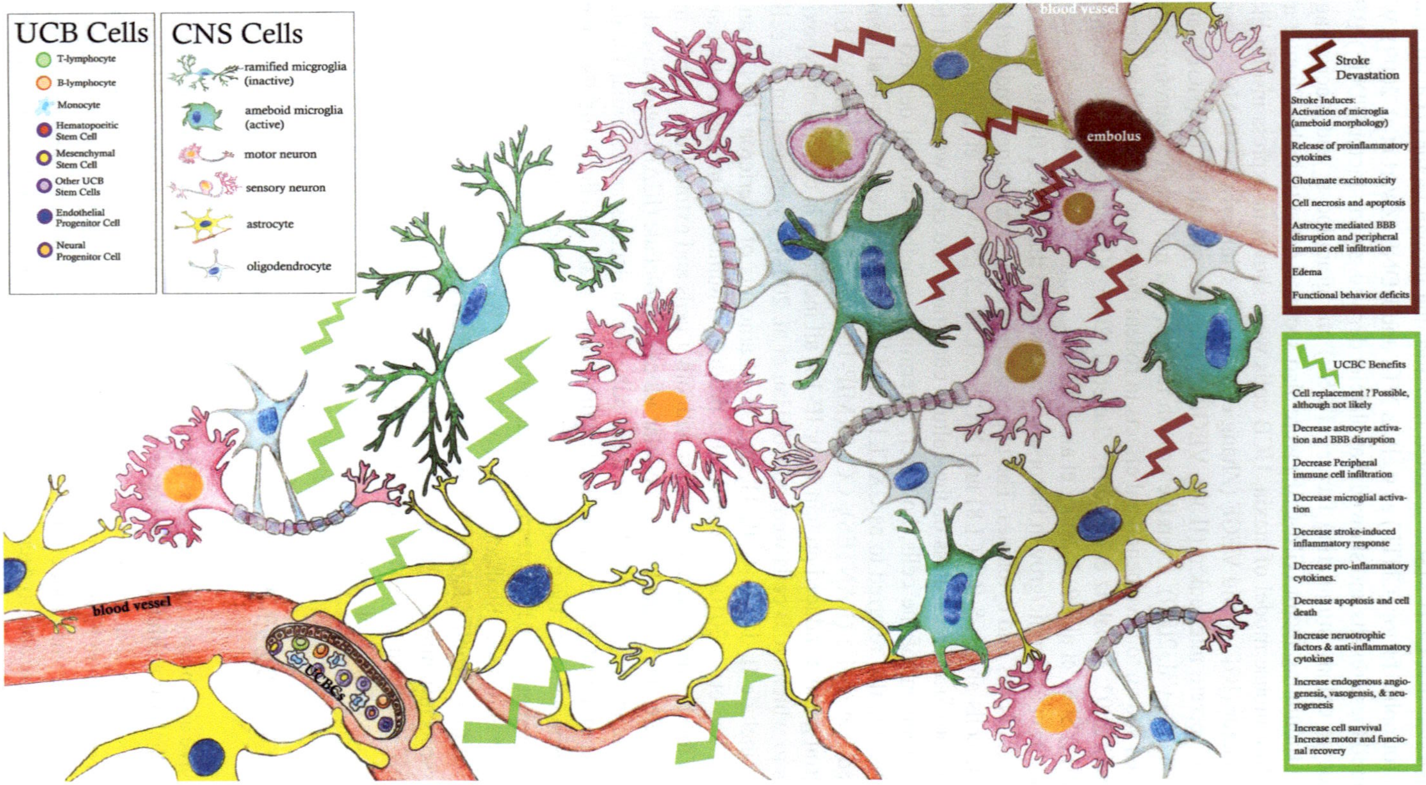

Fig. 5.1 CNS effects of cord blood administration

cortical neuronal cultures or co-cultures of cortical neurons and UCB cells. In the co-cultures, neuronal death as a result of glutamate exposure decreased, apoptosis decreased and there was an upregulation of Akt survival pathways in the cultured neurons. We do observe decreased apoptosis in the brains of MCAO rats treated with UCB cells (Newcomb et al. 2006). Even in cultures under normal conditions, the UCB cells can increase neuronal survival and neurite outgrowth (Chen et al. 2010).

Hall and fellow researchers found UCB cells have a protective effect on oligodendrocytes in vitro (Hall et al. 2009a). Using the lactic acid dehydrogenase (LDH) assay to determine oligodendrocyte survival, researchers found that UCB cells increased oligodendrocyte survival post-ischemia and also reduced caspase-3 activation, indicating a reduction in apoptosis. Further studies using microarray and quantitative real time polymerase chain reaction (qRT-PCR) assays demonstrated that soluble factors secreted by the UCB cells increased expression of oligodendrocyte repair, proliferation, and survival genes Insig1, Mt3, Prdx4, Stmn2, MOG, and Vcan (Rowe et al. 2010). These UCB effects were mediated by Akt signal transduction pathways, since addition of an Akt inhibitor to the culture medium eliminated the effects of the UCB cells (Rowe et al. 2012); UCB induced peroxiredoxin 4 (Prdx4) expression. Prdx4 expression, an endogenous antioxidant often involved in the reduction of oxidative stress, was prevented when Akt was inhibited. The potential of UCB cells to protect neuroglial cells post-ischemia suggests that UCB cell treatments salvage CNS structure and potentially function as well.

5.4.4 *Vascular Repair*

In the first study to suggest that the cells could induce vascular repair, IV administration of CD34+ UCB cells 48 h after distal MCAO resulted in the proliferation of endothelial cells and growth of new blood vessels in the ischemic zone around the infarct (Taguchi et al. 2004). Further, if an anti-angiogenic agent (Endostatin) was administered after UCB injection, stroke-induced neurogenesis was inhibited and cortical volume was decreased. More recently, AC133 (CD133)+ UCB cells were examined. There is also some evidence that endothelial progenitor cells express CD34, CD133, and VEGFR-2 (Peichev et al. 2000). Cell culture studies had shown that these cells could be expanded in vitro and produce tube-like structures reminiscent of capillaries (Janic et al. 2010). In a transient MCAO model with AC133+ cells delivered 24 h post-stroke, the cells increased expression of von Willebrand factor around the infarct; however, there was no change in cerebral blood flow (Iskander et al. 2013). In another study, EPCs and smooth muscle progenitors were isolated from CD34+ UCB cells on the basis of clone morphology and expanded in vitro. Twenty four hours after distal MCAO in the mouse, UCB-derived smooth muscle progenitors, EPCs or a combination of both were administered via the tail vein (Nih et al. 2012). All UCB cells increased vascular density in the cortex around the lesion and enhanced proliferation of CD31+ cells. Consistent with the earlier studies, endostatin eliminated the effect.

This phenomenon has also been observed in other models of ischemic injury. UCB derived progenitors have been shown to stimulate angiogenesis in models of hind limb and myocardial ischemia, restoring some degree of tissue perfusion (Finney et al. 2006; Ma et al. 2006). In none of these studies has there been any evidence that the UCB derived cells have incorporated into the new vessels. Just as the cells cannot be repairing the infarcted tissue through incorporation into existing neural circuitry, the cells must be inducing their effects through the expression of trophic factors or cytokines.

5.4.5 Host Immune Modulation

5.4.5.1 Sequelae of Stroke

One key sequelae of a stroke is the activation of microglia, the immune cells of the brain, and subsequent induction of inflammation. Inflammation begins with the onset of an ischemic stroke, but unlike excitotoxicity it is a more protracted process that does not peak until hours or days later (Dirnagl and Priller 2005). There has been considerable debate as to whether this response has neuroprotective or neurodegenerative effects. The general answer appears to be that inflammation is both depending upon whether it is examined early or late after the injury. In a review of the field, Kriz (Kriz 2006) suggested that after a stroke there is an early proliferative, neuroprotective response followed by a later pro-inflammatory response. When microglia or macrophage are eliminated early after stroke, infarct size is actually increased (Lalancette-Hebert et al. 2007; Smirkin et al. 2010). During this early proliferative phase of microglial activation after transient MCAO, these cells express mRNA for growth factors such as FGF2, GDNF, HGF, IGF-1, PDGF-A and VEGF (Smirkin et al. 2010). In contrast, the later inflammatory response is characterized by the massive influx of peripheral immune cells into the infarcted hemisphere through a permeable blood brain barrier. Many immune cells including neutrophils, T cells, B cells, and macrophages have been detected in the stroke-injured brain (Schroeter et al. 1994; Stevens et al. 2002). Recent evidence suggests that the majority of these cells originate in the spleen (Offner et al. 2006a; Vendrame et al. 2006); it is a reservoir of peripheral macrophages, T cells and B cells. It decreases in size in response to stroke and neurotoxic injury (Benner et al. 2004; Offner et al. 2006a). Removal of the spleen reduces infarct size (Ajmo et al. 2008). There is an increase in circulating macrophages and a reduction in B cells in the spleen in response to stroke (Offner et al. 2006a; Seifert et al. 2012). Further, spleen-derived T cells, monocytes and natural killer cells are observed in the brain at 48 and 96 h post-MCAO (Seifert et al. 2012). With activation of local microglia and the infiltration of peripheral immune cells, it is not surprising that cytokine and chemokine expression increases in the infarcted hemisphere (Offner et al. 2006b; Hurn et al. 2007; Chang et al. 2011).

5.4.5.2 Cord Blood Effects on CNS Inflammation

Microglia are important phagocytic glial cells of the CNS. They are the innate immune cells of the brain, responsible for antigen presentation and much of the inflammatory response. Microglia are known to secrete many different cytokines, chemokines, matrix metalloproteinases, free radicals which contribute to the pro-inflammatory state after cerebral ischemia (Rock et al. 2004). Although helpful under normal conditions, microglia cells can wreak havoc perpetuating inflammation and secondary damage post-stroke. Jiang and fellow researchers found that co-culturing microglia in an indirect culture system with UCB cells decreased microglial viability (Jiang et al. 2010). The researchers observed that UCB cell co-culture also suppressed expression of IL-1β. To further examine the specific cellular effects UCB cells have on microglial cells in vitro, researchers separately co-cultured CD11b+ monocytes, CD8+ T-cells, and CD19+ B-cells isolated from the UCB MNC fraction with microglia under OGD and examined microglial viability. The CD11b+ monocytes and CD19+ B-cells both demonstrated decreased microglial viability, whereas CD8+ T-cells induced the opposite effect. If these effects are representative of how the UCB cells interact with microglia in vivo, it would suggest that the UCB cells decrease the inflammatory response post stroke.

Astrocytes can also contribute to the inflammatory response through the production of cytokines. UCB cells also influence the function of astrocytes. When UCB MNCs were grown in an indirect co-culture system with astrocytes, IL-6 increased significantly in the normoxic cultures, but not in the cultures exposed to OGD (Jiang et al. 2011). Under hypoxic/ischemic conditions, the presence of UCB cells induced an increase in IL-10 concentration. The authors also examined cell populations isolated from the UCB MNC – B cells (CD19+), CD8+ T cells, and monocytes (CD11b+). Of these fractions, only the CD8+ T cells exhibited an effect on astrocyte survival during hypoxic conditions and slightly increased hypoxic IL-6 and IL-1β expression. UCB-derived CD8+ T cells decreased astrocyte survival under OGD compared to non-treated cells, but T cells had no effect under normoxic conditions.

In a more representative model of the nervous system, Hall et al employed the organotypic slice culture model to examine the effects of UCB MNCs on inflammation in the intact nervous system (Hall et al. 2009b). Slices were indirectly co-cultured with UCB MNCs for 48-h of exposure to OGD. The UCB cells significantly decreased OGD-ischemic damage in vitro. UCB cells also reduced the microglial production of nitric oxide (NO), a free radical that can exacerbate secondary ischemic damage and apoptosis. These data suggest that the UCB cells can mediate CNS inflammation, even though the cells were not directly in contact with microglia in the slice.

In our first study to examine the effects of UCB cells on inflammation in the brain, we showed that administering UCB MNCs IV 24 h after permanent MCAO surgery significantly lowered the number of CD45+/CD11b+ leukocytes and CD45+/B220+ B cells compared to rats with MCAO only (Vendrame et al. 2005). The UCB cells had no effect on infiltration of T-cells or neutrophils into the infarcted hemisphere. There was also an alteration in the cytokine expression within the post-MCAO CNS; pro-inflammatory markers, TNF-α, IL2, and IL-1β were significantly reduced. Interestingly, these changes in myeloid cells occur very quickly after UCB administration. Leonardo and colleagues examined changes seen in CNS CD11b+ pro-inflammatory cells when UCB cells were administered at the optimal dose and time: 10^6 cells at 48 h post-MCAO (Leonardo et al. 2010). Within 3 h of the UCB treatment, isolectin binding was reduced and CD11b+ cells demonstrated a ramified (resting) morphology, typical of a non-inflammatory state.

Not only are changes observed in the inflammatory state in the brain, but the cells appear to modulate the peripheral immune response. Vendrame et al observed that when UCB MNCs were administered via the penile vein at 24 h post-MCAO and the rats euthanized at 48 h post-MCAO, the cells restored spleen size to that of non-stroked animals (Vendrame et al. 2006). Further analysis indicated that during MCAO, spleens lost approximately 28 % of their CD8+ T cell population when compared to sham controls, decreasing the CD8+/CD4+ T cell ratio in the spleen, a phenomenon that was reversed with the administration of UCB cells post-MCAO. Splenocyte cytokine profiles after MCAO were also altered by UCB cell administration: TNF-α decreased while INF-γ and IL-10 significantly increased with UCB administration.

5.4.5.3 Peripheral Effects of Cord Blood

Altered cytokine expression and cell evacuation likely contribute to the MCAO-induced changes seen in the spleen (see Fig. 5.2). The administration of UCB cells at 24 h post MCAO seems to ameliorate these splenic changes (Vendrame et al. 2006). In a second study that examined which of the cells within the UCB MNC fraction contributed to the UCB effect on spleen, Golden et al observed that after MCAO, splenic T-cell dramatically increased, and after administration of UCB cells, regardless of whether all the constituent cell populations were present or not, there was a normalizing decrease in the splenic T-cells and a similar trend in circulating T cells (Golden et al. 2012). Perhaps more significant were the changes in circulating and splenic macrophage/monocytes. After stroke, there were significant decreases in these cells in blood and spleen. UCB cell administration resulted in further significant reductions in circulating monocytes/macrophage populations, although the UCB cells had little effect on the number of splenic macrophage. Removing the CD14+ monocytes from the MNC led to a further reduction in splenic macrophage, while removing CD133+ cells restored splenic numbers to normal. However, only the MNC fraction suppressed T cell proliferation.

5.5 Can Cord Blood Cells be Used to Treat Other Forms of Stroke?

5.5.1 Neonatal Models

In addition to providing benefit in adult models of cerebral ischemia, recent reports suggest that the cells are also reparative in models of neonatal hypoxia-ischemia, reducing damage (Xia et al. 2010), improving reflex development (Pimentel-Coelho

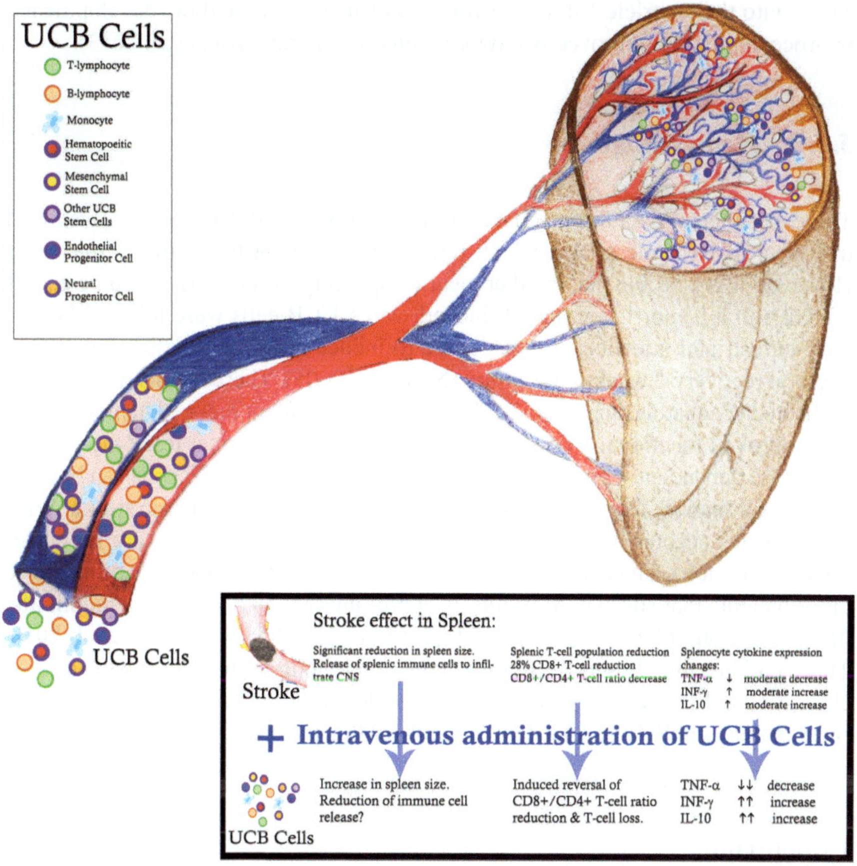

Fig. 5.2 Splenic effects of cord blood administration

et al. 2009), causing reorganization of cognitive maps (Geissler et al. 2011), improving functional recovery (Yasuhara et al. 2010b) and reducing spastic paresis (Meier et al. 2006). Rats treated with UCB cells performed better on the Morris water maze test of learning and memory than did untreated animals after hypoxia/ischemia (de Paula et al. 2012). The putative mechanisms underlying the recovery in neonate are similar to those observed in the adult. For example, the cells were shown to protect progenitors in the subventricular zone as well as mature cortical neurons to ultimately reduce brain damage (Bae et al. 2012; Wang et al. 2013b); Bae et al suggested the cells likely produced these effects by stimulating microglial activity early after the injury. Consistent with these results, the cells were shown to decrease apoptosis in this model and increase angiogenesis (Rosenkranz et al. 2012). The cells appear to also increase GDNF, BDNF and NGF in the brains of treated pups (Yasuhara et al. 2010a).

The beneficial effects of UCB administration were also observed in another neonatal model. One of the complications with premature birth is intraventricular hemorrhage that can lead to post-hemorrhage hydrocephalus; administering UCB

MSCs into the ventricle 1 day after this hemorrhage prevented the development of hydrocephalus and improved behavioral outcomes in rats (Ahn et al. 2013).

5.5.2 *Hemorrhage*

The experimental data that we have discussed in the adult have all involved UCB treatment of ischemic stroke, which represents the majority of strokes that occur. However, it is possible that the therapeutic capability of the cells would be equally effective in a hemorrhagic model. Indeed, when UCB cells were injected IV 24 h after intrastriatal injection of collagenase to induce hemorrhage, they induced behavioral recovery 2 weeks after injury (Nan et al. 2005). These results were recently replicated (Seghatoleslam et al. 2013). The UCB treatment reduced hematoma area and improved functional recovery on limb placement and corner tests; these effects were dose dependent. These initial studies demonstrate that UCB cells can be an efficacious treatment for intracerebral hemorrhage as well as ischemic stroke. While hemorrhages account for approximately 15% of all strokes, there are no current treatment options. The current standard of care is supportive with rehabilitation for those patients that survive. Mortality is approximately 50% in the 1st month after the incident and 60% by the end of 1 year (Juvela 1995; MacLellan et al. 2010). Those patients that do survive usually have severe long-term disabilities (van Asch et al. 2010). If further study bears out the results of these initial studies, a UCB derived therapy could have long-term benefit for this patient population and society.

Conclusions

The evidence presented suggests that the UCB cells have a great deal of potential as a future treatment for stroke, both ischemic and hemorrhagic, in young and adult alike. By virtue of the fact that they are living cells producing a variety of different factors, these cells have been shown to have pleiotropic effects. However, the stroke field is rife with examples of promising pre-clinical experimental treatments that have not panned out in clinical trial. There are studies which have not reproduced these positive outcomes in both adult models (Makinen et al. 2006; Nystedt et al. 2006; Zawadzka et al. 2009; Riegelsberger et al. 2011) and neonatal models (De Paula et al. 2009). There are no obvious differences between these studies and the studies presented earlier showing therapeutic benefit although extensive protocols for cell preparation and storage are not provided in most reports. It is imperative that we understand why these differences exist and their implications for the long-term success of this cell therapy.

There are also other questions that will need to be addressed as we move toward the clinic. The cells appear to be efficacious across the lifespan, but it is too early to draw that conclusion. Almost all of the studies in the adult are in younger adults and not the aged. While the cells have a demonstrated ability to increase proliferation in the subgranular zone of the hippocampus and increase density of dendritic spines (Bachstetter et al. 2008; Shahaduzzaman et al. 2013) in the aged rat, they have not yet been shown to improve stroke outcome in an aged animal.

All the studies discussed here have used human cells in a rodent model of stroke. Xenografting has its own immunologic issues. It is not particularly surprising that the UCB cells are rarely observed surviving in the host even when very stringent immune suppression protocols are employed; it is, perhaps, more surprising that the cells still seem to have such profound effects. When these cells are administered to a human stroke patient, however, it is not clear how much immune suppression or HLA matching will be required. These studies are best addressed in a larger animal model. There is one report in which 114 patients suffering from with 13 varied neurologic injuries or diseases all received allogeneic UCB MSC transplants with no preconditioning, HLA matching or immune suppression (Yang et al. 2010). All hematologic and immunologic parameters and serum chemistries were in the normal range for these patients. The worst adverse events reported were 3 % of the patients experienced headaches and 1 % developed a fever. There was no report on whether any of the patients experiences any benefit from the treatment and the number of patients with any specific neurologic condition was small. Further studies are still needed.

Even given these caveats, the bulk of the evidence suggests that UCB cells may have therapeutic benefit. Multiple groups using different models of cerebral ischemia and hemorrhage, different UCB cell preparations, different routes, different cell doses, differences in timing of cell delivery and different tests of motor and cognitive function have demonstrated efficacy of this experimental treatment. Cautious optimism is warranted as preclinical testing and future clinical trials move forward.

References

Abbas AK, Lichtman AH, Pillai SCaMI (2011) Cellular and molecular immunology: with STUDENT CONSULT Online Access, 7th edn. Elsvier Health Sciences, Amsterdam

Ahn SY, Chang YS, Sung DK, Sung SI, Yoo HS, Lee JH, Oh WI, Park WS (2013) Mesenchymal stem cells prevent hydrocephalus after severe intraventricular hemorrhage. Stroke 44:497–504

Ajmo CTJ, Vernon DO, Collier L, Hall AA, Garbuzova-Davis S, Willing A, Pennypacker KR (2008) The spleen contributes to stroke-induced neurodegeneration. J Neurosci Res 86:2227–2234

Ali H, Al-Mulla F (2012) Defining umbilical cord blood stem cells. Stem Cell Discovery 2:15–23

Arien-Zakay H, Lecht S, Bercu MM, Tabakman R, Kohen R, Galski H, Nagler A, Lazarovici P (2009) Neuroprotection by cord blood neural progenitors involves antioxidants, neurotrophic and angiogenic factors. Exp Neurol 216:83–94

Asahara T, Kawamoto A (2004) Endothelial progenitor cells for postnatal vasculogenesis. Am J Physiol Cell Physiol 287:C572–C579

Auerbach AD (1995) Fanconi anemia. Dermatol Clin 13:41–49

Auffray C, Sieweke MH, Geissmann F (2009) Blood monocytes: development, heterogeneity, and relationship with Dendritic cells. Annu Rev Immunol 27:669–692

Bacchetta R, Bigler M, Touraine JL, Parkman R, Tovo PA, Abrams J, de Waal MR, de Vries JE, Roncarolo MG (1994) High levels of interleukin 10 production in vivo are associated with tolerance in SCID patients transplanted with HLA mismatched hematopoietic stem cells. J Exp Med 179:493–502

Bachstetter AD, Pabon MM, Cole MJ, Hudson CE, Sanberg PR, Willing AE, Bickford PC, Gemma C, Bachstetter AD, Pabon MM, Cole MJ, Hudson CE, Sanberg PR, Willing AE, Bickford PC, Gemma C (2008) Peripheral injection of human umbilical cord blood stimulates neurogenesis in the aged rat brain. BMC Neurosci 9:22

Bae S-H, Kong T-H, Lee H-S, Kim K-S, Hong KS, Chopp M, Kang M-S, Moon J (2012) Long-lasting paracrine effects of human cord blood cells on damaged neocortex in an animal model of cerebral palsy. Cell Transplant 21:2497–2515

Ballen K (2005) New trends in umbilical cord blood transplantation. Blood 105:3786–3782

Ballen KK, Gluckman E, Broxmeyer HE (2013) Umbilical cord blood transplantation: the first 25 years and beyond. Blood 122:491–498

Belge KU, Dayyani F, Horelt A, Siedlar M, Frankenberger M, Frankenberger B, Espevik T, Ziegler-Heitbrock L (2002) The proinflammatory CD14+ CD16+ DR++ monocytes are a major source of TNF. J Immunol 168:3536–3542

Benner EJ, Mosley RL, Destache CJ, Lewis TB, Jackson-Lewis V, Gorantla S, Nemachek C, Green SR, Przedborski S, Gendelman HE (2004) Therapeutic immunization protects dopaminergic neurons in a mouse model of Parkinson's disease. Proc Natl Acad Sci U S A 101:9435–9440

Bertolini F, Lazzari L, Lauri E, Corsini C, Castelli C, Gorini F, Sirchia G (1995) Comparative study of different procedures for the collection and banking of umbilical cord blood. J Hematother 4:29–36

Bicknese AR, Goodwin HS, Quinn CO, Henderson VC, Chien SN, Wall DA (2002) Human umbilical cord blood cells can be induced to express markers for neurons and glia. Cell Transplant 11:261–264

Bieback K, Kern S, Klüter H, Eichler H (2004) Critical parameters for the isolation of mesenchymal stem cells from umbilical cord blood. Stem Cells 22:625–634

Binder CJ, Silverman GJ (2005) Natural antibodies and the autoimmunity of atherosclerosis. Springer Semin Immunopathol 26:385–404

Boltze J, Reich DM, Hau S, Reymann KG, Strassburger M, Lobsein D, Wagner D-C, Kamprad M, Stahl T (2012a) Assessment of neuroprotective effects of human umbilical cord blood mononuclear cell subpopulations in vitro and in vivo. Cell Transplant 21:723–737

Boltze J, Schmidt UR, Reich DM, Kranz A, Reymann KG, Strassburger M, Lobsien D, Wagner DC, Forschler A, Schabitz WR (2012b) Determination of the therapeutic time window for human umbilical cord blood mononuclear cell transplantation following experimental stroke in rats. Cell Transplant 21:1199–1211

Borlongan CV, Hadman M, Davis Sanberg C, Sanberg PR (2004) CNS entry of peripherally injected umbilical cord blood cells is not required for neuroprotection in stroke. Stroke 35:2385–2389

Bracci-Laudiero L, Celestino D, Starace G, Antonelli A, Lambiase A, Procoli A, Rumi C, Lai M, Picardi A, Ballatore G, Bonini S, Aloe L (2003) CD34-positive cells in human umbilical cord blood express nerve growth factor and its specific receptor TrkA. J Neuroimmunol 136:130–139

Brichard B, Varis I, Latinne D, Deneys V, de Bruyere M, Leveugle P, Cornu G (2001) Intracellular cytokine profile of cord and adult blood monocytes. Bone Marrow Transplant 27:1081–1086

Broxmeyer HE, Douglas GW, Hangoc G, Cooper S, Bard J, English D, Arny M, Thomas L, Boyse EA (1989) Human umbilical cord blood as a potential source of transplantable hematopoietic stem/progenitor cells. Proc Natl Acad Sci U S A 86:3828–3832

Broxmeyer HE, Hangoc G, Cooper S, Ribeiro RC, Graves V, Yoder M, Wagner J, Vadhan-Raj S, Benninger L, Rubinstein P, Broun ER (1992) Growth characteristics and expansion of human umbilical cord blood and estimation of its potential for transplantation in adults. Proc Natl Acad Sci U S A 89:4109–4113

Broxmeyer HE, Lee MR, Hangoc G, Cooper S, Prasain N, Kim YJ, Mallet C, Ye Z, Witting S, Cornetta K, Cheng L, Yoder MC (2011) Hematopoietic stem/progenitor cells, generation of induced pluripotent stem cells, and isolation of endothelial progenitors from 21-to 23.5-year cryopreserved cord blood. Blood 117:4773–4777

Bühnemann C, Scholz A, Bernreuther C, Malik CY, Braun H, Schachner M, Reymann KG, Dihné M (2006) Neuronal differentiation of transplanted embryonic stem cell-derived precursors in stroke lesions of adult rats. Brain 129:3238–3248

Buzanska L, Machaj EK, Zablocka B, Pojda Z, Domanska-Janik K (2002) Human cord blood-derived cells attain neuronal and glial features in vitro. J Cell Sci 115:2131–2138

Bużańska L, Jurga M, Stachowiak EK, Stachowiak MK, Domańska-Janik K (2006) Neural stem-like cell line derived from a nonhematopoietic population of human umbilical cord blood. Stem Cells Dev 15:391–406

Chang L, Chen Y, Li J, Liu Z, Wang Z, Chen J, Cao W, Xu Y (2011) Cocaine-and amphetamine-regulated transcript modulates peripheral immunity and protects against brain injury in experimental stroke. Brain Behav Immun 25:260–269

Chen J, Sanberg PR, Li Y, Wang L, Lu M, Willing AE, Sanchez-Ramos J, Chopp M (2001) Intravenous administration of human umbilical cord blood reduces behavioral deficits after stroke in rats. Stroke 32:2682–2688

Chen N, Newcomb J, Garbuzova-Davis S, Sanberg CD, Sanberg PR, Willing AE (2010) Human umbilical cord blood cells have trophic effects on young and aging Hippocampal neurons in vitro. Aging Dis 1:173–190

Chua SJ, Bielecki R, Wong CJ, Yamanaka N, Rogers IM, Casper RF (2009) Neural progenitors, neurons and oligodendrocytes from human umbilical cord blood cells in a serum-free, feeder-free cell culture. Biochem Biophys Res Comm 379:217–221

D'Arena G, Musto P, Cascavilla N, Di Giorgio G, Fusilli S, Zendoli F, Carotenuto M (1998) Flow cytometric characterization of human umbilical cord blood lymphocytes: immunophenotypic features. Haematologica 83:197–203

De Paula S, Santos- Vitola S, Greggio S, De Paula D, Billig-Mello P, Mistrello-Lubianca J, Xavier LL, FIORI HH, Dacosta JC (2009) Hemispheric Brain Injury and Behavioral Deficits Induced by Severe Neonatal Hypoxia-Ischemia in Rats Are Not Attenuated by Intravenous Administration of Human Umbilical Cord Blood Cells. Pediatr Res 65:631–635

De Paula S, Greggio S, Marinowic DR, Machado DC, DaCosta JC (2012) The dose-response effect of acute intravenous transplantation of human umbilical cord blood cells on brain damage and spatial memory deficits in neonatal hypoxia-ischemia. Neurosci 210:431–441

Dhandapani KM, Brann DW (2003) Transforming growth factor-beta: a neuroprotective factor in cerebral ischemia. Cell Biochem Biophys 39:13–22

Dirnagl U, Priller J (2005) Focal cerebral ischemia: the multifaceted role of glial cells. In: Kettenmann H, Ransom BR (eds) Neuroglia, 2nd edn. Oxford University, New York, pp 511–520

Edwards JC, Cambridge G (2006) B-cell targeting in rheumatoid arthritis and other autoimmune diseases. Nat Rev Immunol 6:394–403

Eggermann J, Kliche S, Jarmy G, Hoffmann K, Mayr-Beyrle U, Debatin KM, Beltinger C (2003) Endothelial progenitor cell culture and differentiation in vitro: a methodological comparison using human umbilical cord blood. Cardiovasc Res 58:478–486

Fan C-G, Zhang Q-J, Tang F-W, Han Z-B, Wang G-S, Han Z-C (2005) Human umbilical cord blood cells express neurotrophic factors. Neurosci Lett 380:322–325

Filias A, Theodorou G, Mouzopoulou S, Varvarigou A, Mantagos S, Karakantza M (2011) Phagocytic ability of neutrophils and monocytes in neonates. BMC Pediatr 11:29

Finney MR, Greco NJ, Haynesworth SE, Martin JM, Hedrick DP, Swan JZ, Winter DG, Kadereit S, Joseph ME, Fu P, Pompili VJ, Laughlin MJ (2006) Direct comparison of umbilical cord blood versus bone marrow-derived endothelial precursor cells in mediating neovascularization in response to vascular ischemia. Biol Blood Marrow Transplant 12:585–593

Geissler M, Dinse HR, Neuhoff S, Kreikemeier K, Meier C (2011) Human umbilical cord blood cells restore brain damage induced changes in rat *Somatosensory* cortex. PLoS ONE 6:e20194

Geissmann F, Auffray C, Palframan R, Wirrig C, Ciocca A, Campisi L, Narni-Mancinelli E, Lauvau G (2008) Blood monocytes: distinct subsets, how they relate to dendritic cells, and their possible roles in the regulation of T-cell responses. Immunol Cell Biol 86:398–408

Gille C, Leiber A, Mundle I, Spring B, Abele H, Spellerberg B, Hartmann H, Poets CF, Orlikowsky TW (2009) Phagocytosis and postphagocytic reaction of cord blood and adult blood Monocyte after infection with green fluorescent protein-labeled *Escherichia coli* and Group B Streptococci. Cytometry Part B 76B:271–284

Gilmore GL, DePasquale DK, Lister J, Shadduck RK (2000) Ex vivo expansion of human umbilical cord blood and peripheral blood CD34+ hematopoietic stem cells. Exp Hematol 28:1297–1305

Gluckman E, Rocha V, Boyer-Chammard A, Locatelli F, Arcese W, Pasquini R, Ortega J, Souillet G, Ferreira E, Laporte JP, Fernandez M, Chastang C (1997) Outcome of cord-blood transplantation from related and unrelated donors. N Eng J Med 337:373–381

Go AS et al (2013) Heart disease and stroke statistics-2013 update: a report from the American Heart Association. Circulation 127:e6–e245

Golden JE, Shahaduzzaman M, Wabnitz A, Green S, Womble TA, Sanberg PR, Pennypacker KR, Willing AE (2012) Human umbilical cord blood cells alter blood and spleen cell populations after stroke. Transl Stroke Res 3:491–499

Ha Y, Choi DH, Yeon DS, Lee JJ, Kim HO, Cho YE (2001) Neural phenotype expression of cultured human cord blood cells in vitro. NeuroReport 12:3523–3527

Ha Y, Lee JE, Kim KN, Cho YE, Yoon DH (2003) Intermediate filament nestin expressions in human cord blood monocytes (HCMNCs). Acta Neurochir (Wien) 145:483–487

Habich A, Jurga M, Markiewicz I, Lukomska B, Bany-Laszewicz U, Domanska-Janik K (2006) Early appearance of stem/progenitor cells with neural-like characteristics in human cord blood mononuclear fraction cultured in vitro. Exp Hematol 34:914–925

Haddad R, Guardiola P, Izac B, Thibault C, Radich J, Delezoide AL, Baillou C, Lemoine FM, Gluckman JC, Pflumio F, Canque B (2004) Molecular characterization of early human T/NK and B-lymphoid progenitor cells in umbilical cord blood. Blood 104:3918–3926

Hall A, Guyer A, Leonardo C, Ajmo C, Collier L, Willing AE, Pennypacker KR (2009a) Umbilical cord blood cells directly suppress ischemic oligodendrocyte cell death. J Neurosci Res 87:333–341

Hall AA, Leonardo CC, Collier LA, Rowe DD, Willing AE, Pennypacker KR (2009b) Delayed treatments for stroke influence neuronal death in rat organotypic slice cultures subjected to oxygen glucose deprivation. Neurosci 164:470–477

Harris DT, Schumacher MJ, Locascio J, Besencon FJ, Olson GB, DeLuca D, Shenker L, Bard J, Boyse EA (1992) Phenotypic and functional immaturity of human umbilical cord blood T lymphocytes. Proc Natl Acad Sci U S A 89:10006–10010

Hau S, Reich DM, Scholz M, Naumann W, Emmich F, Kamprad M, Boltze J (2008) Evidence for neuroprotective properties of human umbilical cord blood cells after neuronal hypoxia in vitro. BMC Neurosci 9:30

Hoyert DL, Xu JQ (2012) Deaths: preliminary data for 2011. In National vital statistics reports. Hyattsville

Hurn PD, Subramanian S, Parker SM, Afentoulis ME, Kaler LJ, Vandenbark AA, Offner H (2007) T- and B-cell-deficient mice with experimental stroke have reduced lesion size and inflammation. J Cereb Blood Flow Metab 27:1798–1805

Ingram DA, Mead LE, Tanaka H, Meade V, Fenoglio A, Mortell K, Yoder MC (2004) Identification of a novel hierarchy of endothelial progenitor cells using human peripheral and umbilical cord blood. Blood 104:2752–2760

Iskander A, Knight RA, Zhang ZG, Ewing JR, Shankar A, Varma NR, Bagher-Ebadian H, Ali MM, Arbab AS, Janic B (2013) Intravenous administration of human umbilical cord blood-derived AC133+ endothelial progenitor cells in rat stroke model reduces infarct volume: magnetic resonance imaging and histological findings. Stem Cells Transl Med 2:703–714

Jang YK, Park JJ, Lee MC, Yoon BH, Yang YS, Yang SE, Kim SU (2004) Retinoic acid-mediated induction of neurons and glial cells from human umbilical cord-derived hematopoietic stem cells. J Neurosci Res 75:573–584

Janic B, Guo AM, Iskander AS, Varma NR, Scicli AG, Arbab AS (2010) Human cord blood-derived AC133+ progenitor cells preserve endothelial progenitor characteristics after long term in vitro expansion. PLoS ONE 5:e9173

Jiang Q, Azuma E, Hirayama M, Iwamoto S, Kumamoto T, Kobayashi M, Yamamoto H, Sakurai M, Komada Y (2001) Functional immaturity of cord blood monocytes as detected by impaired response to hepatocyte growth factor. Ped Intl 43:334–339

Jiang L, Newman M, Saporta S, Chen N, Sanberg C, Sanberg PR, Willing AE (2008) MIP-1α and MCP-1 Induce migration of human umbilical cord blood cells in models of stroke. Curr Neurovasc Res 5:118–124

Jiang L, Womble T, Saporta S, Chen N, Sanberg CD, Sanberg PR, Willing AE (2010) Human umbilical cord blood cells depress the microglial inflammatory response in vitro. Stem Cells Dev 19:221–227

Jiang L, Saporta S, Chen N, Sanberg CD, Paul Sanberg PR, Alison Willing AE (2011) The effect of human umbilical cord blood cells on survival and cytokine production by post-ischemic astrocytes in vitro. Stem Cell Rev 6:523–531

Jurga M, Markiewicz I, Sarnowska A, Habich A, Kozlowska H, Lukomska B, Buzanska L, Domanska-Janik K (2006) Neurogenic potential of human umbilical cord blood: neural-like stem cells depend on previous long-term culture conditions. J Neurosci Res 83:627–637

Juvela S (1995) Risk factors for impaired outcome after spontaneous intracerebral hemorrhage. Arch Neurol 52:1193–1200

Kawada H, Takizawa S, Takanashi T, Morita Y, Fujita J, Fukuda K, Takagi S, Okano H, Ando K, Hotta T (2006) Administration of hematopoietic cytokines in the subacute phase after cerebral infarction is effective for functional recovery facilitating proliferation of intrinsic neural stem/progenitor cells and transition of bone marrow-derived neuronal cells. Circulation 113:701–710

Khakoo AY, Finkel T (2005) Endothelial progenitor cells. Annu Rev Med 56:79–101

Kogler G, Radke TF, Lefort A, Sensken S, Fischer J, Sorg RV, Wernet P (2005) Cytokine production and hematopoiesis supporting activity of cord blood-derived unrestricted somatic stem cells. Exp Hematol 33:573–583

Koh SH, Kim KS, Choi MR, Jung KH, Park KS, Chai YG, Kim SH (2008) Implantation of human umbilical cord-derived mesenchymal stem cells as a neuroprotective therapy for ischemic stroke in rats. Brain Res 1229:233–248

Kozłowska H, Jabłonka J, Janowski M, Jurga M, Kossut M, Domańska-Janik K (2007) Transplantation of a novel human cord blood-derived neural-like stem cell line in a rat model of cortical infarct. Stem Cells Dev 16:481–488

Kriz J (2006) Inflammation in ischemic brain injury: timing is important. Crit Rev Neurobiol 18:145–157

Lalancette-Hebert M, Gowing G, Simard A, Weng YC, Kriz J (2007) Selective ablation of proliferating microglial cells exacerbates ischemic injury in the brain. J Neurosci 27:2596–2605

Le T, Leung L, Carroll WL, Schibler KR (1997) Regulation of interleukin—10 gene expression: possible mechanisms accounting for its upregulation and for maturational differences in its expression by blood mononuclear cells. Blood 89:4112–4119

Leonardo C, Hall AA, Collier LA, Ajmo CT Jr, Willing AE, Pennypacker KR (2010) HUCB cell therapy blocks the morphological change and recruitment of CD11b-expressing, isolectin-binding proinflammatory cells after MCAO. J Neurosci Res 88:1213–1222

Lim JY, Park SI, Oh JH, Kim SM, Jeong CH, Jun JA, Lee KS, Oh W, Lee JK, Jeun SS, Lim JY, Park SI, Oh JH, Kim SM, Jeong CH, Jun JA, Lee K-S, Oh W, Lee J-K, Jeun S-S (2008) Brain-derived neurotrophic factor stimulates the neural differentiation of human umbilical cord blood-derived mesenchymal stem cells and survival of differentiated cells through MAPK/ERK and PI3 K/Akt-dependent signaling pathways. J Neurosci Res 86:2168–2178

Liu CH, Hwang SM (2005) Cytokine interactions in mesenchymal stem cells from cord blood. Cytokine 32:270–279

Liu E, Tu W, Law HKW, Lau Y-L (2001) Decreased yield, phenotypic expression and function of immature monocyte-derived dendritic cells in cord blood. Br J Haematol 113:240–246

Ma N, Ladilov Y, Moebius JM, Ong L, Piechaczek C, Dávid Á, Kaminski A, Choi Y-H, Li W, Egger D, Stamm C, Steinhoff G (2006) Intramyocardial delivery of human CD133+ cells in a SCID mouse cryoinjury model: bone marrow vs. cord blood-derived cells. Cardiovasc Res 71:158–169

MacLellan CL, Silasi G, Auriat AM, Colbourne F (2010) Rodent models of intracerebral hemorrhage. Stroke 41:95–98

Makinen S, Kekarainen T, Nystedt J, Liimatainen T, Huhtala T, Narvanen A, Laine J, Jolkkonen J (2006) Human umbilical cord blood cells do not improve sensorimotor or cognitive outcome following transient middle cerebral artery occlusion in rats. Brain Res 1123:207–215

Martins AA, Paiva A, Morgado JM, Gomes A, Pais ML (2009) Quantification and immunophenotypic characterization of bone marrow and umbilical cord blood mesenchymal stem cells by multicolor flow cytometry. Transplant Proc 41:943–946

McGuckin CP, Pearce D, Forraz N, Tooze JA, Watt SM, Pettengell R (2003) Multiparametric analysis of immature cell populations in umbilical cord blood and bone marrow. Eur J Haematol 71:341–350

McGuckin CP, Forraz N, Baradez MO, Navran S, Zhao J, Urban R, Tilton R, Denner L (2005) Production of stem cells with embryonic characteristics from human umbilical cord blood. Cell Prolif 38:245–255

Meier C, Middlelanis J, Wasielewski B, Neuhoff S, Roth-Haerer A, Gantert M, Dinse HR, Dermietzel R, Jensen A (2006) Spastic paresis after perinatal brain damage in rats is reduced by human cord blood mononuclear cells. Pediatr Res 59:244–249

Morgado JM, Pratas R, Laranjeira P, Henriques A, Crespo I, Regateiro F, Paiva A (2008) The phenotypical and functional characteristics of cord blood monocytes and CD14 (-/low)/CD16 (+) dendritic cells can be relevant to the development of cellular immune responses after transplantation. Transpl Immunol 19:55–63

Nahrendorf M, Swirski FK, Aikawa E, Stangenberg L, Wurdinger T, Figueiredo JL, Libby P, Weissleder R, Pittet MJ (2007) The healing myocardium sequentially mobilizes two monocyte subsets with divergent and complementary functions. J Exp Med 204:3037–3047

Nan Z, Grande A, Sanberg CD, Sanberg PR, Low WC (2005) Infusion of human umbilical cord blood amerlieorates neurologic deficits in rats with hemorrhagic brain injury. Ann NY Acad Sci 1049:84–96

Newcomb JD, Ajmo CT, Davis Sanberg C, Sanberg PR, Pennypacker KR, Willing AE (2006) Timing of cord blood treatment after experimental stroke determines therapeutic efficacy. Cell Transplant 15:213–223

Newman MB, Willing AE, Manresa JJ, Davis-Sanberg C, Sanberg PR (2005) Stroke induced migration of human umbilical cord blood: time course and cytokines. Stem Cells Dev 14:576–586

Newman MB, Willing AE, Manresa JJ, Davis Sanberg C, Sanberg PR (2006) Cytokines produced by cultured human umbilical cord blood (HUCB) cells: implications for brain repair. Exp Neurol 199:201–208

Nih LR, Deroide N, Lere-Dean C, Lerouet D, Soustrat M, Levy BI, Silvestre JS, Merkulova-Rainon T, Pocard M, Margaill I, Kubis N (2012) Neuroblast survival depends on mature vascular network formation after mouse stroke: role of endothelial and smooth muscle progenitor cell co-administration. Eur J Neurosci 35:1208–1217

NINDS_rtPA_Stroke_Study_Group (1995) Tissue plasminogen activator for acute ischemic stroke. N Eng J Med 333:1581–1587

Nystedt J, Makinen S, Laine J, Jolkkonen J (2006) Human cord blood CD34+ cells and behavioral recovery following focal cerebral ischemia in rats. Acta Neurobiol Exp (Wars) 66:293–300

Offner H, Subramanian S, Parker SM, Afentoulis ME, Vandenbark AA, Hurn PD (2006a) Experimental stroke induces massive, rapid activation of the peripheral immune system. J Cereb Blood Flow Metab 26:654–665

Offner H, Subramanian S, Parker SM, Afentoulis ME, Vandenbark AA, Hurn PD (2006b) Experimental stroke induces massive, rapid activation of the peripheral immune system. J Cereb Blood Flow Metab 26:654–665

Ong LM, Fan X, Chu PPY, Gay FPH, Bari S, Ang JML, Li Z, Chen J, Lim SK, Bunte RM, Hwang WYK (2012) Cotransplantation of ex vivo expanded and unexpanded cord blood units in immunodeficient mice using insulin growth factor binding protein-2-augmented mesenchymal cell cocultures. Biol Blood Marrow Transplant 18:674–682

Park KS, Kim YS, Kim JH, Choi BK, Kim SH, Oh SH, Ahn YR, Lee MS, Lee MK, Park JB, Kwon CH, Joh JW, Kim KW, Kim SJ (2009) Influence of human allogenic bone marrow and cord blood-derived mesenchymal stem cell secreting trophic factors on ATP (adenosine—5'-triphosphate)/ADP (adenosine—5'-diphosphate) ratio and insulin secretory function of isolated human islets from cadaveric donor. Transplant Proc 41:3813–3818

Peichev M, Naiyer AJ, Pereira D, Zhu Z, Lane WJ, Williams M, Oz MC, Hicklin DJ, Witte L, Moore MAS, Rafii S (2000) Expression of VEGFR-2 and AC133 by circulating human CD34 (+) cells identifies a population of functional endothelial precursors. Blood 95:952–958

Pimentel-Coelho PM, Magalhães ES, Lopes LM, deAzevedo LC, Santiago MF, Mendez-Otero R (2009) Human cord blood transplantation in a neonatal rat model of hypoxic-ischemic brain damage: functional outcome related to neuroprotection in the striatum. Stem Cells Dev 19:351–358

Riegelsberger U-M, Deten A, Pösel C, Zille M, Kranz A, Boltze J, Wagner D-C (2011) Intravenous human umbilical cord blood transplantation for stroke: Impact on infarct volume and caspase-3-dependent cell death in spontaneously hypertensive rats. Exp Neurol 227:218–223

Rock RB, Gekker G, Hu S, Sheng WS, Cheeran M, Lokensgard JR, Peterson PK (2004) Role of microglia in central nervous system infections. Clin Microbiol Rev 17:942–964 (table of contents)

Rosenkranz K, Meier C (2011) Umbilical cord blood cell transplantation after brain ischemia-from recovery of function to cellular mechanisms. Ann Anat 193:371–379

Rosenkranz K, Kumbruch S, Tenbusch M, Marcus K, Marschner K, Dermietzel R, Meier C (2012) Transplantation of human umbilical cord blood cells mediated beneficial effects on apoptosis, angiogenesis and neuronal survival after hypoxic-ischemic brain injury in rats. Cell Tissue Res 348:429–438

Rowe DD, Leonardo CC, Hall AA, Shahaduzzaman MD, Collier LA, Willing AE, Pennypacker KR (2010) Cord blood administration induces oligodendrocyte survival through alterations in gene expression. Brain Res 1366:172–188

Rowe DD, Leonardo CC, Recio JA, Collier LA, Willing AE, Pennypacker KR (2012) Human umbilical cord blood cells protect oligodendrocytes from brain ischemia through Akt signal transduction. J Biol Chem 287:4177–4187

Sanberg PR, Willing AE, Garbuzova-Davis S, Saporta S, Liu G, Sanberg CD, Bickford PC, Klasko SK, El-Badri NS (2005) Umbilical cord blood-derived stem cells and brain repair. Ann NY Acad Sci 1049:67–83

Sanchez-Ramos JR, Song S, Kamath SG, Zigova T, Willing A, Cardozo-Pelaez F, Stedeford T, Chopp M, Sanberg PR (2001) Expression of neural markers in human umbilical cord blood. Exp Neurol 171:109–115

Schabitz WR, Kollmar R, Schwaninger M, Juettler E, Bardutzky J, Scholzke MN, Sommer C, Schwab S (2003) Neuroprotective effect of granulocyte colony-stimulating factor after focal cerebral ischemia. Stroke 34:745–751

Schroeter M, Jander S, Witte OW, Stoll G (1994) Local immune responses in the rat cerebral cortex after middle cerebral artery occlusion. J Neuroimmunol 55:195–203

Seghatoleslam M, Jalali M, Nikravesh MR, Hamidi Alamdari D, Hosseini M, Fazel A (2013) Intravenous administration of human umbilical cord blood-mononuclear cells dose-dependently relieve neurologic deficits in rat intracerebral hemorrhage model. Ann Anat 195:39–49

Seifert HA, Hall AA, Chapman CB, Collier LA, Willing AE, Pennypacker KR (2012) A transient decrease in spleen size following stroke corresponds to splenocyte release into systemic circulation. J Neuroimmune Pharmacol 7:1017–1024

Shahaduzzaman M, Golden JE, Green S, Gronda AE, Adrien E, Ahmed A, Sanberg PR, Bickford PC, Gemma C, Willing AE (2013) A single administration of human umbilical cord blood T cells produces long-lasting effects in the aging hippocampus. Age (Dordr) 35:2071–2087

Shang J, Deguchi K, Ohta Y, Liu N, Zhang X, Tian F, Yamashita T, Ikeda Y, Matsuura T, Funakoshi H, Nakamura T, Abe K (2011) Strong neurogenesis, angiogenesis, synaptogenesis, and antifibrosis of hepatocyte growth factor in rats brain after transient middle cerebral artery occlusion. J Neurosci Res 89:86–95

Shen J, Ishii Y, Xu G, Dang TC, Hamashima T, Matsushima T, Yamamoto S, Hattori Y, Takatsuru Y, Nabekura J, Sasahara M (2012) PDGFR-beta as a positive regulator of tissue repair in a mouse model of focal cerebral ischemia. J Cereb Blood Flow Metab 32:353–367

Smirkin A, Matsumoto H, Takahashi H, Inoue A, Tagawa M, Ohue S, Watanabe H, Yano H, Kumon Y, Ohnishi T, Tanaka J (2010) Iba1+/NG2+ macrophage-like cells expressing a variety of neuroprotective factors ameliorate ischemic damage of the brain. J Cereb Blood Flow Metab 30:603–615
Solomon EP, Berg LR, Martin DW (1999) Biology. Saunders College Publishing, Orlando
Stevens SL, Bao J, Hollis J, Lessov NS, Clark WM, Stenzel-Poore MP (2002) The use of flow cytometry to evaluate temporal changes in inflammatory cells following focal cerebral ischemia in mice. Brain Res 932:110–119
Sun W, Buzanska L, Domanska-Janik K, Salvi RJ, Stachowiak MK (2005) Voltage-sensitive and ligand-gated channels in differentiating neural stem-like cells derived from the nonhematopoietic fraction of human umbilical cord blood. Stem Cells 23:931–945
Swirski FK, Nahrendorf M, Etzrodt M, Wildgruber M, Cortez-Retamozo V, Panizzi P, Figueiredo JL, Kohler RH, Chudnovskiy A, Waterman P, Aikawa E, Mempel TR, Libby P, Weissleder R, Pittet MJ (2009) Identification of splenic reservoir monocytes and their deployment to inflammatory sites. Science 325:612–616
Taguchi A, Soma T, Tanaka H, Kanda T, Nishimura H, Yoshikawa H, Tsukamoto Y, Iso H, Fujimori Y, Stern DM, Naritomi H, Matsuyama T (2004) Administration of CD34+ cells after stroke enhances neurogenesis via angiogenesis in a mouse model. J Clin Invest 114:330–338
Van Asch CJ, Luitse MJ, Rinkel GJ, van der Tweel I, Algra A, Klijn CJ (2010) Incidence, case fatality, and functional outcome of intracerebral haemorrhage over time, according to age, sex, and ethnic origin: a systematic review and meta-analysis. Lancet Neurol 9:167–176
Van Furth R, Cohn ZA (1968) The origin and kinetics of mononuclear phagocytes. J Exp Med 128:415–435
Van Voorhis WC, Valinsky J, Hoffman E, Luban J, Hair LS, Steinman RM (1983) Relative efficacy of human monocytes and dendritic cells as accessory cells for T cell replication. J Exp Med 158:174–191
Vanneaux V, El-Ayoubi F, Delmau C, Driancourt C, Lecourt S, Grelier A, Cras A, Cuccuini W, Soulier J, Lataillade JJ, Lebousse-Kerdiles MC, Oury JF, Sibony O, Marolleau JP, Benbunan M, Uzan G, Larghero J (2010) In vitro and in vivo analysis of endothelial progenitor cells from cryopreserved umbilical cord blood: are we ready for clinical application? Cell Transplant 19:1143–1155
Vendrame M, Cassady CJ, Newcomb J, Butler T, Pennypacker KR, Zigova T, Davis Sanberg C, Sanberg PR, Willing AE (2004) Infusion of human umbilical cord blood cells in a rat model of stroke dose-dependently rescues behavioral deficits and reduces infarct volume. Stroke 35:2390–2395
Vendrame M, Gemma C, De Mesquita D, Collier L, Bickford PC, Davis Sanberg C, Sanberg PR, Pennypacker KR, Willing AE (2005) Anti-inflammatory effects of human cord blood cells in a rat model of stroke. Stem Cells Dev 14:595–604
Vendrame M, Gemma C, Pennypacker KR, Bickford PC, Davis Sanberg C, Sanberg PR, Willing AE (2006) Cord blood rescues stroke-induced changes in splenocyte phenotype and function. Exp Neurol 199:191–200
Walczak P, Chen N, Hudson J, Zigova T, Sanchez-Ramos J, Sanberg PR, Willing AE (2007) Long-term cultured human umbilical cord neural-like cells transplanted into the striatum of NOD SCID mice. Brain Res Bull 14:155–163
Walton M, Brown R, Gupta G, Joneckis C, Llave R (1996) Activase for Acute Ischemic stroke. Summary Basis for Approval. www.fda.gov
Wang J, Tang Y, Zhang W, Zhao H, Wang R, Yan Y, Xu L, Li P (2013a) Insulin-like growth factor-1 secreted by brain microvascular endothelial cells attenuates neuron injury upon ischemia. Febs J 280:3658–3668
Wang X-L, Zhao Y-S, Hu M-Y, Sun Y-Q, Chen Y-X, Bi X-H (2013b) Umbilical cord blood cells regulate endogenous neural stem cell proliferation via hedgehog signaling in hypoxic ischemic neonatal rats. Brain Res 1518:26–35
Weber C, Belge KU, von Hundelshausen P, Draude G, Steppich B, Mack M, Frankenberger M, Weber KS, Ziegler-Heitbrock HW (2000) Differential chemokine receptor expression and function in human monocyte subpopulations. J Leukocyte Biol 67:699–704

Wedgwood JF, Weinberger BI, Hatam L, Palmer R (1997) Umbilical cord blood lacks circulating B lymphocytes expressing surface IgG or IgA. Clin Immunol Immunopathol 84:276–282

Whedon MB, Wujcik D (eds) (1997) Blood and marrow stem cell transplantation: principles, practice, and nursing insights, 2nd edn. Jones & Bartlett Learning, Sudbury

Willing AE, Lixian J, Milliken M, Poulos S, Zigova T, Song S, Hart C, Sanchez-Ramos J, Sanberg PR (2003) Intravenous versus intrastriatal cord blood administration in a rodent model of stroke. J Neurosci Res 73:296–307

Willing AE, Garbuzova-Davis SN, Zayko O, Derasari HM, Rawls AE, James CR, Mervis RF, Sanberg CD, Kuzmin-Nichols N, Sanberg PR (In Press) Repeated administrations of human umbilical cord blood cells improve disease outcomes in a mouse model of Sanfilippo Syndrome Type III B. Cell Transplant

Womble TA, Green S, Shahaduzzaman M, Grieco J, Sanberg PR, Pennypacker KR, Willing AE (2014) Monocytes are essential for the neuroprotective effect of human cord blood cells following middle cerebral artery occlusion in rat. Mol Cell Neurosci 59:76–84

Xia G, Hong X, Chen X, Lan F, Zhang G, Liao L (2010) Intracerebral transplantation of mesenchymal stem cells derived from human umbilical cord blood alleviates hypoxic ischemic brain injury in rat neonates. J Perinat Med 38:215–221

Yang W-Z, Zhang Y, Wu F, Min W-P, Minev B, Zhang M, Luo X-L, Ramos F, Ichim TE, Riordan NH, Hu X (2010) Safety evaluation of allogeneic umbilical cord blood mononuclear cell therapy for degenerative conditions. J Transl Med 8:75

Yasuhara T, Hara K, Maki M, Xu L, Yu G, Ali MM, Masuda T, Yu SJ, Bae EK, Hayashi T, Matsukawa N, Kaneko Y, Kuzmin-Nichols N, Ellovitch S, Cruz EL, Klasko SK, Sanberg CD, Sanberg PR, Borlongan CV (2010a) Mannitol facilitates neurotrophic factor up-regulation and behavioural recovery in neonatal hypoxic-ischaemic rats with human umbilical cord blood grafts. J Cell Mol Med 14:914–921

Yasuhara T, Hara K, Maki M, Xu L, Yu G, Ali MM, Masuda T, Yu SJ, Bae EK, Hayashi T, Matsukawa N, Kaneko Y, Kuzmin-Nichols N, Ellovitch S, Cruz EL, Klasko SK, Sanberg CD, Sanberg PR, Borlongan CV (2010b) Mannitol facilitates neurotrophic factor up-regulation and behavioural recovery in neonatal hypoxic-ischaemic rats with human umbilical cord blood grafts. J Cell Mol Med 14:914–921

Zawadzka M, Lukasiuk K, Machaj EK, Pojda K, Kaminska B (2009) Lack of migration and neurological benefits after infusion of umbilical cord blood cells in ischemic brain injury. Acta Neurobiol Exp 69:46–51

Zhang MJ, Sun JJ, Qian L, Liu Z, Zhang Z, Cao W, Li W, Xu Y (2011) Human umbilical mesenchymal stem cells enhance the expression of neurotrophic factors and protect ataxic mice. Brain Res 1402:122–131

Zhao Y, Wang H, Mazzone T (2006) Identification of stem cells from human umbilical cord blood with embryonic and hematopoietic characteristics. Exp Cell Res 312:2454–2464

Zhu W, Mao Y, Zhao Y, Zhou L-F, Wang Y, Zhu J-H, Zhu Y, Yang G-Y (2005) Transplantation of vascular endothelial growth factor-transfected neural stem cells into the rat brain provides neuroprotection after transient focal cerebral ischemia. Neurosurgery 57:325–333 (discussion 325–333)

Ziegler-Heitbrock HW, Strobel M, Kieper D, Fingerle G, Schlunck T, Petersmann I, Ellwart J, Blumenstein M, Haas JG (1992) Differential expression of cytokines in human blood monocyte subpopulations. Blood 79:503–511

Zigova T, Song S, Willing AE, Hudson JE, Newman MB, Saporta S, Sanchez-Ramos J, Sanberg PR (2002) Human umbilical cord blood cells express neural antigens after transplantation into the developing rat brain. Cell Transplant 11:265–274

Zivin JA (1985) Cyproheptadine reduces or prevents ischemic central nervous system damage. Neurol 35:584–587

Zwart I, Hill AJ, Al-Allaf F, Shah M, Girdlestone J, Sanusi ABR, Mehmet H, Navarrete R, Navarrete C, Jen L-S (2009) Umbilical cord blood mesenchymal stromal cells are neuroprotective and promote regeneration in a rat optic tract model. Exp Neurol 216:439–448

Chapter 6
The Role of Endothelial Progenitor Cells in Stroke

Tomás Sobrino, Francisco Campos and José Castillo

Contents

Abstract Stroke is a devastating disease that is a leading cause of death and disability in developed countries. However, therapeutic options are notably limited, so is mandatory to investigate repairing processes after stroke in order to develop therapeutic strategies able to promote brain repair processes. In this context, therapeutic angiogenesis and vasculogenesis hold promise to improve the prognosis of patients with stroke. In this regard, it is well established that circulating endothelial progenitor cells (EPCs) have been suggested to be a marker of vascular risk and endothelial function. Moreover, low EPC number has been found in patients with

J. Castillo (✉) · T. Sobrino · F. Campos
Department of Neurology, Clinical Neurosciences Research Laboratory,
Hospital Clínico Universitario, c/ Travesa da Choupana s/n, Health Research
Institute of Santiago de Compostela (IDIS), University of Santiago de Compostela,
15706 Santiago de Compostela, Spain
e-mail: jose.castillo.sanchez@sergas.es

T. Sobrino
e-mail: tomas.sobrino.moreiras@sergas.es

F. Campos
e-mail: francisco.campos.perez@sergas.es

L.-R. Zhao, J. H. Zhang (eds.), *Cellular Therapy for Stroke and CNS Injuries*,
Springer Series in Translational Stroke Research, DOI 10.1007/978-3-319-11481-1_6

cerebrovascular diseases. Besides, EPC levels have been associated with good neurological and functional outcome as well as reduced infarct growth in patients with acute ischemic stroke. Finally, experimental and clinical studies indicate that EPC might mediate endothelial cell regeneration and neovascularization. Therefore, EPC-based therapy could be an excellent therapeutic option in stroke. Currently, clinical trials for evaluating EPC treatment in ischemic stroke are ongoing. In this chapter, we discuss the present status of knowledge about the possible therapeutic role of EPCs in stroke, molecular mechanisms, and the future perspectives and strategies for their use in clinical practice.

Abbreviations

Ang-1	Angiopoietin 1
Cdc42	Cell division control protein 42 homolog
CFU-EC	Early outgrowth colony forming unit-endothelial cell
eF2	Elongation factor 2
Enos	Endothelial nitric oxide synthase
EPCs	Endothelial progenitor cells
EPO	Erythropoietin
G-CSF	Granulocyte colony-stimulating factor
ERp29	Endoplasmic reticulum protein 29
HIF-1	Hypoxia-inducible factor 1
HSP-72	72 kilodalton heat shock protein
ICH	Intracerebral hemorrhage
IGF-1	Insulin-like growth factor 1
IS	Ischemic stroke
mKitL	membrane bound Kit ligand
MMP-9	Matrix metalloproteinase 9
NO	Nitric oxide
PRDX1	Peroxiredoxin 1
r-tPA	Recombinant tissue plasminogen activator
SCF	Stem cell factor
SDF-1α	stromal cell-derived factor 1α
VEGF	Vascular endothelial growth factor
VEGFR2	Vascular endothelial growth factor receptor 2

6.1 Introduction

Stroke is the second to third most common cause of death in adults, and more than a third of people who survive a stroke will have severe disability (Lloyd-Jones et al. 2010). Most of the strokes, about 80 %, are ischemic strokes (IS). IS is caused by an occlusion of a cerebral artery, which prevents blood flow to reach brain parenchyma. The remaining 20 % of strokes are intracerebral hemorrhages (ICH), caused by the rupture of a cerebral blood vessel. The ICH is usually more severe with greater rates of mortality and disability than IS. Therapeutic options currently are focusing on recanalization therapies for acute IS, mainly through intravenous or intra-arterial fibrinolyisis and thrombectomy (Mokin et al. 2014), but limitations restrict their use to a small proportion of patients. Moreover, no specific pharmacological treatments exist in order to improve the ICH prognosis, and treatment options for this disease are more limited than in IS, based primarily on surgical treatment.

Ischemic and hemorrhagic injury is a long and dynamic process involving a variety of mechanisms at different times. Stroke triggers many processes, including accumulation of excitatory amino acids, oxidative stress, alterations in gene expression, mitochondrial injury, brain edema, inflammation, and apoptosis, but also brain plasticity and endogenous repair mechanisms. Unfortunately, neurons at risk survive for only a few hours, and there is limited opportunity for effective therapeutic intervention. In this regard, although a wide range of neuroprotective substances has been effective in experimental models of stroke, they have repeatedly failed in clinical trials because of toxicity or loss of effectiveness (Tymianski 2013). Therefore, the development of strategies to increase plasticity and endogenous brain repair mechanisms in order to improve functional outcome in stroke are essential (Rodríguez-González et al. 2007). In this context, therapeutic angiogenesis and vasculogenesis hold promise to improve the prognosis of patients with stroke. In this regard, it is well established that circulating endothelial progenitor cells (EPCs) have been suggested to be a marker of vascular risk and endothelial function (Hill et al. 2003; Vasa et al. 2001; Werner et al. 2005). Moreover, low EPC number has been found in patients with cerebrovascular diseases (Ghani et al. 2005). Besides, EPC levels have been associated with good neurological and functional outcome as well as reduced infarct growth in patients with acute ischemic stroke (Sobrino et al. 2007; Chu et al. 2008; Yip et al. 2008; Navarro-Sobrino et al. 2010; Bogoslovsky et al. 2010; Paczkowska et al. 2013; Martí-Fàbregas et al. 2013). Finally, EPCs have been related to endothelial cell regeneration and neovascularization after tissue ischemia (Zhang et al. 2002; Mao et al. 2014). Therefore, EPC-based therapy could be an excellent therapeutic option in stroke. In this chapter, we discuss the present status of knowledge about the possible therapeutic role of EPCs in stroke, molecular mechanisms, and the future perspectives and strategies for their use in clinical practice.

6.2 Rationale for Therapeutic Use of EPCs to Treat Stroke

It is well established that circulating endothelial progenitor cells (EPCs) have been suggested to be a marker of vascular risk and endothelial function (Hill et al. 2003; Vasa et al. 2001; Werner et al. 2005). The number of circulating EPC has been reported to be decreased in patients with vascular risk factors such as smoking habit, hypercholesterolemia, diabetes and hypertension (Vasa et al. 2001;

Sobrino et al. 2007; Zhao et al. 2013), many of which have been identified as prognostic markers of poor outcome following stroke. Moreover, low EPC number has been found in patients with cerebrovascular diseases (Ghani et al. 2005), and EPC levels have been associated with good neurological and functional outcome as well as reduced infarct growth in patients with acute IS and ICH (Sobrino et al. 2007, 2011a; Chu et al. 2008; Yip et al. 2008; Navarro-Sobrino et al. 2010; Bogoslovsky et al. 2010; Paczkowska et al. 2013; Martí-Fàbregas et al. 2013). However, existing evidence supports that EPCs not only work as biomarker but also might offer a new therapeutic strategy for stroke (Lapergue et al. 2007). In this regard, experimental and human studies indicate that EPC might mediate endothelial cell regeneration and neovascularization (Asahara et al. 1997; Werner et al. 2003; Kong et al. 2004; Werner and Nickenig 2006), and that EPC participate in the cerebral neovascularization present in adult brain after ischemia (Zhang et al. 2002; Mao et al. 2014). These protective vascular effects result from EPC proliferation. On the other hand, as stated above, EPCs have been suggested to maintain endothelial protection/repair and neovascularization and angiogenesis. Today it is known that angiogenesis is coupled with neurogenesis following ischemic injury (Thored et al. 2007). The underlying mechanisms include that the regenerated blood vessels provide nutritive blood flow and that EPCs, by secreting factors such as SDF-1 and VEGF, create a microenvironment for neural regeneration and survival (Imitola et al. 2004; Schänzer et al. 2004). Furthermore, neuroblasts migrate along these regenerated vessels to achieve neurogenesis in peri- infarct area. Consequently, suppression of angiogenesis substantially reduces migration of neuroblasts from the subventricular zone to the ischemic region (Zhang and Chopp 2009). Therefore, EPC-based therapy might be an excellent therapeutic option in stroke.

6.3 EPC-Based Cellular Therapy for Stroke

EPC levels have been associated with good neurological and functional outcome as well as reduced infarct growth in patients with acute IS (Sobrino et al. 2007; Chu et al. 2008; Yip et al. 2008; Navarro-Sobrino et al. 2010; Bogoslovsky et al. 2010; Paczkowska et al. 2013; Martí-Fàbregas et al. 2013) and ICH (Sobrino et al. 2011a; Paczkowska et al. 2013). Likewise, EPCs have been related to endothelial cell regeneration and neovascularization after tissue ischemia (Zhang et al. 2002; Mao et al. 2014). Therefore, EPC-based therapy could be an excellent therapeutic opportunity for stroke. The aim of cellular therapy is to restore brain function by replacing dead cells with new ones through transplantation or stimulation of endogenous stem or precursor cells (Hurtado et al. 2006). There is growing evidence that the adult stem cell system, including EPCs, is more flexible than previously thought and may be an excellent therapeutic option for stroke (Rodriguez-González et al. 2007). In this regard, it has been suggested that resident pools of adult stem cells, such as EPCs, can be used in two ways: (a) by isolating, harvesting and growing them in vitro and then administering them locally or systemically; (b) or by endogenous stimulating them (see factors associated to EPC increase in Table 15.1).

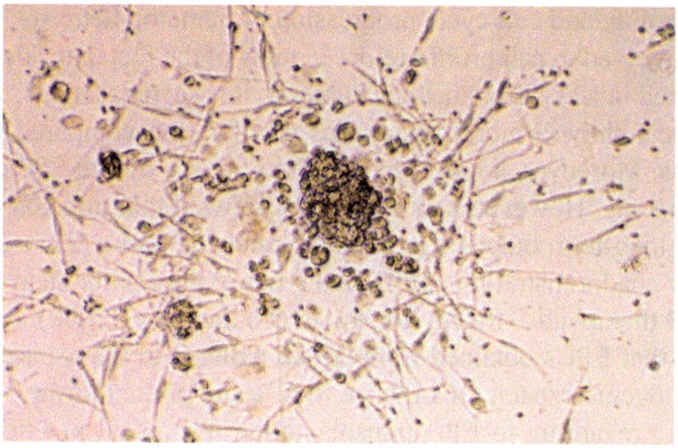

Fig. 6.1 Micrograph of an early outgrowth colony forming unit—endothelial cell (*CFU-EC*) isolated from an ischemic stroke patient in a phase-contrast microscope (×100). A CFU-EC consisted of a central cluster of rounded cells with elongated sprouting cells at the periphery

6.3.1 Exogenous Administration of EPCs

There are numerous concerns about the cell therapy with exogenous EPC transplantation. The optimal starting time point for administration of EPCs following stroke may be critical for their therapeutic efficacy. However, there are few studies on this important issue. Based on the capacity of EPCs to secrete several growth factors with protective effects on the brain, their transplantation in the early phase of stroke may have better efficacy. By contrast, it has also been suggested that oxidative stress and inflammation in the acute phase of stroke may limit the function on survival of transplanted EPCs (Locatelli et al. 2009). Therefore, preclinical and clinical studies are needed to evaluate the best time point after stroke onset for EPC transplantation.

Another important issue is about the autologous or allogencic transplantation of EPCs, as well as the source of obtaining EPCs from stroke patients or healthy subjects. A recent proteomic study has analyzed differences in protein expression of early outgrowth colony forming unit-endothelial cell (CFU-EC) (Fig. 6.1) from IS patients and healthy subjects (see Brea et al. 2011 for review). Remarkably, the proteomic analysis revealed a greater expression of cell division control protein 42 homolog (CdC42) and endoplasmic reticulum protein 29 (ERp29) in EPCs from healthy subjects, and a greater expression of elongation factor 2 (eF2) and peroxiredoxin 1 (PRDX1) in EPCs obtained from IS patients. It has been reported that PRDX1 expression dramatically increases during processes such as spontaneous differentiation of human embryonic stem cells, targeted differentiation of neural progenitor cells and differentiation of human neural stem cell line respect to proliferating cells. eF2 is also up-regulated 4–7 days after differentiation of the human neuronal stem cell line, ReNcell VM. Therefore, these findings could be indicating that EPCs from IS patients are in a more advanced differentiation state than EPCs isolated from control subjects. On the other hand, CdC-42 and ERp29 were found to be up-regulated in EPCs from healthy subjects. Cdc42 regulates adhesion, migration, homing, and cell cycle progression of hematopoietic stem cells. ERp29 seems to be involved in cell proliferation. In view of the fact that Cdc42 and ERp29 are up-regulated in EPCs from healthy subjects, it is seems that EPCs isolated from healthy subjects show a more capacity of proliferation compared to EPCs from stroke patients. Moreover, it has been proposed the use of late EPCs as an optimal EPC-based therapy. However, another studies showed that infusion of early EPCs enhanced the long-term outcome in animal models of stroke (Zhao et al. 2013). In fact, currently coadministration of different types of progenitor/stem cells may constitute a novel therapeutic strategy for stroke (Foubert et al. 2008). Data from other studies show that EPCs obtained from stroke patients in the subacute phase have greater vasculogenic capacity than those from acute phase (Navarro-Sobrino et al. 2010). Finally, regarding to EPC transplantation in clinical practice, intravenous infusion should be the optimal route because intra-arterial infusion may be inconvenient and could provoke embolism. Likewise, intracerebral injection of EPCs is complex and might cause intracerebral hemorrhage and parenchymal damage. Although, it remains to be determined whether administration of autologous or allogeneic EPCs in the subacute period is more effective, it is tempting to postulate, based on the above data, that early and late EPCs obtained from stroke patients in the subacute phase could be the most suitable source of EPCs for cell therapy in stroke by using intravenous administration. However, larger clinical studies are needed to evaluate this hypothesis.

6.3.2 Endogenous Stimulation of EPCs

Restoring blood flow supply after ischemia and re-endothelization after hemorrhage may contribute to cell survival and tissue repair. Formation of new blood vessels in the adult brain after stroke is not only mediated by angiogenesis but also involves vasculogenesis mediated by EPCs, which are involved in processes of re-endothelization and repair of vascular endothelium in response to vascular trauma or tissue ischemia, promoted by biochemical factors that activate its proliferation. They have been described that EPCs migrate through the peripheral blood from bone marrow to sites of neovascularization where EPCs are able to differentiate into mature endothelial cells. Recruitment and incorporation of EPCs into ischemic or hemorrhagic tissues requires a coordinated multistep process including mobilization, chemoattraction, adhesion, migration, tissue invasion and in situ differentiation (Fig. 6.2) (Rodríguez-González et al. 2007). Many molecular and physiological-pathological factors, as well as drugs, are involved in these processes (For review: Arenillas et al. 2007; Brea et al. 2009; Sobrino et al. 2011a, b, 2012a, b; Zhao et al. 2013) (Table 6.1). For example, the activity of matrix metalloproteinase 9 (MMP-9), which causes a massive release of stem cell factor (SCF) and activation of membrane bound Kit ligand (mKitL) that favors the recruitment of progenitor cells, among which are EPC, from bone marrow. Likewise, active MMP-9 induces the release of cytokines that cause the mobilization of quiescent EPC (Rafii et al. 2002). Moreover, EPC release and mobilization

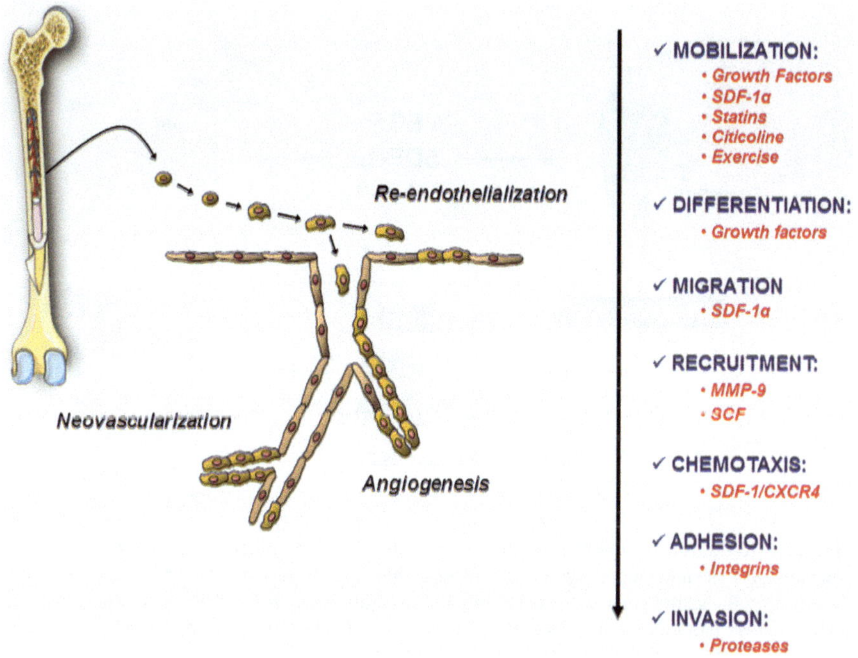

Fig. 6.2 Factors involved in the mobilization, differentiation, migration, recruitment, chemotaxis, adhesion and invasion of EPCs in stroke. These mechanisms are necessary in order that EPCs exert their beneficial functions such as re-endothelialization, angiogenesis and neovascularization

are regulated by vascular endothelial growth factor (VEGF), stromal-cell-derived factor 1 (SDF-1), granulocyte colony-stimulating factor (G-CSF), erythropoietin (EPO), angiopoietin 1, endothelial NO synthase (eNOS), exercise, estrogens and several drugs such as statins, EPO or citicoline. In fact, clinical studies in IS and ICH patients demonstrated that serum levels of VEGF, SDF-1α and active MMP-9 increase in response to cerebral ischemia or ICH within the first 72 h from symptom onset, and that the magnitude of this increase is directly related to an EPC increment (Bogoslovsky et al. 2011a, b; Sobrino et al. 2011a, 2012b). On the other hand, the fact that serum levels of molecular markers at 24 h from stroke onset correlated with EPC increment during the 1st week, but not at admission, and that EPC increment during the 1st week, but not EPC counts at baseline, has been associated with better neurological outcome and reduced infarct growth supports the hypothesis that cerebral ischemia induces the activation of molecular pathways of EPC mobilization focused on promoting endogenous processes of vascular and neurorepair (Fig. 6.3). Furthermore, similar results were found in ICH patients (Sobrino et al. 2009, 2011a; Paczkowska et al. 2013). It has been reported in ICH patients a strong correlation between VEGF and SDF-1α serum levels and circulating concentrations of bone marrow-derived progenitor cells (BMPCs) at day 7 (Sobrino et al. 2011a). Given that the EPC is a subtype of BMPCs, it is tempting to hypothesize that similar molecular and cellular mechanisms are involved in the two major subtypes of stroke (ischemic and hemorrhagic stroke).

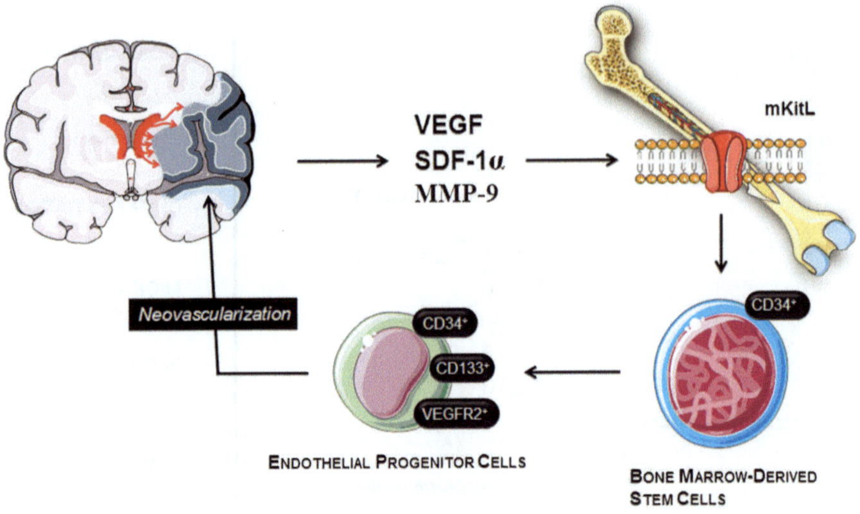

Fig. 6.3 Mechanism of endothelial progenitor cells (*EPC*) mobilization in stroke. Stroke is a potent inducer of endogenous repair mechanisms, which are initially activated by a massive expression of growth factors such as vascular endothelial growth factor (*VEGF*), and other molecules such as matrix metalloproteinase 9 (*MMP-9*) and stromal cell-derived factor 1α (*SDF-1α*). These molecular factors, especially MMP-9 activity, induce the mobilization of bone marrow-derived stem cells, including EPCs, through membrane bound Kit ligand (*mKitL*) activation. These EPCs may migrate into the areas of brain injury, mediating repair mechanisms. Likewise, these mechanisms also enhance the endogenous neurogenesis

On the other hand, because EPCs express functional CXCR4 and VEGF receptors (Salcedo et al. 1999; Yamaguchi et al. 2003), an interaction between the SDF-1α/CXCR4 pathway and VEGF might form a positive-feedback loop which would increase the therapeutic effects of EPCs in cerebral neovascularization. Consequently, growth factors and SDF-1α might be an effective therapy in IS and ICH because of their role mediating the mobilization of circulating EPCs, as well as in restoring endothelium integrity and decreasing brain edema, inflammation, and perihematoma cell death (Krizanac-Bengez et al. 2004).

As a clinical implication, the number of circulating EPCs is inversely correlated with vascular risk factors such as diabetes, hypertension, hypercholesterolemia or smoking (Hill et al. 2003; Vasa et al. 2001; Werner et al. 2005), and also with molecular markers of endothelial dysfunction and inflammation such as homocysteine or C-reactive protein. Therefore, it can be rationally speculated that environment of circulation is essential for the living and functionality of EPCs, which would raise the perspective on the demand in managing risk factors of stroke (Zhao et al. 2013).

Finally, it has also been demonstrated that several drugs can modulate endogenous EPC behavior (Table 6.1). Statins treatment during acute phase leads to an increase in EPCs in IS patients (Sobrino et al. 2012a; Martí-Fàbregas et al. 2013). In view of the fact that statin treatment during the acute phase increases circulating EPC and statin withdrawal is associated with poor outcome in IS patients

Table 6.1 Factors involved in the release, mobilization and recruitment of EPCs. (Arenillas et al. 2007; Brea et al. 2009; Sobrino et al. 2011a, b, 2012a, b; Zhao et al. 2013)

Release and/or Mobilization	
Chemokines/growth factors	
Hypoxia-inducible factor 1 (HIF-1)	>
Stromal cell derived SDF-1	>
Vascular endothelial growth factor (VEGF)	>
Insulin-like growth factor 1 (IGF-1)	>
Granulocyte-colony stimulating factor (G-CSF)	>
Angiopoietin 2	>
Drugs	
Statins	>
Angiotensin II type 1 receptor blockers	>
Angiotensin-converting enzyme inhibitor	>
Erythropoietin (EPO)	>
Berberine	>
Morphine	<
Citicoline	>
Recombinant tissue plasminogen activator (r-tPA)	>
PPAR-γ agonist	>
Proteins/hormones	
Estrogens	>
Nitric oxide (NO) and eNOS (endothelial NO Synthase)	>
Aldosterone	<
Angiotensin II	<
Endostatin	<
Heme-oxigenase 1	>
Matrix metalloproteinase 9 (MMP-9)	>
Physiological/pathological factors	
Exercise and physical training	>
Wound	>
Ischemic events	>
Aging	<
Obesity	<
Smoking	<
Hypertension	<
Diabetes	<
Hypercholesterolemia	<
Homocysteine	<

> Increase, < Decrease

Table 6.1 (continued)

Recruitment
Chemokines/growth factors
SDF-1/CXCR4
CCL5/CCR5
CXCL1
CXCL7/CXCR2
VEGF/VEGFR2
IL-8/Gro
IGF2/IGFR2
Other molecular factors
Caspase-8
Hyaluronic acid
Thrombin
CD9
Alpha6 integrin subunit

(Blanco et al. 2007), the positive effects of statin treatment during the acute phase on functional outcome in ischemic stroke could be mediated by EPC. Moreover, patients treated with statins showed also higher serum levels of VEGF, active MMP-9 and nitric oxide (NO)x at 24 h (Sobrino et al. 2012a). Statins induce the production of NO by eNOS, the expression of angiogenic factors such as VEGF, and the mobilization and proliferation of EPC (Endres 2005), so these mechanisms may be interrelated. G-CSF is one of the early drugs discovered to be able to enhance EPC mobilization into the circulation after venous administration (Powell et al. 2005). Afterwards, other drugs such as Angiotensin II type 1 receptors blocker, Angiotensin-converting enzyme inhibitors, EPO, berberine, citicoline, recombinant tissue plasminogen activator (r-tPA) and PPAR-γ agonist have been shown to increase the number and functional activity of EPCs in vitro and in vivo (Arenillas et al. 2007; Rodríguez-González et al. 2007; Sobrino et al. 2011b, 2012a, b; Zhao et al. 2013). As these drugs are commonly used in clinical treatment of vascular diseases, all these clinical data may help to interpret the beneficial effects of these drugs on top of their known pharmacological actions. However, further studies are needed in order to facilitate the discovery of new drugs targeting EPCs.

6.4 Promising Strategies Related to EPCs

As a promising strategy for cellular-based therapies for stroke, induced pluripotent stem cells (iPSC) technology (Takahashi and Yamanaka 2006; Takahashi et al. 2007), which enables the reprogramming of a wide variety of cell types isolated from humans into embryonic stem cell-like pluripotent cells, offers a novel strategy for the patient-specific derivation of a lineage-specific cells from iPSC, such as

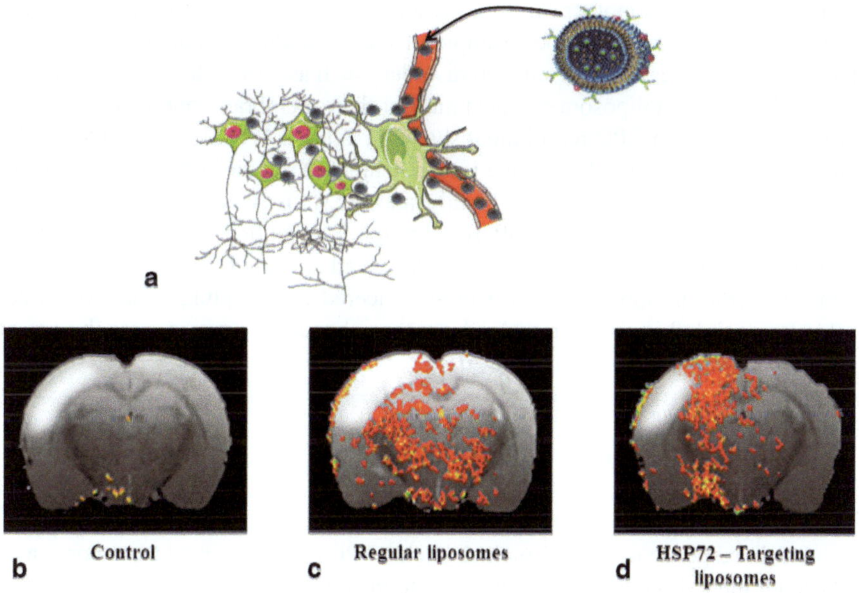

Fig. 6.4 Schematic representation of theranostics feasibility inside the brain parenchyma despite the challenge of crossing the blood brain barrier (**a**). Representative MR images of ischemic brains from rats treated (*i.v.*) with 1 ml of saline (**b**), regular (*non-vectorized*) liposomes (**c**) or anti-HSP72 vectorized liposomes (**d**) showing the in vivo distribution of liposomes at 24 h after treatment. These anti-HSP72 immunoliposomes, containing in their membrane one or more of the proteins involved in EPC recruitment shown in Table 6.1, could be used to locally increase the number of EPCs in the ischemic area after their exogenous transplantation or endogenous stimulation

EPCs (Choi et al. 2009; Park et al. 2010; Xu et al. 2012; Yoo et al. 2013). Moreover, the therapeutic potential of transplanted human iPSC-derived EPCs (hiPSC-EPCs) has been shown in animal disease models of hind-limb ischemia (Park et al. 2010; Rufaihah et al. 2011; Lai et al. 2013). Therefore, it is tempting to postulate that hiPSC-EPCs may represent a strategy for patient-specific EPC therapies in stroke.

Moreover, new strategies are necessary in order to increase the local number of EPCs in the ischemic or hemorrhagic areas. In this regard, nanomedicine may be useful to achieve this goal. A recent study has demonstrated the potential role of superparamagnetic iron oxide nanoparticles (SPION)-loaded EPCs by using a magnetic guidance to the ischemic tissue in animal models of cerebral ischemia. The authors demonstrate ex vivo cellular viability and maintained function following SPION load as well as successful guidance of the EPCs to the target site via magnetic resonance imaging (MRI) (Carenza et al. 2014). On the other hand, another recent study from our group (Agulla et al. 2014) has report a new theranostic nano-platform vectorized towards peri-infarct tissue, the key target for the treatment of cerebral ischemia. Anti-HSP72 (72 kDa heat shock protein) stealth immunoliposomes containing MRI probes were used to allocate the peri-infarct region in vivo and to achieve a superior therapeutic effect in comparison to other non-targeted drug delivery means (Fig. 6.4). Thus, despite the challenge of crossing the blood-brain barrier, this study demonstrates that theranostics inside the brain parenchyma is feasible and represents a good example of the potential that nanotechnology offers for the treatment of neurological disorders such as stroke. In this regard, these anti-HSP72 immunoliposomes, containing in their membrane one or more of the proteins involved in EPC recruitment shown in Table 6.1, could be used to locally increase the number of EPCs in the ischemic area after their exogenous transplantation or endogenous stimulation. Finally, another strategy in order to increase the local number of EPCs may be offered through interventional therapy for stroke, such as percutaneous transluminal angioplasty and stenting (PTAS) and thrombectomy. In addition, application of a bio-engineered EPC-capture stent, which accelerates re-endothelialization and reduces thrombogenicity, may reduce the rate of restenosis after PTAS (Larsen et al. 2012).

Conclusions

In brief, in response to stimuli such as stroke, EPCs are mobilized from bone marrow to peripheral blood and may participate in endothelial cell repair-regeneration and in tissue neovascularization processes. In this context, experimental and human studies have shown that neovascularization is present in the adult brain exposed to ischemia and that EPCs participate in cerebral neovascularization processes. Finally, we and others have observed that a higher increment in the number of circulating EPCs is associated with a better outcome in patients with stroke. Taken together, these findings suggest that EPCs may mediate neurorepair processes after stroke, and that exogenous supplementation or endogenous stimulation of EPCs have a great therapeutic potential for stroke. However, larger clinical trials are needed to evaluate the safety and efficacy of EPC transplantation for treating stroke. Furthermore, how to improve the strategies in order to maximize the endogenous stimulation of EPCs deserves also further studies.

References

Agulla J, Brea D, Campos F, Sobrino T, Argibay B, Al-Soufi W, Blanco M, Castillo J, Ramos-Cabrer P (2014) In vivo theranostics at the peri-infarct region in cerebral ischemia. Theranostics 4:90–105

Arenillas JF, Sobrino T, Castillo J, Dávalos A (2007) The role of angiogenesis in damage and recovery from ischemic stroke. Curr Treat Options Cardiovasc Med 9:205–212

Asahara T, Murohara T, Sullivan A, Silver M, van der Zee R, Li T, Witzenbichler B, Schatteman G, Isner JM (1997) Isolation of putative progenitor endothelial cells for angiogenesis. Science 275:964–967

Blanco M, Nombela F, Castellanos M, Rodriguez-Yáñez M, García-Gil M, Leira R, Lizasoain I, Serena J, Vivancos J, Moro MA, Dávalos A, Castillo J (2007) Statin treatment withdrawal in ischemic stroke: a controlled randomized study. Neurology 69:904–910

Bogoslovsky T, Chaudhry A, Latour L, Maric D, Luby M, Spatz M, Frank J, Warach S (2010) Endothelial progenitor cells correlate with lesion volume and growth in acute stroke. Neurology 75:2059–2062

Bogoslovsky T, Spatz M, Chaudhry A, Maric D, Luby M, Frank J, Warach S (2011a) Circulating CD133+ CD34+ progenitor cells inversely correlate with soluble ICAM-1 in early ischemic stroke patients. J Transl Med 9:145

Bogoslovsky T, Spatz M, Chaudhry A, Maric D, Luby M, Frank J, Warach S, NINDS Natural History of Stroke Investigators (2011b) Stromal-derived factor-1[alpha] correlates with circulating endothelial progenitor cells and with acute lesion volume in stroke patients. Stroke 42:618–625

Brea D, Sobrino T, Ramos-Cabrer P, Castillo J (2009) Reorganisation of the cerebral vasculature following ischaemia. Rev Neurol 49:645–654

Brea D, Rodríguez-González R, Sobrino T, Rodríguez-Yañez M, Blanco M, Castillo J (2011) Proteomic analysis shows differential protein expression in endothelial progenitor cells between healthy subjects and ischemic stroke patients. Neurol Res 33:1057–1063

Carenza E, Barceló V, Morancho A, Levander L, Boada C, Laromaine A, Roig A, Montaner J, Rosell A (2014) In vitro angiogenic performance and in vivo brain targeting of magnetized endothelial progenitor cells for neurorepair therapies. Nanomedicine 10:225–234

Choi KD, Yu J, Smuga-Otto K, Salvagiotto G, Rehrauer W, Vodyanik M, Thomson J, Slukvin I (2009) Hematopoietic and endothelial differentiation of human induced pluripotent stem cells. Stem Cells 27:559–567

Chu K, Jung KH, Lee ST, Park HK, Sinn DI, Kim JM, Kim DH, Kim JH, Kim SJ, Song EC, Kim M, Lee SK, Roh JK (2008) Circulating endothelial progenitor cells as a new marker of endothelial dysfunction or repair in acute stroke. Stroke 39:1441–1447

Endres M (2005) Statins and stroke. J Cereb Blood Flow Metab 25:1093–1110

Foubert P, Matrone G, Souttou B, Leré-Déan C, Barateau V, Plouët J, Le Ricousse-RoussanneS, Lévy BI, Silvestre JS, Tobelem G (2008) Coadministration of endothelial and smooth muscle progenitor cells enhances the efficiency of proangiogenic cell-based therapy. Circ Res 103:751–760

Ghani U, Shuaib A, Salam A, Nasir A, Shuaib U, Jeerakathil T, Sher F, O'Rourke F, Nasser AM, Schwindt B, Todd K (2005) Endothelial progenitor cells during cerebrovascular disease. Stroke 36:151–153

Hill JM, Zalos G, Halcox JP, Schenke WH, Waclawiw MA, Quyyumi AA, Finkel T (2003) Circulating endotelial progenitor cells, vascular function, and cardiovascular risk. N Engl J Med 348:593–600

Hurtado O, Pradillo JM, Alonso-Escolano D, Lorenzo P, Sobrino T, Castillo J, Lizasoain I, Moro MA (2006) Neurorepair versus neuroprotection in stroke. Cerebrovasc Dis 21(Suppl 2):54–63

Imitola J, Raddassi K, Park KI, Mueller FJ, Nieto M, Teng YD, Frenkel D, Li J, Sidman RL, Walsh CA, Snyder EY, Khoury SJ (2004) Directed migration of neural stem cells to sites of CNS injury by the stromal cell-derived factor 1alpha/CXC chemokine receptor 4 pathway. Proc Natl Acad Sci U S A 101:18117–18122

Kong D, Melo LG, Gnecchi M, Zhang L, Mostoslavsky G, Liew CC, Pratt RE, Dzau VJ (2004) Cytokine-induced mobilization of circulating endothelial progenitor cells enhances repair of injured arteries. Circulation 110:2039–2046

Krizanac-Bengez L, Mayberg MR, Janigro D (2004) The cerebral vasculature as a therapeutic target for neurological disorders and the role of shear stress in vascular homeostatis and pathophysiology. Neurol Res 26:846–853

Lai WH, Ho JC, Chan YC, Ng JH, Au KW, Wong LY, Siu CW, Tse HF (2013) Attenuation of hindlimb ischemia in mice with endothelial-like cells derived from different sources of human stem cells. PLoS ONE 8:e57876

Lapergue B, Mohammad A, Shuaib A (2007) Endothelial progenitor cells and cerebrovascular diseases. Prog Neurobiol 83:349–362

Larsen K, Cheng C, Tempel D, Parker S, Yazdani S, den Dekker WK, Houtgraaf JH, de Jong R, Swager-ten Hoor S, Ligtenberg E, Hanson SR, Rowland S, Kolodgie F, Serruys PW, Virmani R, Duckers HJ (2012) Capture of circulatory endothelial progenitor cells and accelerated

re-endothelialization of a bio-engineered stent in human *ex vivo* shunt and rabbit denudation model. Eur Heart J 33:120–128
Lloyd-Jones D et al (2010) Heart disease and stroke statistics-2010 update: a report from the American Heart Association. Circulation 121:e46–e215
Locatelli F, Bersano A, Ballabio E, Lanfranconi S, Papadimitriou D, Strazzer S, Bresolin N, Comi GP, Corti S (2009) Stem cell therapy in stroke. Cell Mol Life Sci 66:757–772
Mao L, Huang M, Chen SC, Li YN, Xia YP, He QW, Wang MD, Huang Y, Zheng L, Hu B (2014) Endogenous endothelial progenitor cells participate in neovascularization via CXCR4/SDF-1 axis and improve outcome after stroke. CNS Neurosci Ther 20:460–468
Martí-Fàbregas J, Crespo J, Delgado-Mederos R, Martínez-Ramírez S, Peña EMarínR, Dinia L, Jiménez-Xarrié E, Fernández-Arcos A, Pérez-Pérez J, Querol L, Suárez-Calvet M, Badimon L (2013) Endothelial progenitor cells in acute ischemic stroke. Brain Behav 3:649–655
Mokin M, Khalessi AA, Mocco J, Lanzino G, Dumont TM, Hanel RA, Lopes DK, Fessler RD 2nd, Ringer AJ, Bendok BR, Veznedaroglu E, Siddiqui AH, Hopkins LN, Levy EI (2014) Endovascular treatment of acute ischemic stroke: the end or just the beginning? Neurosurg Focus 36:E5
Navarro-Sobrino M, Rosell A, Hernandez-Guillamon M, Penalba A, Ribó M, Alvarez-Sabín J, Montaner J (2010) Mobilization, endothelial differentiation and functional capacity of endothelial progenitor cells after ischemic stroke. Microvasc Res 80:317–323
Paczkowska E, Gołąb-Janowska M, Bajer-Czajkowska A, Machalińska A, Ustianowski P, Rybicka M, Kłos P, Dziedziejko V, Safranow K, Nowacki P, Machaliński B (2013) Increased circulating endothelial progenitor cells in patients with haemorrhagic and ischaemic stroke: the role of endothelin-1. J Neurol Sci 325:90–99
Park SW, Jun Koh Y, Jeon J, Cho YH, Jang MJ, Kang Y, Kim MJ, Choi C, Sook Cho Y, Chung HM, Koh GY, Han YM (2010) Efficient differentiation of human pluripotent stem cells into functional CD34+ progenitor cells by combined modulation of the MEK/ERK and BMP4 signaling pathways. Blood 116:5762–5772
Powell TM, Paul JD, Hill JM, Thompson M, Benjamin M, Rodrigo M, McCoy JP, Read EJ, Khuu HM, Leitman SF, Finkel T, Cannon RO 3rd (2005) Granulocyte colony stimulating factor mobilizes functional endothelial progenitor cells in patients with coronary artery disease. Arterioscler Thromb Vasc Biol 25:296–301
Rafii S, Heissig B, Hattori K (2002) Efficient mobilization and recruitment of marrow-derived endothelial and hematopoietic stem cells by adenoviral vectors expressing angiogenic factors. Gene Ther 9:631–641
Rodríguez-González R, Hurtado O, Sobrino T, Castillo J (2007) Neuroplasticity and cellular therapy in cerebral infarction. Cerebrovasc Dis 24(Suppl 1):167–180
Rufaihah AJ, Huang NF, Jamé S, Lee JC, Nguyen HN, Byers B, De A, Okogbaa J, Rollins M, Reijo-Pera R, Gambhir SS, Cooke JP (2011) Endothelial cells derived from human iPSCS increase capillary density and improve perfusion in a mouse model of peripheral arterial disease. Arterioscler Thromb Vasc Biol 31:e72–79
Salcedo R, Wasserman K, Young HA, Grimm MC, Howard OM, Anver MR, Kleinman HK, Murphy WJ, Oppenheim JJ (1999) Vascular endothelial growth factor and basic fibroblast growth factor induce expression of CXCR4 on human endothelial cells: in vivo neovascularization induced by stromal-derived factor-1alpha. Am J Pathol 154:1125–1135
Schänzer A, Wachs FP, Wilhelm D, Acker T, Cooper-Kuhn C, Beck H, Winkler J, Aigner L, Plate KH, Kuhn HG (2004) Direct stimulation of adult neural stem cells in vitro and neurogenesis in vivo by vascular endothelial growth factor. Brain Pathol 14:237–248
Sobrino T, Hurtado O, Moro MA, Rodríguez-Yáñez M, Castellanos M, Brea D, Moldes O, Blanco M, Arenillas JF, Leira R, Dávalos A, Lizasoain I, Castillo J (2007) The increase of circulating endothelial progenitor cells after acute ischemic stroke is associated with good outcome. Stroke 38:2759–2764
Sobrino T, Arias S, Rodríguez-González R, Brea D, Silva Y, de la ONP, Agulla J, Blanco M, Pumar JM, Serena J, Dávalos A, Castillo J (2009) High serum levels of growth factors are associated with good outcome in intracerebral hemorrhage. J Cereb Blood Flow Metab 29:1968–1974

Sobrino T, Arias S, Pérez-Mato M, Agulla J, Brea D, Rodríguez-Yáñez M, Castillo J (2011a) Cd34+ progenitor cells likely are involved in the good functional recovery after intracerebral hemorrhage in humans. J Neurosci Res 89:979–985

Sobrino T, Rodríguez-González R, Blanco M, Brea D, Pérez-Mato M, Rodríguez-Yáñez M, Leira R, Castillo J (2011b) CDP-choline treatment increases circulating endothelial progenitor cells in acute ischemic stroke. Neurol Res 33:572–577

Sobrino T, Blanco M, Pérez-Mato M, Rodríguez-Yáñez M, Castillo J (2012a) Increased levels of circulating endothelial progenitor cells in patients with ischaemic stroke treated with statins during acute phase. Eur J Neurol 19:1539–1546

Sobrino T, Pérez-Mato M, Brea D, Rodríguez-Yáñez M, Blanco M, Castillo J (2012b) Temporal profile of molecular signatures associated with circulating endothelial progenitor cells in human ischemic stroke. J Neurosci Res 90:1788–1793

Takahashi K, Yamanaka S (2006) Induction of pluripotent stem cells from mouse embryonic and adult fibroblast cultures by defined factors. Cell 126:663–676

Takahashi K, Tanabe K, Ohnuki M, Narita M, Ichisaka T, Tomoda K, Yamanaka S (2007) Induction of pluripotent stem cells from adult human fibroblasts by defined factors. Cell 131:861–872

Thored P, Wood J, Arvidsson A, Cammenga J, Kokaia Z, Lindvall O (2007) Long-term neuroblast migration along blood vessels in an area with transient angiogenesis and increased vascularization after stroke. Stroke 38:3032–3039

Tymianski M (2013) Novel approaches to neuroprotection trials in acute ischemic stroke. Stroke 44:2942–2950

Vasa M, Fichtlscherer S, Aicher A, Adler K, Urbich C, Martin H, Zeiher AM, Dimmeler S (2001) Number and migratory activity of circulating endothelial progenitor cells inversely correlate with risk factors for coronary artery disease. Circ Res 89:e1–e7

Werner N, Nickenig G (2006) Influence of cardiovascular risk factors on endothelial progenitor cells: limitations for therapy? Arterioscler Thromb Vasc Biol 26:257–266

Werner N, Junk S, Laufs U, Link A, Walenta K, Bohm M, Nickenig G (2003) Intravenous transfusion of endothelial progenitor cells reduces neointima formation after vascular injury. Circ Res 93:e17–24

Werner N, Kosiol S, Schiegl T, Ahlers P, Walenta K, Link A, Böhm M, Nickenig G (2005) Circulating endothelial progenitor cells and cardiovascular outcomes. N Eng J Med 353:999–1007

Xu Y, Liu L, Zhang L, Fu S, Hu Y, Wang Y, Fu H, Wu K, Xiao H, Liu S, Yu X, Zheng W, Feng B, Huang H (2012) Efficient commitment to functional CD34+ progenitor cells from human bone marrow mesenchymal stem-cell-derived induced pluripotent stem cells. PLoS ONE 7:e34321

Yamaguchi J, Kusano KF, Masuo O, Kawamoto A, Silver M, Murasawa S, Bosch-Marce M, Masuda H, Losordo DW, Isner JM, Asahara T (2003) Stromal cell-derived factor-1 effects on ex vivo expanded endothelial progenitor cell recruitment for ischemic neovascularization. Circulation 107:1322–1328

Yip HK, Chang LT, Chang WN, Lu CH, Liou CW, Lan MY, Liu JS, Youssef AA, Chang HW (2008) Level and value of circulating endothelial progenitor cells in patients after acute ischemic stroke. Stroke 39:69–74

Yoo CH, Na HJ, Lee DS, Heo SC, An Y, Cha J, Choi C, Kim JH, Park JC, Cho YS (2013) Endothelial progenitor cells from human dental pulp-derived iPS cells as a therapeutic target for ischemic vascular diseases. Biomaterials 34:8149–8160

Zhang ZG, Chopp M (2009) Neurorestorative therapies for stroke: underlying mechanisms and translation to the clinic. Lancet Neurol 8:491–500

Zhang ZG, Zhang L, Jiang Q, Chopp M (2002) Bone marrow-derived endothelial progenitor cells participate in cerebral neovascularization after focal cerebral Ischemia in the adult mouse. Circ Res 90:284–288

Zhao YH, Yuan B, Chen J, Feng DH, Zhao B, Qin C, Chen YF (2013) Endothelial progenitor cells: therapeutic perspective for ischemic stroke. CNS Neurosci Ther 19:67–75

Chapter 7
Endothelial Progenitor Cell Therapy in Stroke

Yaning Li, Yuanyuan Ma, Yongting Wang and Guo-Yuan Yang

Contents

G.-Y. Yang (✉) · Y. Li · Y. Wang
Neuroscience and Neuroengineering Research Center, Med-X Research Institute
and School of Biomedical Engineering, Jiao Tong University,
1954 Hua Shan Road, 200030 Shanghai, China
e-mail: gyyang0626@gmail.com

Y. Ma · G.-Y. Yang
Department of Neurology, School of Medicine, Ruijin Hospital,
Jiao Tong University, 200025 Shanghai, China

L.-R. Zhao, J. H. Zhang (eds.), *Cellular Therapy for Stroke and CNS Injuries*,
Springer Series in Translational Stroke Research, DOI 10.1007/978-3-319-11481-1_7

Abstract Stroke is a major cause of death globally, which induces irreversible neuronal and endothelial cell death. Endothelial progenitor cell (EPC) based therapeutics result in neovascularization and the improvement of vascular perfusion, which benefits clinical stroke patients. Although EPC transplantation in experimental stroke models shows functional improvement, EPC therapy in clinical stroke patients continues to face an arduous task. In this chapter, we give a brief introduction of EPCs including the source of EPCs, methods of isolation and identification of EPCs, the therapeutic potential for stroke, and signaling in modulating EPC function. Furthermore, we summarize the molecular mechanisms of EPC action after transplantation either through differentiating into mature endothelial cells to replace damaged cells or by enhancing trophic/regenerative support for endogenous repair processes. We discuss the routes of transplantation and the modifying methods for EPC safety and efficacy *in vivo.* Finally, we discuss the pros and cons for the application of EPCs for transplantation in clinical patients. Though EPC-based therapy is a potential treatment for stroke and holds promise for vascular regeneration, this field needs more study to uncover and resolve unsolved problems.

Abbreviations

acLDL	Acetylated low density lipoprotein
ACE2	Angiotensin-converting enzyme 2
BBB	Blood-brain barrier
BDNF	Brain-derived neurotrophic factor
bFGF	Basic fibroblast growth factor
BM	Bone marrow
CACs	Circulating angiogenic cells
CFU-Hil	Colony forming unit-Hill
ECFC	Endothelial colony forming cell
ECs	Endothelial cells
EGF	Epidermal growth factor
eNOS	Endothelial NO synthase
EPCs	Endothelial progenitor cells
EPO	Erythropoietin
FACS	Fluorescence activated cell
G-CSF	Granulocyte colony-stimulating factor
GDNF	Glial cell line-derived neurotrophic factor
GM-CSF	Granulocyte-macrophage colony-stimulating factor
HGF	Hepatocyte growth factor
HIF−Iα	Hypoxia-inducible factor−1α
HMGB−1	High-mobility group box 1
HSCs	Hematopoietic stem cells
ICAM−1	Intercellular adhesion molecule 1
IGF−1	Insulin-like growth factor−1
IL−8	Interleukin−8
MCP−1	Monocyte chemoattractant protein−1

MMP−9	Matrix metalloproteinase 9
MNCs	Mononuclear cells
MRI	Magnetic resonance imaging
NPCs	Neural progenitor cells
PB	Periblood blood
PDGF	Platelet-derived growth factor
PlGF	Placental growth factor
SDF−1α	Stromal derived factor−1α
TGF-b2	Transforming growth factor b2
tPA	Recombinant tissue plasminogen activator
UCB	Umbilical cord blood
UEA−E	Ulex europaeus agglutinin 1
VEGF	Vascular endothelial growth factor

7.1 Stroke

Stroke is a multifactorial disease, and ischemic and hemorrhagic stroke are the leading causes of death globally (Hassan and Markus 2000; Rubattu et al. 2000). A variety of risk factors have been illustrated to relate with stroke incidence including cerebrovascular diseases, aging, smoking, hypertension, diabetes, hypercholesterolemia, and lack of exercise. (Hankey 2006; Allen and Bayraktutan 2008; Flynn et al. 2008; Karam et al. 2008). The brain is very vulnerable to ischemic insult because it is sensitive to a lack of oxygen and glucose. Neurological dysfunction usually occurs within minutes after stroke onset. However, the deterioration of the brain may continue in the following minutes, hours or even days.

Ischemic stroke (>70% of strokes) is the most common type of stroke in clinical stroke patients. After ischemic stroke onset, a process of pathophysiological events are triggered, including energy failure, loss of cell ion homeostasis, the release of excitatory amino acid and reactive oxygen species, increase of intracellular calcium, disruption of the blood-brain barrier (BBB), activation of glial cells, and the infiltration of leukocytes (Bliss et al. 2007; Moskowitz et al. 2010). These interrelated and coordinated events result in ischemic cell necrosis, which exhibits non-selective damage of all cells including neurons, astrocytes, oliogdendrocytes, microglia and endothelial cells (Broughton et al. 2009). The size and location of these infarcts are determinants of the long-term functional deficits (Sims and Muyderman 2010). The ischemic penumbra area represents the region in which there is a chance for recovery via post-stroke therapy (Ginsberg 1997).

The only effective treatment for ischemic stroke patients is to administer recombinant tissue plasminogen activator (tPA). However, very few patients are lucky enough to receive tPA treatment because tPA has a very narrow time window (<4.5 h). Stem cell therapy has been proposed as a potential treatment for ischemic stroke in recent years, especially after putative progenitor endothelial cells have been isolated from bone marrow (BM) and identified as CD34 positive (Asahara et al. 1997). This kind of cells is named endothelial progenitor cells (EPCs) and is capable of contributing to the formation of new vessels by differentiating into mature endothelial cells (ECs) or supporting/promoting the endogenous repair process. EPCs can also serve as a marker during stroke occurrence and prognosis (Chu et al. 2008), and preclinical studies have shown EPC transplantation improves functional recovery by promoting neurogenesis and angiogenesis or provide trophic/protective factors through paracrine effects. Several clinical studies are currently investigating the safety and efficacy of EPC transplantation. EPC transplantation in stroke represents a promising therapeutic approach, although it is still in its infancy.

7.2 Endothelial Progenitor Cells (EPCs)

7.2.1 Discovery of EPCs

EPCs are BM mononuclear cells (MNCs), which were first isolated from peripheral blood (PB) by Asahara in 1997 (Asahara et al. 1997). This novel technique opened a new field for the treatment of vascular disease. Increasing evidence showed that EPCs could be mobilized to the PB after ischemic stroke and restore the damaged vessels via vasculogenesis (Asahara et al. 1999; Takahashi et al. 1999; Kalka et al. 2000a; Shintani et al. 2001; Murayama et al. 2002; Asahara and Kawamoto 2004; Zhan et al. 2013). Considering these cells' lack of unique markers, and that they share similar surface antigens with some hematopoietic lineages and mature ECs, such as CD31/KDR (VEGFR-2)/CD34/VE-cadherin/E-selectin (Rafii 2000; Khakoo and Finkel 2005), it is difficult to precisely define EPCs. However, it has been generally accepted that EPCs exist in circulating blood and possess angiogenic capability and the potential to differentiate into ECs, which contribute to the process of vasculogenesis and the maintenance of homeostasis in vascular ECs (Asahara et al. 1997; Shi et al. 1998; Asahara et al. 1999; Lin et al. 2000; Rafii 2000; Cesari et al. 2012). EPCs may be mixed with the circulating ECs in peripheral circulation, which may partially differentiate into mature ECs. EPCs play a more important role in promoting postnatal vasculogenesis compared with circulating ECs (Kalka et al. 2000a). Therefore, EPCs and circulating ECs are two different cells.

7.2.2 Source of EPCs

EPCs can be divided into two types based on their origin: hematopoietic and non-hematopoietic EPCs. Hematopoietic EPCs originating from BM are considered a subtype of hematopoietic stem cells (HSCs). Non-hematopoietic cells could be isolated from PB, umbilical cord blood (UCB), and tissue samples (Asahara et al. 2011). Although the origin of non-hematopoietic cells is unclear, this type of cell is likely derived from organ blood vessels and tissue stem cells (Alev et al. 2011).

In addition, increased studies have reported additional sources of non-hematopoietic cells. For example, the myogenic-EPCs in the interstitial spaces of skeletal muscle contribute to skeletal muscle growth (Tamaki et al. 2002); EPCs could also exist in the boundary between smooth muscle and the adventitial layer of human vascular walls (Zengin et al. 2006). Other sources of EPCs include the liver and intestine (Aicher et al. 2007), dental pulp-derived iPS cells (Yoo et al. 2013), the kidney (Sirker et al. 2009) and adipose tissue (Planat-Benard et al. 2004). Therefore, it is plausible that EPCs could be found in other sources, further study is needed to investigate the mysterious origin of EPCs.

7.2.3 Methods for the Isolation and Culturing of EPCs

Actually, it is a challenge, and controversial work, to isolate and identify EPCs from the PB, because of these cells' lack of unique and specific surface markers. Currently three methods are mainly used to isolate EPCs from the PB. The first and perhaps simplest method is to collect low density MNCs via density barrier centrifugation, and then plate these cells on fibronectin coated dishes with culture medium containing a variety of growth factors and fetal serum. After 4–5 days, remove non-adherent cells (Asahara et al. 2000; Vasa et al. 2001; Tepper et al. 2002). The remaining adherent cells present the early EPCs with spindle shape (Fig. 7.1). The second method is based on cell surface antigens, using a technology known as fluorescence activated cell (FACS) analysis to distinguish EPCs from other cells in PB (Yoder 2009; Kirton and Xu 2010; Basile and Yoder 2014). Although there are no specific antigens to isolate and identify EPCs, some have been accepted as fundamental elements representing the EPC phenotype, which are CD34, CD133 (AC133), and KDR (VEGFR−2) (Peichev et al. 2000). Subsequently, different combinations of these antigens have been used to isolate EPCs from PB, UCB and fetal liver (Timmermans et al. 2009). However, recent studies provide opposing evidence that cells expressing that the three antigens mentioned above should not represent EPCs, but stand for hematopoietic progenitors, because no observed vessel structure formed *in vivo* (Timmermans et al. 2007). Other surface antigens have also been used to identify EPCs, such as CXCR4, CD31, CD144, CD105, CD106, CD117, and CD45 (Basile and Yoder 2014). However, all of these surface antigens, including CD34, CD133 and KDR, do not only emerge on EPCs, but are also expressed on other cells, for example HSCs (Hirschi et al. 2008; Alaiti et al. 2010; Fadini et al. 2012; Yoder 2012), leading to unpersuasive results when isolating and identifying EPCs. Therefore, novel specific markers need to be found to identify true EPCs. The last method includes two colony-forming assays *in vitro*, which are based on cluster formation. One is called colony forming unit-Hill (CFU-Hil) assay. Briefly, the MNCs isolated from the PB are plated on the fibronectin coated dishes for 48 h, and the non-adherent cells are collected to culture again, clusters occurring after 4–9 days, which are named CFU-Hil EPCs (Hill et al. 2003). These cells have similar characteristics to early EPCs (Fadini et al. 2012), express endothelial and hematopoietic cell markers, and fail to form vessels *in vivo*. The other method

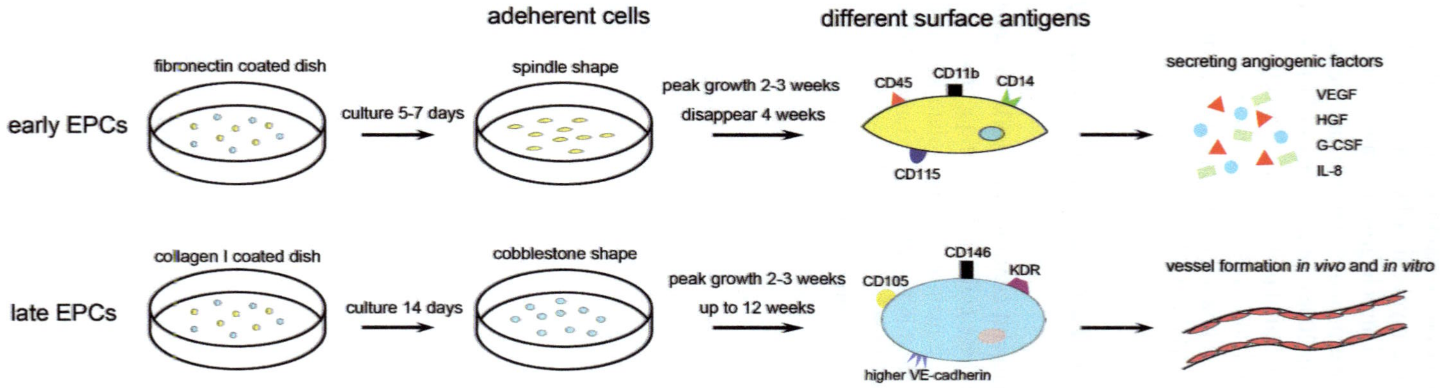

Fig. 7.1 An overview of isolation methods and characteristics of early and late EPCs

Table 7.1 Characteristics of early and late EPCs

	Early EPCs	Late EPCs	Ref
Source	PB	BM, PB/CB, vessel well	(Ingram et al. 2004; Kirton and Xu 2010)
Culture time	5–7 days	>14 days	(Ahrens et al. 2011; Zhao et al. 2013)
Morphology	Spindle	Cobblestone	(Hur et al. 2004)
Survival fate	Peak growth 2–3 weeks, disappear 4 weeks	Peak growth 2–3 weeks, up to 12 weeks	(Zhao et al. 2013)
In vitro	Low proliferative ability, no tube-like structure formation	High proliferative ability, capillary-like formation	(Hur et al. 2004)
In vivo	No vessel formation	Vessel formation	(Hur et al. 2004)
Function	Cannot regenerate a damaged endothelium	Physically contribute to vascular regeneration	(Hur et al. 2004; Yoder et al. 2007; Kirton and Xu 2010)
Neuro-vasculogenesis	Release angiogenic cytokines	Differentiate into endothelial cells	(Rehman et al. 2003; Hur et al. 2004)
Same surface antigens	CD34/KDR (VEGFR-2)/CD31/ CD114/vWF	CD34/KDR (VEGFR-2)/CD31/ CD114/vWF	(Kirton and Xu 2010)
Different surface antigens	CD45/CD14/CD11b/ CD115	CD105/CD146, higher VE-cadherin/ KDR	(Hur et al. 2004; Ingram et al. 2004; Kirton and Xu 2010; Fadini et al. 2012)

BM bone marrow, *PB* peripheral blood, *CB* cord blood

is endothelial colony forming cells (ECFCs) assay. Plate the isolated MNCs on collagen I coated dishes, and adherent cells form colonies 2–3 weeks later. These cells are named ECFCs, which are known as late EPCs (Ingram et al. 2004; Kirton and Xu 2010). ECFCs express antigens like primary ECs, have a huge potential to form colonies, and are able to form vessels *in vivo* and *in vitro* (Yoder 2009) (Fig. 7.1).

7.2.4 Classification of EPCs

According to their culture characteristics and functions, circulation EPCs can be classified into two different populations: early EPCs, which are also called circulating angiogenic cells (CACs), and late EPCs, which are also known as ECFCs (Hur et al. 2004) (Table 7.1). The early EPCs emerge 5–7 days after isolation of MNCs from the PB and disappear at 4 weeks. They have spindle shape, can be stained with Ulex europaeus agglutinin 1 (UEA−1) and take up acetylated low-density lipoprotein (acLDL) (Hur et al. 2004; Hirschi et al. 2008). They express EC markers and keep hematopoietic antigen expression (Kirton and Xu 2010). Early EPCs cannot form vessels *in vivo*, but contribute to angiogenesis by secreting angiogenic cytokines (Gehling et al. 2000; Lin et al. 2000; Vaughan and O'Brien 2012). Late EPCs form a monolayer of cobblestone shaped cells 2–4 weeks after plating, have huge potential to proliferate, and can be maintained for up to 12 weeks. Similarly, these cells can also be stained with UEA−1, take up acLDL and express the same markers as early EPCs, such as CD34/KDR (VEGFR−2)/CD31, but they lack the expression of antigens like CD14, CD133, CD45 and CD115 (Hur et al. 2004; Ingram et al. 2004; Kirton and Xu 2010). More importantly, late EPCs are able to form vessels *in vitro* and *in vivo* (Lin et al. 2000; Grant et al. 2002). Late EPCs are thought to be the true EPCs and show greater sensitivity to vascular endothelial growth factor (VEGF), basic fibroblast growth factor (bFGF), and placental growth factor (PlGF) (Bompais et al. 2004; Pasquier 2010). Therefore, these two types of EPCs have different morphologies, proliferative abilities and survival rates but both of them display vasculogenic capacity *in vivo* (Hur et al. 2004) (Fig. 7.1).

7.3 Therapeutic Potential of EPCs for Stroke

7.3.1 A Biomarker of Diseases

There is no doubt that EPCs exist in adult PB (Asahara et al. 1997), promote vascular repair after ischemia, and attenuate the progression of arteriosclerosis (Medina et al. 2012). In the past years, a lot of studies have demonstrated that the number and functional stage of circulation EPCs are associated with arteriosclerosis, hypertension, diabetes, and metabolic syndrome (Vasa et al. 2001; Hill et al. 2003; Werner et al. 2005; Liao et al. 2010; Mandraffino et al. 2011; Devaraj and Jialal 2012; Flammer et al. 2012). Based on this evidence, levels of circulation EPCs can be used as novel biomarkers. More importantly, the levels of EPCs also have a close relationship with ischemic stroke, studies have shown that lower levels of circulation EPCs indicate poor outcomes among ischemic stroke patients (Ghani et al. 2005; Sobrino et al. 2007; Chu et al. 2008; Yip et al. 2008; Tsai et al. 2014) (Table 7.2).

7.3.2 Protection of Blood Brain Barrier (BBB)

As we all know, the BBB is comprised of brain microvascular ECs, basement membrane, astrocytes and pericytes, all of these parts are now called the neurovascular unit (Wong et al. 2013). The integrity of the BBB plays an important role in maintaining the homeostasis of the brain. Once destroyed, the balance of the brain's microenvironment is disrupted, leading to a series of pathological processes, including the swelling of endothelial cells, an increase in vascular permeability, inflammatory cell infiltration and tissue edema. As mentioned above, EPCs have the potential

Table 7.2 EPCs as a marker for diseases or good outcomes

Diseases	No. patient	EPC phenotype	EPC number	Marker for	Ref
IS	30	$CD34^+/KDR^+$	↓	Increase burden of carotid atherosclerosis	(Lau et al. 2007)
IS	48	CFU-EC	↑	Good functional outcome; reduced infarct growth	(Sobrino et al. 2007)
Stroke	75-acute 45-chronic	CFU	↓	Endothelial dysfunction or repair in acute stroke	(Chu et al. 2008)
Acute IS	138	$CD31^+/CD34^+$ $CD62E^+/CD34^+$ $KDR^+/CD34^+$	↑in IS ↓in severe neurological impairment	↓EPC: severe neurological impairment and adverse clinical outcomes	(Yip et al. 2008)
CM	214	$CD34^+/VEGFR\text{-}2^+CD34^+/CD133^+$	↓	A pathophysiologic feature of CM	(Gyan et al. 2009)
Acute IS	100	$CD133^+/KDR^+$	↓	Mobilization of EPCs in response to stroke stress	(Zhou et al. 2009)
Acute IS	17	$CD34^+/CD133^+/VEGFR\text{-}2^+$	↓	Larger lesion volumes; acute phase stroke severity	(Bogoslovsky et al. 2010)
IS	42	$CD133^+/CD34^+/KDR^+/CD45^-$	↑	Enhanced angiogenic function in the subacute phase of stroke	(Navarro-Sobrino et al. 2010)
IS	17	$CD133^+/CD34^+$	↓	Tissue injury and stroke severity in early ischemia	(Bogoslovsky et al. 2011a)
CA	27	$CD133^+/CD34^+$	↓	Contributes to the pathophysiological process of aneurysm formation	(Wei et al. 2011a)
Ruptured CA	14	$CD133^+/CD34^+$	↑	Prognostic marker for the outcomes of ruptured CA	(Wei et al. 2011b)
Hypertension	32	$(CD31^+/CD34^+/KDR^+/CD45^-$	↓	Relates to radiological cerebral small vessel disease manifestations	(Rouhl et al. 2012)

Table 7.2 (continued)

Diseases	No. patient	EPC phenotype	EPC number	Marker for	Ref
IAS	108	$CD34^+/CD133^+/VEGFR\text{-}2^+$	↑	Independent markers of IAS	(Liu et al. 2013c)
IS	146	$CD34^+/CD133^+/VEGFR\text{-}2^+$	↑at day 7	A better functional outcome	(Marti-Fabregas et al. 2013)
Acute IS	55	$CD133^+/CD34^+$; $KDR^+/CD34^+$	↓	Poor 6-month outcome in patients with AIS	(Tsai et al. 2014)

CFU-EC early outgrowth colony-forming unit-endothelial cell, *IS* ischemic stroke, *CA* cerebral aneurysm, *CM* cerebral malaria, *IAS* intracranial arterial stenosis to differentiate into ECs and promote vascular repair (Ponio et al. 2014), and to support the integrity and function of the BBB (Kaneko et al. 2012). However, how EPCs beneficially influence the BBB is still a mystery. Therefore, more work is needed to elucidate the protective mechanism of EPCs on the BBB after stroke.

7.3.3 Promotion of Neovascularization After Stroke

It has been widely accepted that neovascularization after stroke is essential and critical for tissue repair and neurological function recovery. Animal and human studies have proved that EPCs participate in neovascularization (Zhang et al. 2002; Fan et al. 2010; Paczkowska et al. 2013), mainly via two approaches: by directly differentiating into ECs and incorporating into the damaged vessels, which is called vasculogenesis; and by indirectly promoting migration and proliferation of pre-existing ECs, which is called angiogenesis, through releasing a variety of angiogenic cytokines (Masuda and Asahara 2003; Tepper et al. 2005; Urbich et al. 2005; Chen et al. 2013b). In addition, these cytokines also enhance EC and neuron survival, and recruit endogenous progenitor cells (Chen et al. 2013b). Because of the unique characteristic and advantage of angiogenic ability, EPCs may be an important agent for the treatment of stroke.

7.3.4 Factors Influence EPCs In Vivo

In the past decade, it had been demonstrated that tissue ischemia and exogenous cytokines could mobilize endogenous circulating EPCs and thereby contribute to neovascularization (Asahara et al. 1999; Takahashi et al. 1999). Subsequently, more and more studies have proven that the levels and functional stages of EPCs are correlated with many diseases and are considered as a biomarker (Table 7.2). Moreover, transplantation of EPCs as a therapeutic strategy is beneficial to the hindlimb and cerebral ischemia (Kalka et al. 2000a; Fan et al. 2010; Moubarik et al. 2011). Recently, several studies showed that a variety of factors could influence the number and function of circulating EPCs *in vivo*. For example, statin treatment for 4 days may increase circulating EPCs levels in acute ischemic stroke patients, probably by nitric oxide (NO)-related mechanisms (Sobrino et al. 2012a). VEGF and stromal derived factor−1α (SDF−1α) are independent factors for the increment of circulating EPCs (Sobrino et al. 2012b). In addition, factors like homocysteine, haptoglobin 1–1, citicoline, cilostazol, systolic blood pressure, total cholesterol, erythropoietin (EPO), high-mobility group box 1 (HMGB−1), and matrix metalloproteinase (MMP−9) are also proven to influence the number of circulating EPCs in humans and animals (Table 7.3). The therapeutic effects of transplantation of EPCs for stroke can be improved by modulating these factors (Morancho et al. 2013).

Table 7.3 Factors that influence EPCs *in vivo*

Factors	EPC source	Route and time for factors of administration	EPC phenotype	Influence	Patients or animals	Ref
Homocysteine	PB	—	$CD34^+/CD31^+$	↑EPCs apoptosis	? patients with hyperhomocysteinemia	(Alam et al. 2009)
Haptoglobin 1-1	PB	—	$CD34^+/KDR^+$	↓EPCs number	42 patients with lacunar stroke	(Rouhl et al. 2009)
BP and total cholesterol	PB	—	$CD133^+/KDR^+$	↓EPCs number	? patients with acute stroke	(Zhou et al. 2009)
SDF-1	PB	—	$CD34^+/CD133^+/$ $VEGF\text{-}R2^+$	↑EPCs number	17 patients with acute IS	(Bogoslovsky et al. 2011b)
Citicoline	PB	Oral citicoline (2000 mg/day)	CFU-EC	↑EPCs number	48 patients with IS	(Sobrino et al. 2011)
Cilostazol	PB	200 mg daily	$CD34^+/CD133^+$	↑EPCs number	20 diabetic patients with leukoaraiosis or asymptomatic old cerebral infarction	(Ueno et al. 2011)
EPO	PB	200 U/kg 1 month after surgery	$CD34^+/CD133^+$	↑EPCs number	rat with CA	(Xu et al. 2011)
EPO	PB	5000 IU Each time, subcutaneously	$CD31^+/CD34^+$, $CD62E^+/CD34^+$, $KDR^+/CD34^+$	↑EPCs number	167 patients with acute IS	(Yip et al. 2011)
HMGB-1	Peri-infarct in brain	—	$CD34^+/Flk1^+$	↑EPCs number	mice with 45 min MCAO	(Hayakawa et al. 2012)
Statins	PB	20 mg atorvastatin/day	CFU-EC	↑EPCs number	48 patients with acute IS	(Sobrino et al. 2012a)
VEGF and SDF-1α	PB	—	CFU-EC	↑EPCs number	48 patients with IS	(Sobrino et al. 2012b)
MMP-9	Spleens	—	$vWF^+/KDR^+/$ $CD133^+$	Delay EPCs release	Animals with MCAO	(Morancho et al. 2013)

BP blood pressure, *PB* peripheral blood, *CFU-EC* Foutgrowth colony forming unit-endothelial cell, *HMGB−M* high-mobility group box 1, *EPO* erythropoietin, *MMP−M* matrix metalloproteinase 9, *IS* ischemic stroke, *CA* cerebral aneurysm, *MCAO* middle cerebral artery occlusion, — no administration, *?* not

7.4 Signaling in Regulating EPC Functions

Studies are investigating a variety of factors that influence EPC proliferation, migration and maturation (Table 7.3). Additional researchers are trying to discover the signaling pathways activated by these factors to influence EPCs. Early EPCs secrete a large number of factors, including VEGF, brain-derived neurotrophic factor (BDNF), bFGF, insulin-like growth factor 1 (IGF−1), and interleukin−8 (IL−8). (He et al. 2005; Moubarik et al. 2011; Rosell et al. 2013), which are pro-angiogenic factors that increase endothelial proliferation, tube formation, migration and MMP secretion in ECs to enhance the invasiveness of EPCs (Carmeliet 2003; Li et al. 2003). MMP−9 is essential for ischemia-induced neovascularization, which modulates the neovascularization of EPCs by increasing the release of cytokines (Huang et al. 2009; Morancho et al. 2013). Integrin-linked kinase is upregulated in ECs and associated with increased intercellular adhesion molecule 1 (ICAM−1) and SDF−1 under hypoxic stress, which recruits EPCs to ischemic tissue (Lee et al. 2006). CD18 and its ligand ICAM−1 also play an essential role in mediating EPC recruitment in infracted hearts (Wu et al. 2006). Activated AKT signals promoted the expression of ICAM−1 on ECs and closely associated with EPC entrapment, which might be important in regulating the process of neovascularization through enhancing EPC migration and trans-endothelial migration (Yoon et al. 2006; Hur et al. 2007).

IL−10 increases EPC survival and mobilization through the activation of STAT3/VEGF signaling cascades (Krishnamurthy et al. 2011). SDF−1 released from the ischemic tissue form a concentration gradient to promote EPC homing through interaction with its receptor CXCR4 (Fan et al. 2010). Deltalike−4 gene modified EPCs show enhanced functional neovascularization in ischemic tissue due to the activation of Notch/Hey1/mTOR/p70S6K signaling pathways (Huang et al. 2013a). Wnt1 is a pro-angiogenic molecule and enhances EPC function in a hepatocyte growth factor (HGF)-dependent manner (Gherghe et al. 2011). HMGB1 secreted by astrocytes after ischemic stroke increases EPC homing involved in neurovascular remodeling and functional recovery (Hayakawa et al. 2012). Other factors have also been reported to influence EPC functions, such as E-selectin, estrogen and β2-adrenergic receptor (Oh et al. 2007; Tan et al. 2012; Galasso et al. 2013). Fully understanding the mechanisms underlying EPC function will help improve the safety and efficiency of EPC transplantation.

7.5 Action Mechanism of EPCs

7.5.1 Cell Replacement

EPCs derived from BM or other tissues have an intrinsic capacity for differentiating into ECs (Asahara et al. 1997; Beltrami et al. 2003; Planat-Benard et al. 2004; Chen et al. 2008; Chen et al. 2012; Nih et al. 2012; Iskander et al. 2013;

Pellegrini et al. 2013). The injured ECs in the brain can be replaced by transplanted EPCs. Granulocyte colony-stimulating factor (G-CSF) mobilizes circulating EPCs to engage 39 % of the total luminal length of the neoendothelium (Takamiya et al. 2006). LacZ-transduced $CD34^+$EPC transplantation leads to about 60 % reendothelialization of balloon-injured rabbit carotid arteries costained with CD31 as early as 4 days after transplantation and this increases to about 70 % at 30 days after transplantation (Griese et al. 2003b). Fluorescence-labeled EPCs are found in the neointima and costaining with vWF is found after 4 weeks in the injured carotid artery of balloon injured New Zealand white rabbits (Hu et al. 2013). Hence, cell replacement is one of the mechanisms of vascular repair by progenitor cells. BM derived EPCs contribute to the microvascular structure of the choroid plexus by differentiating into ECs during cerebral ischemia in adult mice (Zhang et al. 2002). 14 days after the transplantation of EPCs in the cerebral ischemia rabbit model, a decrease in the number of apoptotic cells and an increase in the microvessel density in the ischemic boundary area has been witnessed, and most of EPCs capable of binding to UEA−1 lectin are incorporated into capillaries (Chen et al. 2008).

However, the extent of incorporation of BM derived cells in cerebral vessels after stroke has varied in previous studies (Hess et al. 2002; Zhang et al. 2002; Machein et al. 2003; Chen et al. 2008; Moubarik et al. 2011). Whereas positive vessels had an average of 34 % endothelial marker expressing BM derived cells (Hess et al. 2002; Zhang et al. 2002), others could not detect endothelial marker expressing cells (Machein et al. 2003; Moubarik et al. 2011).

7.5.2 Enhanced Trophic/Regenerative Support for Endogenous Repair Processes

Neovascularization is not solely the result of the incorporation of EPCs in newly formed vessels; the release of trophic factors in a paracrine manner may also influence neovascularization. Cultured PB MNCs secrete high levels of VEGF, HGF, G-CSF and granulocyte-macrophage colony-stimulating factor (GM-CSF) (Rehman et al. 2003). More and more researchers are paying close attention to the trophic effects of EPCs. *In vitro*, early EPCs cultivated from different sources have shown marked expression and the release of angiogenic cytokines including G-CSF, GM-CSF, VEGF, platelet-derived growth factor (PDGF), epidermal growth factor (EGF), FGF, HGF, IL−8, transforming growth factor β2 (TGF-β2), IGF−1 and etc (He et al. 2004; Hur et al. 2004; Yoon et al. 2005). The release of these growth factors in turn may influence the classical process of angiogenesis, particularly the proliferation and migration as well as the survival of mature ECs (Folkman 1995; Urbich and Dimmeler 2004). EPCs can also exert a strong mitogenic effect on mature ECs and enhance the angiogenic capacity of outgrowth of ECs via secretion of IL−8 with/without VEGF (He et al. 2005; Yoon et al. 2005).

In cerebral arteries, the paracrine effect of EPCs promotes vasoprotection by increasing prostacyclin production and the intracellular concentration of cAMP

(Santhanam et al. 2007). EPCs from stroke patients present higher levels of pro-angiogenic factors at early stages, which decrease in mature ECs when they become more similar to mature microvascular ECs (Navarro-Sobrino et al. 2013). 24 h after the administration of EPCs expressing GFP, they are found to express endothelial NO-synthase (eNOS) and distribute in the brain parenchyma and around the endothelial layer of pial arteries in the ischemic lesions (Ohta et al. 2006). EPC transplantation can also induce humoral effects, which are sustained by host tissues, decrease apoptosis and augment proliferation of cells. Transplantation of EPCs enhances the mobilization of endogenous EPCs and HSCs mainly by upregulation of humoral VEGF, FGF−2, IGF, HGF, angiopoietin−1 and SDF−1 (Cho et al. 2007).

Studies have shown that vascular niche can support neurogenesis in the subventricular zone and the dentate gyrus by secreting growth factors associated with neurogenesis, such as VEGF or BDNF (Leventhal et al. 1999; Palmer et al. 2000). In an experimental stroke study, neovascularization related to neurogenesis, and also to the migration of neural progenitor cells (NPCs), along the newly formed vessels (Thored et al. 2007). Thus, administered EPCs may enhance the proliferation of endogenous NPCs in the brain (Rouhl et al. 2008). EPCs injected 24 h after MCAO were found in the injured area and improved functional recovery, which was linked to a reduction in ischemia-induced apoptosis and a stimulation of ischemia-induced angiogenesis and neurogenesis (Moubarik et al. 2011). Transplantation of BM-derived EPCs exerts potent neuroprotective functions against cerebral ischemia/reperfusion injury in rats, and the protective effects may be associated with decreased expression of Bax, caspase−3 and p67phox and the increasing expression of Bcl−2 and manganese superoxide dismutase (MnSOD), which promotes anti-oxidative and anti-apoptotic properties (Qiu et al. 2013).

7.6 Transplantation of EPCs in Ischemic Stroke Animals

7.6.1 Transplantation Routes for EPCs

The optimal transplantation route for EPCs following ischemic stroke may be important for the therapeutic efficacy. The two routes mostly used for the transplantation of EPCs in stroke are intracerebral and intravascular injections. They each have their own advantages and disadvantages. EPCs intracerebrally injected into the peri-infarct area may be immediately involved in incorporating newly formed vessels or secreting trophic factors to support endogenous repair processes, especially in permanent ischemic stroke to bypass the occlusion of blood vessels. However, invasive injury to the brain raises safety issues.

Intravascular injection either through veins or arteries has minimal invasive injury potential for systematical cell distribution, as well as the far-flung secretion of neuroprotective, pro-angiogenic and immunomodulatory factors (Misra et al. 2012). Intravenously grafted cells can follow a chemokine generated gradient formed by the injured brain and penetrate through the BBB (Guzman et al. 2008), and grafted cells do not have to be near the lesion to be effective (Borlongan et al. 2004). However, very few cells have been found to integrate into the infarct area. The majority of cells became stuck in the lung, liver, and spleen after intravenous administration. Intra-arterial delivery, in contrast, overpasses the peripheral filtering organs, leading to higher cell engraftment to the brain (Li et al. 2010; Zhang et al. 2012), and greater efficacy (Kamiya et al. 2008; Pendharkar et al. 2010). There is a concern that intra-artery transplanted cells can stick together and cause microemboli, including lethal pulmonary emboli or a reduction in cerebral blood flow, which is associated with microstrokes (Walczak et al. 2008).

In preclinical experimental stroke, intravascular injections are usually used, and they are applied through the tail vein (Zhang et al. 2002; Chen et al. 2012; Nih et al. 2012; Chen et al. 2013d; Decano et al. 2013; Qiu et al. 2013), femoral vein (Moubarik et al. 2011; Pellegrini et al. 2013), jugular vein (Fan et al. 2010; Li et al. 2013), and internal carotid artery (Ohta et al. 2006) (Table 7.4). Despite the different routes used for EPC transplantation, decreased infarct volume, improved neurobehavioral outcomes, increased angiogenesis and neurogenesis, attenuation of endothelial dysfunction, even anti-apoptosis effects have been observed during study. These studies may benefit from both functions of EPCs during cell replacement and enhanced trophic/regenerative support for endogenous repair processes.

However, when considering application in clinical trials, the routes of transplantation of EPCs should be standardized to ease administration. Several clinical studies have been carried out as illustrated in Table 7.5. These studies look into the safety and efficiency of routes for EPC transplantation in human patients, and there is still a lot of work to do in this field.

7.6.2 Modification of EPCs

7.6.2.1 Gene Modification

Considering the paracrine-mediated mechanisms of EPCs, the enhancement of their secretion of trophic factors capacity by the overexpression of related genes would be valuable to magnify the efficacy of EPC therapies in stroke treatment (Chen et al. 2013a). EPCs have been modified by a variety of genes before transplantation and have been reported to enhance functional recovery, these genes include VEGF (Asahara 2007; Gou et al. 2011; Yang et al. 2012), HGF (Song et al. 2009), IGF−1 (Sen et al. 2010), paraoxonase−1 (Wang et al. 2010), CXCR4 (Chen et al. 2012), SDF−1 (Schuh et al. 2012), NO (Chen et al. 2013c), home oxygenase−1 (Long et al. 2013), hypoxia-inducible factor−1α (HIF−1α) (Jiang et al. 2008) and Deltalike−4 (Huang et al. 2013a). There are two major methods for gene transfer systems, viral and nonviral. The most widely used viral vectors for gene transfer are adenovirus and retrovirus. Nonviral methods include the introduction of naked DNA into the target cells and the use of liposomes (Vale et al. 2001).

Table 7.4 Preclinical EPC transplantation studies

Animal models	EPC source	EPC phenotype	Modification	Co-transplantation or not?	Dose	Route and administration time	Outcomes	Mechanisms	Ref
Embolic stroke in mice	BM	NM	No	No	2×10^6	IV via tail vein, 4 weeks before stroke	↑neovascularization	↑angiogenesis and vasculogenesis	(Zhang et al. 2002)
tMCAO in rats	BM	NM	No	No	2.5×10^5	IA, immediately after stroke	↓infarct volume and neurological deficits	↑eNOS, VEGF, IGF-1; ↓endothelial dysfunction	(Ohta et al. 2006)
pMCAO in rabbits	BM	UEA-1, acLDL	No	No	3×10^7	IV, 1 day after ischemia	↓infarct volume and neurological score	↓tunel$^+$ cells; ↑vessel density↑angiogenesis	(Chen et al. 2008)
1 h tMCAO in mice	PB	KDR$^+$/VE-cadherin$^+$/vWF$^+$/Tie–2$^+$; UEA, acLDL	No	No	1×10^6	IV via left jugular vein, 1 h after tMCAO	↓ischemic infarct volume and neurological deficits	↑angiogenesis	(Fan et al. 2010)
tMCAo in rats	PB	CFU-EC	No	No	4×10^6	IV via femoral vein, 1 day after MCAO	↓apoptosis and reactive astrogliosis; ↑capillary density and neurogenesis	↑IGF-1, VEGF secretion	(Moubarik et al. 2011)
pMCAO in type 2 diabetic mice	BM	CD34$^+$/VEGFR-2$^+$/Lectin$^+$; acLDL	CXCR4 gene modified	No	2×10^5	IV via tail vein, 2 h after MCAO	↑CBF; ↓infarct volume and neurological score	↑angiogenesis and neurogenesis	(Chen et al. 2012)
Distal MCAO in mice	UCB	CD34$^+$	No	SMPCs, 2.5×10^5	2.5×10^5	IV via tail vein, 1 day after MCAO	↑angiogenesis, vascular remodeling	maintenance of neurogenesis and neuroblast migration	(Nih et al. 2012)

Table 7.4 (continued)

Animal models	EPC source	EPC phenotype	Modification	Co-transplantation or not?	Dose	Route and administration time	Outcomes	Mechanisms	Ref
TBI in rats	Spleens	NM	Labeled with SPIO	No	10^6	IV via tail vein, at 6–12 h after TBI	↑CBF	↑microvascular density	(Chen et al. 2013d)
Stroke-prone Rats	PB	$CD45^-/CD34^+/KDR^+$	No	No	NM	IA via tail vein, —	Attenuated progression and delayed stroke onset	Maintain vascular healthy	(Decano et al. 2013)
tMCAO in rat	UCB	$CD133^+$	No	No	1×10^7	IV, 1 day after tMCAO	↓infarct volume	↑endogenous proliferation, angiogenesis, and neurogenesis	(Iskander et al. 2013)
tMCAO in rat	UCB	CFU-EC	No	EPO (2500 UI/kg per day ×3)	5×10^6	IV via femoral vein, 1 day after tMCAO		↓apoptosis; ↑angiogenesis and neurogenesis	(Pellegrini et al. 2013)
tMCAO in rat	BM	$CD31^+/CD34^+/CD133^+/Flk\text{-}1^+$	No	No	10^6	IV via tail vein, immediately and 12 h after reperfusion	↓I/R injury	↓caspase-3 activity, Bax and NF-κB, ↑Bcl-2 expression; anti-oxidative and anti-apoptotic properties	(Qiu et al. 2013)

BM bone marrow; *PB* peripheral blood; *UCB* umbilical cord blood; *t/pMCAO* transient/permanent middle cerebral occlusion; *TBI* traumatic brain injury; *CFU-EC* outgrowth colony forming unit-endothelial cell; *SPIO* super-paramagnetic iron oxide; *IV* intravenous, *IA* intra-artery; *acLDL* acetylated low density lipoprotein; *EPO* erythropoietin; *SMPCs* smooth muscle progenitor cells, *CBF* cerebral blood flow; *I/R* ischemia/reperfusion; *NM* not mentioned

Table 7.5 Studies on clinical EPC transplantation

Phase	Patients' symptoms	No. patients	EPC source	EPC markers	Dose	Route	Outcome measures	Status	Location	NTC identifier
I/II	18–80 years, ≤7 days of after stroke; NIHSS≥7 at day 7	NM	BM		2.5×10^6/kg	IV	Number of adverse events; Changes in functional outcomes	Recruiting	China	NCT01468064
I	35–75 years, ≥6 and ≤60 months after stroke	NM	UCB	$CD34^+$	$2–8 \times 10^6$/patient	IC	Change from baseline in NIHSS and brain Image	Not yet recruiting	China	NCT01438593
II	35–70 years, ≥6 and ≤60 months after stroke, NIHSS (9–20)	30	PB	$CD34^+$	$2–8 \times 10^6$/patient	IC	NIHSS, European stroke scale, Barthel index and Mini-Mental State Examination, MRI and CT scans	Completed	China	NCT00950521
I/II	30–80 years, ≤7 days of onset, NIHSS≥8	10	BM	$CD34^+$	NM	IA	Safety, Modified Rankin Score, NIHSS	Recruiting	United Kingdom	NCT00535197
—	≥30 years, ≥6 months and ≤5 years after stroke, NIHSS (9–20)	36	PB	$CD34^+$	Plus G-CSF	IC	Not mentioned	Unknown	China	NCT01239602

BM bone marrow; *PB* peripheral Blood; *UCB* umbilical cord blood; *NIHSS* the national institutes of health stroke scale; *IV* intravenous; *IC* intracerebral; *IA* intra-artery; *MRI* magnetic resonance imaging; *CT* computed tomography;—Not mentioned

EPCs modified by VEGF gene show significantly enhanced neovascularization, even when ten times fewer cells were infused (Asahara 2007), and promote vascular regeneration of ischemic flaps (Yi et al. 2006). In ischemic hindlimb model, transfection of VEGF or heme oxygenase−1genes into EPCs significantly increased the number of differentiated ECs, blood perfusion levels and neovascularization compared to the bare EPCs (Yang et al. 2012; Long et al. 2013). Transfection of EPCs with other genes, such as IGF−1 (Sen et al. 2010), SDF−1 (Schuh et al. 2012), NO (Chen et al. 2013c) and Deltalike−4 (Huang et al. 2013a) genes, to treat ischemic myocardial injury show cell protective and myocardial regeneration effects and functional neovascularization recovery. EPCs modified by paraoxonase−1 genes are potentially valuable in the treatment of atherosclerosis (Wang et al. 2010). Transfection of HGF genes enhances EPC function and improves EPC transplantation efficiency by decreasing neointima formation and increasing reendothelialization for balloon-induced arterial injury (Song et al. 2009).

Some investigators tried to transfect multiple genes into EPCs by using retroviruses to encode both tPA and heparin. Local transplantation of engineered EPCs in a balloon-injured carotid artery model attenuates reendothelialization of angioplasty-injured arteries, but fails to inhibit neointima proliferation (Griese et al. 2003a). In experimental stroke models, only one study found that the transplantation of *CXCR4* gene-modified EPCs reduces cerebral ischemic damage and promotes repair in diabetic mice, and that modified EPCs show better therapeutic effects for ischemic stroke than unmodified EPCs (Chen et al. 2012).

Until now, there have been no clinical trials using gene-modified EPC therapy for the treatment of stroke. It is important to confirm the safety and efficacy of delivering exogenous genes into patients by modifying EPCs. The main concern is the possibility of tumorigenesis after gene delivery. Although exogenous genes are transferred into EPCs rather than to host cells, viral vectors may increase the risk of genotoxicity by insertional mutagenesis and the activation of adjacent oncogenes. To avoid malignant transformation in clinical patients, the vector should be designed for self-inactivation and only contain nonviral, physiologic promoter/enhancer elements (Payen and Leboulch 2012). Second, the therapeutic genes may serve different functions during different pathological stages. For example, SDF−1 plays a key role in promoting angiogenesis and neurogenesis during development (Mithal et al. 2012; Virgintino et al. 2013) and can recruit EPCs towards ischemic lesions for reendovascularization (Fan et al. 2010). Blocking SDF−1/CXCR4 interaction suppresses inflammatory responses and reduces brain infarction in the acute phase of ischemic stroke (Huang et al. 2013b; Ruscher et al. 2013), which indicates that SDF−1 is an inflammation initiator and exaggerates the BBB leakage and ischemic lesions. Whether SDF−1-overexpressing cells could exhibit a similar deterioration effect is unknown, but such studies are fundamental in calling attention to the administration paradigm of EPC gene modified therapy. Third, most completed and ongoing clinical trials employ autologous EPCs for transplantation; the exogenous gene expression in EPCs is time consuming and unavoidably delays cell transplantation. Further studies should be carried out on the effects of delivering gene-modified EPCs in a later period after stroke or the transplantation ex vivo of expanded EPCs from allogenic sources, which allows for transformation and *in vitro* expansion of EPCs before transplantation (Chen et al. 2013a).

7.6.2.2 Preconditioning EPCs

In addition to the exogenous gene modification of EPCs, investigators have been trying to manipulate endogenous mechanisms for optimizing the therapeutic potential of cell-based stroke therapy by pre-treating EPCs before transplantation. Various factors seem to influence the number of EPCs and their functions, both in experimental stroke models and in clinical patients (Table 7.3). The hypoxia induced by HIF−1α and trophic growth factors such as BDNF, glial cell line-derived neurotrophic factor (GDNF), VEGF and its receptor FIK−1, EPO and its receptor EPOR, SDF−1 and its receptor CXCR4, enhance EPC proliferation, mobilization and the homing to ischemic lesions involved in the repairing process (Kalka et al. 2000b; Vale et al. 2001; Yamaguchi et al. 2003; Bennis et al. 2012). Increased HIF−1α and its downstream genes play central roles in hypoxia-induced defense responses (Ogle et al. 2012). Ischemic preconditioning increases EPC numbers to attenuate partial nephrectomy-induced ischemia/reperfusion injury (Liu et al. 2013a).

VEGF is an important humoral factor for EPC mobilization/differentiation, which is supported by the correlation between the increase in VEGF serum concentration and the increase in circulating EPCs (Sobrino et al. 2012a). EPO stimulates normal EPC-mediated endothelial turnover and improves cardiac microvascularization and function in the presence of ischemia (Westenbrink et al. 2008). Pretreatment of EPCs with EPO before transplantation enhances their angiogenic potential (Bennis et al. 2012). SDF−1 pretreatment during EPC expansion stimulates the adhesion of EPCs to ECs and augments the efficiency of EPC-based cell therapy for ischemic diseases (Zemani et al. 2008). The hormone melatonin stimulates the protective effect of EPCs in acute ischemic kidney injury (Patschan et al. 2012). Exposure to sub-lethal hypoxia can significantly increase the tolerance and regenerative properties of stem/progenitor cells *in vitro* and after transplantation for other cell types (Francis and Wei 2010; Wei et al. 2013; Yu et al. 2013).

Estradiol preserves the integrity of ischemic tissue by augmenting the mobilization and incorporation of EPCs into sites of neovascularization by the eNOS-mediated augmentation of MMP−9 expression in the BM (Iwakura et al. 2006). Angiotensin-converting enzyme 2 (ACE2) improves EPC functions, by regulating eNOS and Nox pathways, enhancing the efficacy of EPC-based therapy for ischemic stroke (Chen et al. 2013b). Other methods, such as the pretreatment of EPCs with extracorporeal shock waves (Lee and Kou 2012) or magnetic bionanoparticles (Li et al. 2013) to enhance the homing and functions of EPCs may also be promising and novel strategies. It is expected that the preconditioning strategy will be further explored due to its potential to enhance the benefits of EPC-based transplantation therapies in stroke therapy (Liu et al. 2013b).

7.7 Pros and Cons in the Application of EPCs in Clinical Trials

EPC transplantation in stroke has pros and cons. As mentioned above, EPCs have shown much potential for stroke therapy either through directly differentiating into mature ECs to replace damaged tissue or by secreting trophic factors to enhance the endogenous repairing processes. Additionally, EPCs can be derived from a variety of sources including PB (Medina et al. 2010), BM (Kwon et al. 2010), cord blood (Li et al. 2013), spleen (Wassmann et al. 2006), adipose tissue (Planat-Benard et al. 2004), and the liver or intestines (Aicher et al. 2007). Ethical limitations are avoided because fetal or embryonic tissues are not necessary sources. A lot of experience in administration of HSCs in clinical treatment of patients with leukemia (Rouhl et al. 2008) shows it is not necessary for autogenous transplanted cells. It allows plenty of time for ex vivo expanded EPCs to be cultured, pretreated or even gene modified, so as to enhance therapeutic capacity when transplanted *in vivo*. However, there are still difficulties that need to be resolved. As EPCs can be cultured by many methods and derived from different sources, and they bear both the characteristic of hematopoietic and endothelial cells, there are no specific markers to identify them and they may also be contaminated by other cell lines like lymphocytes, macrophages or other dendritic cells (Ishikawa and Asahara 2004). Gene expression profiles may also change during EPC culturing (Gremmels et al. 2011). Whether the exogenous gene modified EPCs increase malignant transformation in clinical patients still needs to be further explored.

7.8 Problems Need to be Clarified for the Treatment of Patients

7.8.1 Evaluation of Clinical Safety

Preclinical studies have shown that EPC transplantation is beneficial for functional outcomes without showing side effects, such as enhancing inflammatory responses or forming teratoma. However, to fully ensure the safety of transplanting EPCs in clinical patients, clinical studies have been carried out. A small pilot study suggested that intravenous infusion of autologous EPCs was safe and improved exercise capacity in children with idiopathic pulmonary arterial hypertension (Zhu et al. 2008). EPC transplantation in 20 patients with acute myocardial infarction showed no incensement in the levels of inflammatory markers or troponin T (a marker for cardiac ischemia) (Assmus et al. 2002). Thus, in this small number of patients, EPCs neither seem to stimulate the inflammatory response nor increase ischemia. A variety of clinical studies has also shown that autologous BM stem cell and mesenchymal stem cell transplantation in stroke patients showed nothing related to abnormal cell growth or tumorigenesis, deteriorated functional outcomes or venous thrombuses (Suarez-Monteagudo et al. 2009; Lee et al. 2010; Honmou et al. 2011; Friedrich et al. 2012; Moniche et al. 2012).

EPCs can secret inflammatory factors such as IL−8 and monocyte chemoattractant protein−1 (MCP−1) (Hur et al. 2004; van der Strate et al. 2007), which might recruit monocytes and macrophages to aggravate ischemia. Currently, several clinical trials (clinicaltrials.gov identifier: NCT00950521; NCT01468064; NCT00535197) are trying to evaluate the safety and efficacy of autologous EPC transplantation in ischemic stroke. EPC transplantation cannot be routinely performed on patients for the treatment of stroke before larger clinical trials further ensure their clinical safety.

7.8.2 Identifying Acceptable Patients for EPC Transplantation

No treatment is appropriate for all stroke patients therapy. Therefore, establishing a criterion for choosing suitable patients for EPC transplantation is vitally important. Stroke patients range in age from 30–80 years old; age should be something to consider because elder patients tend to be suffering from other diseases such as hypertension, diabetes mellitus and dyslipidemic syndromes. Patients with these syndromes show endothelial dysfunction and decreased EPC numbers (Rouhl et al. 2012). It might be difficult for this kind of patient to receive autologous EPCs, which may lose their functional therapeutic effects after transplantation. Methods of modifying EPCs to increase their vasculogenic potential or allogenic EPCs from healthy people may provide options.

Studies have shown that the pathology of stroke in young and aged rats are not identical. For example, after intracerebral hemorrhage, aged rats showed a wider spread of activated microglia/macrophages around the parenchyma and higher astrocyte activity than young rats (Wasserman et al. 2008). Another study showed that aging mice had significantly less edema formation after stroke (Liu et al. 2009). In addition, EPCs are critical components of tumor angiogenesis (Nolan et al. 2007); therefore, EPC transfusion to patients with tumors should be avoided. Considering estrogen has the capacity to promote EPC proliferation (Tan et al. 2012), men and women may respond differently to EPC-based treatments. Whether EPC treatment would have the same efficacy in males and females needs to be considered further.

Infarct location and volume are other factors for determining a patient's suitability for cell transfusion. Preclinical studies in EPC transplantation in stroke (Table 7.4) and different animal models may result in different infarct location and volume in striatum, cortex or both. EPC transplantation shows improvement in functional recovery; however, we cannot exclude the possibility that it may not be as effective in clinical patients. Scoring patients with different lesions, which are usually determined by magnetic resonance imaging (MRI), choosing suitable patients and accordingly giving the appropriate EPC treatment is necessary.

7.8.3 *Time, Dose, Route and Type of EPCs for Transplantation*

7.8.3.1 Time of EPC Transplantation

Preclinical studies provide various time points to deliver EPCs (Table 7.4). However, the optimal time for transplantation after a stroke is still unclear. After stroke onset, the microenvironment in the brain changes dramatically (Moskowitz et al. 2010). The optimal timing of delivery depends on EPC mechanisms of action, which could replace the damaged cells and promote the endogenous repair process by paracrine effects. If the treatment strategy focuses on cell survival and later cells integrate into the damaged tissue to replace the dead cells, cell survival is extremely important and transplanting during the recovery phase of stroke to avoid inflammation could be beneficial. Otherwise, if the treatment acts to enhance the endogenous repair process or protective mechanisms by paracrine effects, acute phase delivery is critical (Hayashi et al. 2003; Carmichael 2006). Preclinical studies of the delivery of EPCs were done either immediately after stroke or from 1 h to 1 day after stroke, which showed functional recovery in animals. However, a systematic analysis of transplantation timing and its effect on functional recovery has not been done.

7.8.3.2 Dose Injection of EPCs

In addition, as we move towards clinical trials, cell dosage becomes an important question to consider. Different cell dosages have been applied during preclinical trials (Table 7.4). Cell dosages influence cell viability after transplantation; fewer cells may not be enough to function as a therapeutic treatment, while an excessive amount of cells may result in side effects such as inflammation, teratoma or microembolus. Ongoing clinical studies are designed to use $2\text{–}8 \times 10^6$ EPCs to treat stroke patients (clinical trials.gov identifier: NCT01468064; NCT01438593; NCT00950521), and the safety and efficacy of EPC therapies are not yet clear.

7.8.3.3 Routes of Administration

Studies have reported functional recovery using the intravenous and intracerebral delivery of EPCs. All routes resulted in cells targeting the lesion, but more cells were found at the lesion with intracerebral delivery than with intravenous delivery (Jin et al. 2005). Preclinical studies choosing intravenous or intra-artery delivery of EPCs have shown functional recovery. Clinical studies plan to apply either intravenous or intracerebral delivery routes. In regards to EPC transplantation in clinical trials, intravenous infusion should be the optimal route because intracerebral injection is invasive and inconvenient, and intra-artery delivery may cause embolisms (Borlongan et al. 2004).

7.8.3.4 Types of Used EPCs

Different types of EPCs play specific roles, with early EPCs protecting damaged tissue by secreting amounts of pro-angiogenic factors, and late EPCs integrate in to host vessels to replace damaged ECs. However, it is still difficult to define EPCs, because they have multiple markers. Investigators are trying to identify, isolate and expand EPCs using their normal markers, such as CD34, KDR, and CD133, but related cell types might bear the same markers. All the ongoing clinical trials use CD34 to identify EPCs, which might not be sufficient. Optimizing the isolation and identification of EPCs from patients is still a critical problem.

7.8.4 Bio-Distribution and Persistence of EPCs

When EPCs are transplanted into ischemic animals or patients, it is crucial to monitor where the EPCs travel and into what cell types they differentiate. This helps us to understand how these cells mediate functional recovery. Therefore, dynamic non-invasive tracking of grafted EPCs *in vivo* is necessary. Optical imaging, MRI and nuclear imaging are potential imaging strategies and MRI is most often used for the dynamic tracking of grafted EPCs *in vivo*. For the tracking of exogenous EPCs *in vivo*, the grafted cells must be labeled with contrast agents *in vitro* before transplantation so that they are distinguishable from the host tissue. Molecular probes such as transferrin have successfully been used to tag putative stem cells followed by high-resolution MRI to track the homing of cells (Weissleder et al. 2000). Currently, gadolinium rhodamine dextran (GRID) and superparamagnetic iron oxide (SPIO) are two groups of commonly used contrast agents. Some studies have been successful in long-term monitoring EPCs using MRI in a rat hindlimb ischemic model (Agudelo et al. 2011; Agudelo et al. 2012). However, little work has been done on the dynamic tracking of EPCs in ischemic stroke and it calls for much attention in order to provide fundamental data for its application in clinical trials.

Conclusions

Preclinical studies have shown great promise for EPC transplantation as a therapy for stroke. Beneficial effects from EPC transplantation have been observed, including functional improvement, increased neovascularization and decreased apoptotic cells. However, there are many fundamental unsolved problems, mentioned above, and relevant clinical trials are needed. A guideline called Stem Cell Therapy as an Emerging Paradigm for Stroke (STEPS) in 2009 (The STEPS Participants, 2009) has been drawn up. To some extent, this guideline can help scientists in all fields collaborate to accelerate the use of EPC transplantation in clinical patients for those who might benefit from EPC therapy. Currently, EPC transplantation for stroke treatment in clinic is only a vision, preclinical studies and clinical research are needed to maximize the therapeutic benefit and minimize the risks of EPCs transplantation in stroke.

References

Agudelo CA, Tachibana Y, Noboru T, Iida H, Yamaoka T (2011) Long-term in vivo magnetic resonance imaging tracking of endothelial progenitor cells transplanted in rat ischemic limbs and their angiogenic potential. Tissue Eng Part A 17:2079–2089

Agudelo CA, Tachibana Y, Hurtado AF, Ose T, Iida H, Yamaoka T (2012) The use of magnetic resonance cell tracking to monitor endothelial progenitor cells in a rat hindlimb ischemic model. Biomaterials 33:2439–2448

Ahrens I, Domeij H, Topcic D, Haviv I, Merivirta RM, Agrotis A, Leitner E, Jowett JB, Bode C, Lappas M, Peter K (2011) Successful in vitro expansion and differentiation of cord blood derived CD34+ cells into early endothelial progenitor cells reveals highly differential gene expression. PLoS ONE 6:e23210

Aicher A, Rentsch M, Sasaki K, Ellwart JW, Fandrich F, Siebert R, Cooke JP, Dimmeler S, Heeschen C (2007) Nonbone marrow-derived circulating progenitor cells contribute to postnatal neovascularization following tissue ischemia. Circ Res 100:581–589

Alaiti MA, Ishikawa M, Costa MA (2010) Bone marrow and circulating stem/progenitor cells for regenerative cardiovascular therapy. Transl Res 156:112–129

Alam MM, Mohammad AA, Shuaib U, Wang C, Ghani U, Schwindt B, Todd KG, Shuaib A (2009) Homocysteine reduces endothelial progenitor cells in stroke patients through apoptosis. J Cereb Blood Flow Metab 29:157–165

Alev C, Ii M, Asahara T (2011) Endothelial progenitor cells: a novel tool for the therapy of ischemic diseases. Antioxid Redox Signal 15:949–965

Allen CL, Bayraktutan U (2008) Risk factors for ischaemic stroke. Int J Stroke 3:105–116

Asahara T (2007) Cell therapy and gene therapy using endothelial progenitor cells for vascular regeneration. Handb Exp Pharmacol 180:181–194

Asahara T, Kawamoto A (2004) Endothelial progenitor cells for postnatal vasculogenesis. Am J physiol Cell Physiol 287:C572–579

Asahara T, Murohara T, Sullivan A, Silver M, van der Zee R, Li T, Witzenbichler B, Schatteman G, Isner JM (1997) Isolation of putative progenitor endothelial cells for angiogenesis. Science 275:964–967

Asahara T, Takahashi T, Masuda H, Kalka C, Chen D, Iwaguro H, Inai Y, Silver M, Isner JM (1999) VEGF contributes to postnatal neovascularization by mobilizing bone marrow-derived endothelial progenitor cells. EMBO J 18:3964–3972

Asahara T, Kalka C, Isner JM (2000) Stem cell therapy and gene transfer for regeneration. Gene Ther 7:451–457

Asahara T, Kawamoto A, Masuda H (2011) Concise review: circulating endothelial progenitor cells for vascular medicine. Stem Cells 29:1650–1655

Assmus B, Schachinger V, Teupe C, Britten M, Lehmann R, Dobert N, Grunwald F, Aicher A, Urbich C, Martin H, Hoelzer D, Dimmeler S, Zeiher AM (2002) Transplantation of Progenitor Cells and Regeneration Enhancement in Acute Myocardial Infarction (TOPCARE-AMI). Circulation 106:3009–3017

Basile DP, Yoder MC (2014) Circulating and tissue resident endothelial progenitor cells. J Cell Physiol 229:10–16

Beltrami AP, Barlucchi L, Torella D, Baker M, Limana F, Chimenti S, Kasahara H, Rota M, Musso E, Urbanek K, Leri A, Kajstura J, Nadal-Ginard B, Anversa P (2003) Adult cardiac stem cells are multipotent and support myocardial regeneration. Cell 114:763–776

Bennis Y, Sarlon-Bartoli G, Guillet B, Lucas L, Pellegrini L, Velly L, Blot-Chabaud M, Dignat-Georges F, Sabatier F, Pisano P (2012) Priming of late endothelial progenitor cells with erythropoietin before transplantation requires the CD131 receptor subunit and enhances their angiogenic potential. J Thromb Haemost 10:1914–1928
Bliss T, Guzman R, Daadi M, Steinberg GK (2007) Cell transplantation therapy for stroke. Stroke 38:817–826
Bogoslovsky T, Chaudhry A, Latour L, Maric D, Luby M, Spatz M, Frank J, Warach S (2010) Endothelial progenitor cells correlate with lesion volume and growth in acute stroke. Neurology 75:2059–2062
Bogoslovsky T, Spatz M, Chaudhry A, Maric D, Luby M, Frank J, Warach S (2011a) Circulating CD133+ CD34+ progenitor cells inversely correlate with soluble ICAM−1 in early ischemic stroke patients. J Transl Med 9:145
Bogoslovsky T, Spatz M, Chaudhry A, Maric D, Luby M, Frank J, Warach S (2011b) Stromal-derived factor-1[alpha] correlates with circulating endothelial progenitor cells and with acute lesion volume in stroke patients. Stroke 42:618–625
Bompais H, Chagraoui J, Canron X, Crisan M, Liu XH, Anjo A, Tolla-Le Port C, Leboeuf M, Charbord P, Bikfalvi A, Uzan G (2004) Human endothelial cells derived from circulating progenitors display specific functional properties compared with mature vessel wall endothelial cells. Blood 103:2577–2584
Borlongan CV, Hadman M, Sanberg CD, Sanberg PR (2004) Central nervous system entry of peripherally injected umbilical cord blood cells is not required for neuroprotection in stroke. Stroke 35:2385–2389
Broughton BR, Reutens DC, Sobey CG (2009) Apoptotic mechanisms after cerebral ischemia. Stroke 40:e331–e339
Carmeliet P (2003) Angiogenesis in health and disease. Nat Med 9:653–660
Carmichael ST (2006) Cellular and molecular mechanisms of neural repair after stroke: making waves. Ann Neurol 59:735–742
Cesari F, Gori AM, Romagnuolo I, Abbate R (2012) [Endothelial progenitor cells and vascular health: effects of lifestyle's modifications]. Monaldi archives for chest disease = Archivio Monaldi per le malattie del torace/ Fondazione clinica del lavoro, IRCCS [and] Istituto di clinica tisiologica e malattie apparato respiratorio, Universita di Napoli, Secondo ateneo, 78:66–72
Chen ZZ, Jiang XD, Zhang LL, Shang JH, Du MX, Xu G, Xu RX (2008) Beneficial effect of autologous transplantation of bone marrow stromal cells and endothelial progenitor cells on cerebral ischemia in rabbits. Neurosci Lett 445:36–41
Chen J, Chen S, Zhang C, Zhang L, Xiao X, Das A, Zhao Y, Yuan B, Morris M, Zhao B, Chen Y (2012) Transfusion of CXCR4-primed endothelial progenitor cells reduces cerebral ischemic damage and promotes repair in db/db diabetic mice. PLoS ONE 7:e50105
Chen C, Wang Y, Yang GY (2013a) Stem cell-mediated gene delivering for the treatment of cerebral ischemia: progress and prospectives. Curr Drug Targets 14:81–89
Chen J, Xiao X, Chen S, Zhang C, Yi D, Shenoy V, Raizada MK, Zhao B, Chen Y (2013b) Angiotensin-converting enzyme 2 priming enhances the function of endothelial progenitor cells and their therapeutic efficacy. Hypertension 61:681–689
Chen X, Gu M, Zhao X, Zheng X, Qin Y, You X (2013c) Deterioration of cardiac function after acute myocardial infarction is prevented by transplantation of modified endothelial progenitor cells overexpressing endothelial NO synthases. Cell Physiol Biochem 31:355–365
Chen X, Yin J, Wu X, Li R, Fang J, Chen R, Zhang B, Zhang W (2013d) Effects of magnetically labeled exogenous endothelial progenitor cells on cerebral blood perfusion and microvasculature alterations after traumatic brain injury in rat model. Acta Radiol 54:313–323
Cho HJ, Lee N, Lee JY, Choi YJ, Ii M, Wecker A, Jeong JO, Curry C, Qin G, Yoon YS (2007) Role of host tissues for sustained humoral effects after endothelial progenitor cell transplantation into the ischemic heart. J Exp Med 204:3257–3269
Chu K, Jung KH, Lee ST, Park HK, Sinn DI, Kim JM, Kim DH, Kim JH, Kim SJ, Song EC, Kim M, Lee SK, Roh JK (2008) Circulating endothelial progenitor cells as a new marker of endothelial dysfunction or repair in acute stroke. Stroke 39:1441–1447

Decano JL, Moran AM, Giordano N, Ruiz-Opazo N, Herrera VL (2013) Analysis of CD45-[CD34+/KDR+] endothelial progenitor cells as juvenile protective factors in a rat model of ischemic-hemorrhagic stroke. PLoS ONE 8:e55222

Devaraj S, Jialal I (2012) Dysfunctional endothelial progenitor cells in metabolic syndrome. Exp Diabetes Res 2012:585018

Fadini GP, Losordo D, Dimmeler S (2012) Critical reevaluation of endothelial progenitor cell phenotypes for therapeutic and diagnostic use. Circ Res 110:624–637

Fan Y, Shen F, Frenzel T, Zhu W, Ye J, Liu J, Chen Y, Su H, Young WL, Yang GY (2010) Endothelial progenitor cell transplantation improves long-term stroke outcome in mice. Ann Neurol 67:488–497

Flammer AJ, Gossl M, Widmer RJ, Reriani M, Lennon R, Loeffler D, Shonyo S, Simari RD, Lerman LO, Khosla S, Lerman A (2012) Osteocalcin positive CD133+/CD34-/KDR+ progenitor cells as an independent marker for unstable atherosclerosis. Eur Heart J 33:2963–2969

Flynn RW, MacWalter RS, Doney AS (2008) The cost of cerebral ischaemia. Neuropharmacology 55:250–256

Folkman J (1995) Angiogenesis in cancer, vascular, rheumatoid and other disease. Nat Med 1:27–31

Francis KR, Wei L (2010) Human embryonic stem cell neural differentiation and enhanced cell survival promoted by hypoxic preconditioning. Cell Death Dis 1:e22

Friedrich MA, Martins MP, Araujo MD, Klamt C, Vedolin L, Garicochea B, Raupp EF, Sartori EAmmarJ, Machado DC, Costa JC, Nogueira RG, Rosado-de-Castro PH, Mendez-Otero R, Freitas GR (2012) Intra-arterial infusion of autologous bone marrow mononuclear cells in patients with moderate to severe middle cerebral artery acute ischemic stroke. Cell Transplant 21(Suppl 1):S13–S21

Galasso G, De Rosa R, Ciccarelli M, Sorriento D, Del Giudice C, Strisciuglio T, De Biase C, Luciano R, Piccolo R, Pierri A, Di Gioia G, Prevete N, Trimarco B, Piscione F, Iaccarino G (2013) Beta2-adrenergic receptor stimulation improves endothelial progenitor cell-mediated ischemic neoangiogenesis. Circ Res 112:1026–1034

Gehling UM, Ergun S, Schumacher U, Wagener C, Pantel K, Otte M, Schuch G, Schafhausen P, Mende T, Kilic N, Kluge K, Schafer B, Hossfeld DK, Fiedler W (2000) In vitro differentiation of endothelial cells from AC133-positive progenitor cells. Blood 95:3106–3112

Ghani U, Shuaib A, Salam A, Nasir A, Shuaib U, Jeerakathil T, Sher F, O'Rourke F, Nasser AM, Schwindt B, Todd K (2005) Endothelial progenitor cells during cerebrovascular disease. Stroke 36:151–153

Gherghe CM, Duan J, Gong J, Rojas M, Klauber-Demore N, Majesky M, Deb A (2011) Wnt1 is a proangiogenic molecule, enhances human endothelial progenitor function, and increases blood flow to ischemic limbs in a HGF-dependent manner. FASEB J 25:1836–1843

Ginsberg MD (1997) The new language of cerebral ischemia. AJNR Am J Neuroradiol 18:1435–1445

Gou X, He WY, Xiao MZ, Qiu M, Wang M, Deng YZ, Liu CD, Tang ZB, Li J, Chen Y (2011) Transplantation of endothelial progenitor cells transfected with VEGF165 to restore erectile function in diabetic rats. Asian J Androl 13:332–338

Grant MB, May WS, Caballero S, Brown GA, Guthrie SM, Mames RN, Byrne BJ, Vaught T, Spoerri PE, Peck AB, Scott EW (2002) Adult hematopoietic stem cells provide functional hemangioblast activity during retinal neovascularization. Nat Med 8:607–612

Gremmels H, Fledderus JO, van Balkom BW, Verhaar MC (2011) Transcriptome analysis in endothelial progenitor cell biology. Antioxid Redox Signal 15:1029–1042

Griese DP, Achatz S, Batzlsperger CA, Strauch UG, Grumbeck B, Weil J, Riegger GA (2003a) Vascular gene delivery of anticoagulants by transplantation of retrovirally-transduced endothelial progenitor cells. Cardiovasc Res 58:469–477

Griese DP, Ehsan A, Melo LG, Kong D, Zhang L, Mann MJ, Pratt RE, Mulligan RC, Dzau VJ (2003b) Isolation and transplantation of autologous circulating endothelial cells into denuded vessels and prosthetic grafts: implications for cell-based vascular therapy. Circulation 108:2710–2715

Guzman R, Choi R, Gera A, De Los Angeles A, Andres RH, Steinberg GK (2008) Intravascular cell replacement therapy for stroke. Neurosurg Focus 24:E15

Gyan B, Goka BQ, Adjei GO, Tetteh JK, Kusi KA, Aikins A, Dodoo D, Lesser ML, Sison CP, Das S, Howard ME, Milbank E, Fischer K, Rafii S, Jin D, Golightly LM (2009) Cerebral malaria is associated with low levels of circulating endothelial progenitor cells in African children. Am J Trop Med Hyg 80:541–546

Hankey GJ (2006) Potential new risk factors for ischemic stroke: what is their potential? Stroke 37:2181–2188

Hassan A, Markus HS (2000) Genetics and ischaemic stroke. Brain 123(Pt 9):1784–1812

Hayakawa K, Pham LD, Katusic ZS, Arai K, Lo EH (2012) Astrocytic high-mobility group box 1 promotes endothelial progenitor cell-mediated neurovascular remodeling during stroke recovery. Proc Natl Acad Sci U S A 109:7505–7510

Hayashi T, Noshita N, Sugawara T, Chan PH (2003) Temporal profile of angiogenesis and expression of related genes in the brain after ischemia. J Cereb Blood Flow Metab 23:166–180

He T, Smith LA, Harrington S, Nath KA, Caplice NM, Katusic ZS (2004) Transplantation of circulating endothelial progenitor cells restores endothelial function of denuded rabbit carotid arteries. Stroke 35:2378–2384

He T, Peterson TE, Katusic ZS (2005) Paracrine mitogenic effect of human endothelial progenitor cells: role of interleukin−8. Am J Physiol Heart Circ Physiol 289:H968–972

Hess DC, Hill WD, Martin-Studdard A, Carroll J, Brailer J, Carothers J (2002) Bone marrow as a source of endothelial cells and NeuN-expressing cells aAfter stroke. Stroke 33:1362–1368

Hill JM, Zalos G, Halcox JP, Schenke WH, Waclawiw MA, Quyyumi AA, Finkel T (2003) Circulating endothelial progenitor cells, vascular function, and cardiovascular risk. N Engl J Med 348:593–600

Hirschi KK, Ingram DA, Yoder MC (2008) Assessing identity, phenotype, and fate of endothelial progenitor cells. Arterioscler Thromb Vasc Biol 28:1584–1595

Honmou O, Houkin K, Matsunaga T, Niitsu Y, Ishiai S, Onodera R, Waxman SG, Kocsis JD (2011) Intravenous administration of auto serum-expanded autologous mesenchymal stem cells in stroke. Brain 134:1790–1807

Hu CH, Ke X, Chen K, Yang DY, Du ZM, Wu GF (2013) Transplantation of human umbilical cord-derived endothelial progenitor cells promotes re-endothelialization of the injured carotid artery after balloon injury in New Zealand white rabbits. Chin Med J (Engl) 126:1480–1485

Huang PH, Chen YH, Wang CH, Chen JS, Tsai HY, Lin FY, Lo WY, Wu TC, Sata M, Chen JW, Lin SJ (2009) Matrix metalloproteinase−9 is essential for ischemia-induced neovascularization by modulating bone marrow-derived endothelial progenitor cells. Arterioscler Thromb Vasc Biol 29:1179–1184

Huang H, Huang F, Huang JP (2013a) Transplantation of bone marrowderived endothelial progenitor cells overexpressing Deltalike4 enhances functional neovascularization in ischemic myocardium. Mol Med Rep 8:1556–1562

Huang J, Li Y, Tang Y, Tang G, Yang GY, Wang Y (2013b) CXCR4 antagonist AMD3100 protects blood-brain barrier integrity and reduces inflammatory response after focal ischemia in mice. Stroke 44:190–197

Hur J, Yoon CH, Kim HS, Choi JH, Kang HJ, Hwang KK, Oh BH, Lee MM, Park YB (2004) Characterization of two types of endothelial progenitor cells and their different contributions to neovasculogenesis. Arterioscler Thromb Vasc Biol 24:288–293

Hur J, Yoon CH, Lee CS, Kim TY, Oh IY, Park KW, Kim JH, Lee HS, Kang HJ, Chae IH, Oh BH, Park YB, Kim HS (2007) Akt is a key modulator of endothelial progenitor cell trafficking in ischemic muscle. Stem Cells 25:1769–1778

Ingram DA, Mead LE, Tanaka H, Meade V, Fenoglio A, Mortell K, Pollok K, Ferkowicz MJ, Gilley D, Yoder MC (2004) Identification of a novel hierarchy of endothelial progenitor cells using human peripheral and umbilical cord blood. Blood 104:2752–2760

Ishikawa M, Asahara T (2004) Endothelial progenitor cell culture for vascular regeneration. Stem Cells Dev 13:344–349

Iskander A, Knight RA, Zhang ZG, Ewing JR, Shankar A, Varma NR, Bagher-Ebadian H, Ali MM, Arbab AS, Janic B (2013) Intravenous administration of human umbilical cord blood-derived AC133+ endothelial progenitor cells in rat stroke model reduces infarct volume: magnetic resonance imaging and histological findings. Stem Cells Transl Med 2:703–714

Iwakura A, Shastry S, Luedemann C, Hamada H, Kawamoto A, Kishore R, Zhu Y, Qin G, Silver M, Thorne T, Eaton L, Masuda H, Asahara T, Losordo DW (2006) Estradiol enhances recovery after myocardial infarction by augmenting incorporation of bone marrow-derived endothelial progenitor cells into sites of ischemia-induced neovascularization via endothelial nitric oxide synthase-mediated activation of matrix metalloproteinase−9. Circulation 113:1605–1614

Jiang M, Wang B, Wang C, He B, Fan H, Shao Q, Gao L, Liu Y, Yan G, Pu J (2008) In vivo enhancement of angiogenesis by adenoviral transfer of HIF−1alpha-modified endothelial progenitor cells (Ad-HIF−1alpha-modified EPC for angiogenesis). Int J Biochem Cell Biol 40:2284–2295

Jin K, Sun Y, Xie L, Mao XO, Childs J, Peel A, Logvinova A, Banwait S, Greenberg DA (2005) Comparison of ischemia-directed migration of neural precursor cells after intrastriatal, intraventricular, or intravenous transplantation in the rat. Neurobiol Dis 18:366–374

Kalka C, Masuda H, Takahashi T, Kalka-Moll WM, Silver M, Kearney M, Li T, Isner JM, Asahara T (2000a) Transplantation of ex vivo expanded endothelial progenitor cells for therapeutic neovascularization. Proc Natl Acad Sci U S A 97:3422–3427

Kalka C, Masuda H, Takahashi T, Gordon R, Tepper O, Gravereaux E, Pieczek A, Iwaguro H, Hayashi SI, Isner JM, Asahara T (2000b) Vascular endothelial growth factor(165) gene transfer augments circulating endothelial progenitor cells in human subjects. Circ Res 86:1198–1202

Kamiya N, Ueda M, Igarashi H, Nishiyama Y, Suda S, Inaba T, Katayama Y (2008) Intra-arterial transplantation of bone marrow mononuclear cells immediately after reperfusion decreases brain injury after focal ischemia in rats. Life Sci 83:433–437

Kaneko Y, Tajiri N, Shinozuka K, Glover LE, Weinbren NL, Cortes L, Borlongan CV (2012) Cell therapy for stroke: emphasis on optimizing safety and efficacy profile of endothelial progenitor cells. Curr Pharm Des 18:3731–3734

Karam JG, Loney-Hutchinson L, McFarlane SI (2008) High-dose atorvastatin after stroke or transient ischemic attack: tThe sStroke pPrevention by aAggressive rReduction in cCholesterol lLevels (SPARCL) iInvestigators. J Cardiometab Syndr 3:68–69

Khakoo AY, Finkel T (2005) Endothelial progenitor cells. Annu Rev Med 56:79–101

Kirton JP, Xu Q (2010) Endothelial precursors in vascular repair. Microvasc Res 79:193–199

Krishnamurthy P, Thal M, Verma S, Hoxha E, Lambers E, Ramirez V, Qin G, Losordo D, Kishore R (2011) Interleukin−10 deficiency impairs bone marrow-derived endothelial progenitor cell survival and function in ischemic myocardium. Circ Res 109:1280–1289

Kwon O, Miller S, Li N, Khan A, Kadry Z, Uemura T (2010) Bone marrow-derived endothelial progenitor cells and endothelial cells may contribute to endothelial repair in the kidney immediately after ischemia-reperfusion. J Histochem Cytochem 58:687–694

Lau KK, Chan YH, Yiu KH, Li SW, Tam S, Lau CP, Kwong YL, Tse HF (2007) Burden of carotid atherosclerosis in patients with stroke: relationships with circulating endothelial progenitor cells and hypertension. J Hum Hypertens 21:445–451

Lee SP, Youn SW, Cho HJ, Li L, Kim TY, Yook HS, Chung JW, Hur J, Yoon CH, Park KW, Oh BH, Park YB, Kim HS (2006) Integrin-linked kinase, a hypoxia-responsive molecule, controls postnatal vasculogenesis by recruitment of endothelial progenitor cells to ischemic tissue. Circulation 114:150–159

Lee JS, Hong JM, Moon GJ, Lee PH, Ahn YH, Bang OY (2010) A long-term follow-up study of intravenous autologous mesenchymal stem cell transplantation in patients with ischemic stroke. Stem Cells 28:1099–1106

Lee TS, Kou YR (2012) Enhancing endothelial progenitor cell therapy for critical limb ischemia by extracorporeal shock wave. Crit Care Med 40:332–333

Leventhal C, Rafii S, Rafii D, Shahar A, Goldman SA (1999) Endothelial trophic support of neuronal production and recruitment from the adult mammalian subependyma. Mol Cell Neurosci 13:450–464

Li A, Dubey S, Varney ML, Dave BJ, Singh RK (2003) IL−8 directly enhanced endothelial cell survival, proliferation, and matrix metalloproteinases production and regulated angiogenesis. J Immunol 170:3369–3376

Li L, Jiang Q, Ding G, Zhang L, Zhang ZG, Li Q, Panda S, Lu M, Ewing JR, Chopp M (2010) Effects of administration route on migration and distribution of neural progenitor cells transplanted into rats with focal cerebral ischemia, an MRI study. J Cereb Blood Flow Metab 30:653–662

Li Q, Tang G, Xue S, He X, Miao P, Li Y, Wang J, Xiong L, Wang Y, Zhang C, Yang GY (2013) Silica-coated superparamagnetic iron oxide nanoparticles targeting of EPCs in ischemic brain injury. Biomaterials 34:4982–4992

Liao YF, Chen LL, Zeng TS, Li YM, Fan Y, Hu LJ, Ling Y (2010) Number of circulating endothelial progenitor cells as a marker of vascular endothelial function for type 2 diabetes. Vasc Med 15:279–285

Lin Y, Weisdorf DJ, Solovey A, Hebbel RP (2000) Origins of circulating endothelial cells and endothelial outgrowth from blood. J Clin Invest 105:71–77

Liu F, Yuan R, Benashski SE, McCullough LD (2009) Changes in experimental stroke outcome across the life span. J Cereb Blood Flow Metab 29:792–802

Liu H, Wu R, Jia RP, Zhong B, Zhu JG, Yu P, Zhao Y, Ge YZ, Wu JP (2013a) Ischemic preconditioning increases endothelial progenitor cell number to attenuate partial nephrectomy-induced ischemia/reperfusion injury. PLoS ONE 8:e55389

Liu X, Ye R, Yan T, Yu SP, Wei L, Xu G, Fan X, Jiang Y, Stetler RA, Chen J (2013b) Cell based therapies for ischemic stroke: fFrom basic science to bedside. Progress Neurobiol 115:92–115ogy.

Liu Z, Ding X, Fang F, Wang R, Chen Y, Ma Y, Zhang G, Kang X (2013c) Higher numbers of circulating endothelial progenitor cells in stroke patients with intracranial arterial stenosis. BMC Neurol 13:161

Long J, Wang S, Zhang Y, Liu X, Zhang H (2013) The therapeutic effect of vascular endothelial growth factor gene- or heme oxygenase−1 gene-modified endothelial progenitor cells on neovascularization of rat hindlimb ischemia model. J Vasc Surg 58(756–765):e752

Machein MR, Renninger S, de Lima-Hahn E, Plate KH (2003) Minor contribution of bone marrow-derived endothelial progenitors to the vascularization of murine gliomas. Brain Pathol 13:582–597

Mandraffino G, Sardo MA, Riggio S, Loddo S, Imbalzano E, Alibrandi A, Saitta C, Cinquegrani M, Mormina EM, Saitta A (2011) Circulating progenitor cells are increased in newly diagnosed untreated hypertensive patients with arterial stiffening but normal carotid intima-media thickness. Hypertens Res 34:876–883

Marti-Fabregas J, Crespo J, Delgado-Mederos R, Martinez-Ramirez S, Pena E, Marin R, Dinia L, Jimenez-Xarrie E, Fernandez-Arcos A, Perez-Perez J, Querol L, Suarez-Calvet M, Badimon L (2013) Endothelial progenitor cells in acute ischemic stroke. Brain Behav 3:649–655

Masuda H, Asahara T (2003) Post-natal endothelial progenitor cells for neovascularization in tissue regeneration. Cardiovasc Res 58:390–398

Medina RJ, O'Neill CL, Sweeney M, Guduric-Fuchs J, Gardiner TA, Simpson DA, Stitt AW (2010) Molecular analysis of endothelial progenitor cell (EPC) subtypes reveals two distinct cell populations with different identities. BMC Med Genomics 3:18

Medina RJ, O'Neill CL, O'Doherty TM, Wilson SE, Stitt AW (2012) Endothelial progenitors as tools to study vascular disease. Stem Cells Int 2012:346735

Misra V, Ritchie MM, Stone LL, Low WC, Janardhan V (2012) Stem cell therapy in ischemic stroke: role of IV and intra-arterial therapy. Neurology 79:S207–212

Mithal DS, Banisadr G, Miller RJ (2012) CXCL12 signaling in the development of the nervous system. J Neuroimmune Pharmacol 7:820–834

Moniche F, Gonzalez A, Gonzalez-Marcos JR, Carmona M, Pinero P, Espigado I, Garcia-Solis D, Cayuela A, Montaner J, Boada C, Rosell A, Jimenez MD, Mayol A, Gil-Peralta A (2012) Intra-arterial bone marrow mononuclear cells in ischemic stroke: a pilot clinical trial. Stroke 43:2242–2244

Morancho A, Hernandez-Guillamon M, Boada C, Barcelo V, Giralt D, Ortega L, Montaner J, Rosell A (2013) Cerebral ischaemia and matrix metalloproteinase−9 modulate the angiogenic function of early and late outgrowth endothelial progenitor cells. J Cell Mol Med 17(12):1543–1553

Moskowitz MA, Lo EH, Iadecola C (2010) The science of stroke: mechanisms in search of treatments. Neuron 67:181–198

Moubarik C, Guillet B, Youssef B, Codaccioni JL, Piercecchi MD, Sabatier F, Lionel P, Dou L, Foucault-Bertaud A, Velly L, Dignat-George F, Pisano P (2011) Transplanted late outgrowth endothelial progenitor cells as cell therapy product for stroke. Stem Cell Rev 7:208–220

Murayama T, Tepper OM, Silver M, Ma H, Losordo DW, Isner JM, Asahara T, Kalka C (2002) Determination of bone marrow-derived endothelial progenitor cell significance in angiogenic growth factor-induced neovascularization in vivo. Exp Hematol 30:967–972

Navarro-Sobrino M, Rosell A, Hernandez-Guillamon M, Penalba A, Ribo M, Alvarez-Sabin J, Montaner J (2010) Mobilization, endothelial differentiation and functional capacity of endothelial progenitor cells after ischemic stroke. Microvasc Res 80:317–323

Navarro-Sobrino M, Hernandez-Guillamon M, Fernandez-Cadenas I, Ribo M, Romero IA, Couraud PO, Weksler BB, Montaner J, Rosell A (2013) The angiogenic gene profile of circulating endothelial progenitor cells from ischemic stroke patients. Vasc Cell 5:3

Nih LR, Deroide N, Lere-Dean C, Lerouet D, Soustrat M, Levy BI, Silvestre JS, Merkulova-Rainon T, Pocard M, Margaill I, Kubis N (2012) Neuroblast survival depends on mature vascular network formation after mouse stroke: role of endothelial and smooth muscle progenitor cell co-administration. Eur J Neurosci 35:1208–1217

Nolan DJ, Ciarrocchi A, Mellick AS, Jaggi JS, Bambino K, Gupta S, Heikamp E, McDevitt MR, Scheinberg DA, Benezra R, Mittal V (2007) Bone marrow-derived endothelial progenitor cells are a major determinant of nascent tumor neovascularization. Genes Dev 21:1546–1558

Ogle ME, Gu X, Espinera AR, Wei L (2012) Inhibition of prolyl hydroxylases by dimethyloxaloylglycine after stroke reduces ischemic brain injury and requires hypoxia inducible factor−1alpha. Neurobiol Dis 45:733–742

Oh IY, Yoon CH, Hur J, Kim JH, Kim TY, Lee CS, Park KW, Chae IH, Oh BH, Park YB, Kim HS (2007) Involvement of E-selectin in recruitment of endothelial progenitor cells and angiogenesis in ischemic muscle. Blood 110:3891–3899

Ohta T, Kikuta K, Imamura H, Takagi Y, Nishimura M, Arakawa Y, Hashimoto N, Nozaki K (2006) Administration of ex vivo-expanded bone marrow-derived endothelial progenitor cells attenuates focal cerebral ischemia-reperfusion injury in rats. Neurosurgery 59:679-686 (discussion 679–686)

Paczkowska E, Golab-Janowska M, Bajer-Czajkowska A, Machalinska A, Ustianowski P, Rybicka M, Klos P, Dziedziejko V, Safranow K, Nowacki P, Machalinski B (2013) Increased circulating endothelial progenitor cells in patients with haemorrhagic and ischaemic stroke: the role of endothelin−1. J Neurol Sci 325:90–99

Palmer TD, Willhoite AR, Gage FH (2000) Vascular niche for adult hippocampal neurogenesis. J Comp Neurol 425:479–494

Pasquier E, Dias S (2010) Endothelial progenitor cells: hope beyond controversy. Curr Cancer Drug Targets 10:914–921

Patschan D, Hildebrandt A, Rinneburger J, Wessels JT, Patschan S, Becker JU, Henze E, Kruger A, Muller GA (2012) The hormone melatonin stimulates renoprotective effects of "early outgrowth" endothelial progenitor cells in acute ischemic kidney injury. Am J Physiol Renal Physiol 302:F1305–1312

Payen E, Leboulch P (2012) Advances in stem cell transplantation and gene therapy in the beta-hemoglobinopathies. Hematology Am Soc Hematol Educ Program 2012:276–283

Peichev M, Naiyer AJ, Pereira D, Zhu Z, Lane WJ, Williams M, Oz MC, Hicklin DJ, Witte L, Moore MA, Rafii S (2000) Expression of VEGFR−2 and AC133 by circulating human CD34(+) cells identifies a population of functional endothelial precursors. Blood 95:952–958

Pellegrini L, Bennis Y, Guillet B, Velly L, Garrigue P, Sabatier F, Dignat-George F, Bruder N, Pisano P (2013) Therapeutic benefit of a combined strategy using erythropoietin and endothelial progenitor cells after transient focal cerebral ischemia in rats. Neurol Res 35:937–947

Pendharkar AV, Chua JY, Andres RH, Wang N, Gaeta X, Wang H, De A, Choi R, Chen S, Rutt BK, Gambhir SS, Guzman R (2010) Biodistribution of neural stem cells after intravascular therapy for hypoxic-ischemia. Stroke 41:2064–2070

Planat-Benard V, Silvestre JS, Cousin B, Andre M, Nibbelink M, Tamarat R, Clergue M, Manneville C, Saillan-Barreau C, Duriez M, Tedgui A, Levy B, Penicaud L, Casteilla L (2004) Plasticity of human adipose lineage cells toward endothelial cells: physiological and therapeutic perspectives. Circulation 109:656–663

Ponio JB, El-Ayoubi F, Glacial F, Ganeshamoorthy K, Driancourt C, Godet M, Perriere N, Guillevic O, Couraud PO, Uzan G (2014) Instruction of circulating endothelial progenitors in vitro towards specialized blood-brain barrier and arterial phenotypes. PLoS ONE 9:e84179

Qiu J, Li W, Feng S, Wang M, He Z (2013) Transplantation of bone marrow-derived endothelial progenitor cells attenuates cerebral ischemia and reperfusion injury by inhibiting neuronal apoptosis, oxidative stress and nuclear factor-kappaB expression. Int J Mol Med 31:91–98

Rafii S (2000) Circulating endothelial precursors: mystery, reality, and promise. J Clin Invest 105:17–19

Rehman J, Li J, Orschell CM, March KL (2003) Peripheral blood "endothelial progenitor cells" are derived from monocyte/macrophages and secrete angiogenic growth factors. Circulation 107:1164–1169

Rosell A, Morancho A, Navarro-Sobrino M, Martinez-Saez E, Hernandez-Guillamon M, Lope-Piedrafita S, Barcelo V, Borras F, Penalba A, Garcia-Bonilla L, Montaner J (2013) Factors secreted by endothelial progenitor cells enhance neurorepair responses after cerebral ischemia in mice. PLoS ONE 8:e73244

Rouhl RP, van Oostenbrugge RJ, Damoiseaux J, Tervaert JW, Lodder J (2008) Endothelial progenitor cell research in stroke: a potential shift in pathophysiological and therapeutical concepts. Stroke 39:2158–2165

Rouhl RP, van Oostenbrugge RJ, Damoiseaux JG, Debrus-Palmans LL, Theunissen RO, Knottnerus IL, Staals JE, Delanghe JR, Tervaert JW, Lodder J (2009) Haptoglobin phenotype may alter endothelial progenitor cell cluster formation in cerebral small vessel disease. Curr Neurovasc Res 6:32–41

Rouhl RP, Mertens AE, van Oostenbrugge RJ, Damoiseaux JG, Debrus-Palmans LL, Henskens LH, Kroon AA, de Leeuw PW, Lodder J, Tervaert JW (2012) Angiogenic T-cells and putative endothelial progenitor cells in hypertension-related cerebral small vessel disease. Stroke 43:256–258

Rubattu S, Giliberti R, Volpe M (2000) Etiology and pathophysiology of stroke as a complex trait. Am J Hypertens 13:1139–1148

Ruscher K, Kuric E, Liu Y, Walter HL, Issazadeh-Navikas S, Englund E, Wieloch T (2013) Inhibition of CXCL12 signaling attenuates the postischemic immune response and improves functional recovery after stroke. J Cereb Blood Flow Metab 33:1225–1234

Santhanam AV, Smith LA, He T, Nath KA, Katusic ZS (2007) Endothelial progenitor cells stimulate cerebrovascular production of prostacyclin by paracrine activation of cyclooxygenase-2. Circ Res 100:1379–1388

Schuh A, Kroh A, Konschalla S, Liehn EA, Sobota RM, Biessen EA, Bot I, Tolga Taha S, Schober A, Marx N, Weber C, Sasse A (2012) Myocardial regeneration by transplantation of modified endothelial progenitor cells expressing SDF-1 in a rat model. J Cell Mol Med 16:2311–2320

Sen S, Merchan J, Dean J, Ii M, Gavin M, Silver M, Tkebuchava T, Yoon YS, Rasko JE, Aikawa R (2010) Autologous transplantation of endothelial progenitor cells genetically modified by adeno-associated viral vector delivering insulin-like growth factor-1 gene after myocardial infarction. Hum Gene Ther 21:1327–1334

Shi Q, Rafii S, Wu MH, Wijelath ES, Yu C, Ishida A, Fujita Y, Kothari S, Mohle R, Sauvage LR, Moore MA, Storb RF, Hammond WP (1998) Evidence for circulating bone marrow-derived endothelial cells. Blood 92:362–367

Shintani S, Murohara T, Ikeda H, Ueno T, Honma T, Katoh A, Sasaki K, Shimada T, Oike Y, Imaizumi T (2001) Mobilization of endothelial progenitor cells in patients with acute myocardial infarction. Circulation 103:2776–2779

Sims NR, Muyderman H (2010) Mitochondria, oxidative metabolism and cell death in stroke. Biochim Biophys Acta 1802:80–91
Sirker AA, Astroulakis ZM, Hill JM (2009) Vascular progenitor cells and translational research: the role of endothelial and smooth muscle progenitor cells in endogenous arterial remodelling in the adult. Clin Sci (Lond) 116:283–299
Sobrino T, Hurtado O, Moro MA, Rodriguez-Yanez M, Castellanos M, Brea D, Moldes O, Blanco M, Arenillas JF, Leira R, Davalos A, Lizasoain I, Castillo J (2007) The increase of circulating endothelial progenitor cells after acute ischemic stroke is associated with good outcome. Stroke 38:2759–2764
Sobrino T, Rodriguez-Gonzalez R, Blanco M, Brea D, Perez-Mato M, Rodriguez-Yanez M, Leira R, Castillo J (2011) CDP-choline treatment increases circulating endothelial progenitor cells in acute ischemic stroke. Neurol Res 33:572–577
Sobrino T, Blanco M, Perez-Mato M, Rodriguez-Yanez M, Castillo J (2012a) Increased levels of circulating endothelial progenitor cells in patients with ischaemic stroke treated with statins during acute phase. Eur J Neurol 19:1539–1546
Sobrino T, Perez-Mato M, Brea D, Rodriguez-Yanez M, Blanco M, Castillo J (2012b) Temporal profile of molecular signatures associated with circulating endothelial progenitor cells in human ischemic stroke. J Neurosci Res 90:1788–1793
Song MB, Yu XJ, Zhu GX, Chen JF, Zhao G, Huang L (2009) Transfection of HGF gene enhances endothelial progenitor cell (EPC) function and improves EPC transplant efficiency for balloon-induced arterial injury in hypercholesterolemic rats. Vascul Pharmacol 51:205–213
Suarez-Monteagudo C et al (2009) Autologous bone marrow stem cell neurotransplantation in stroke patients. an open study. Restor Neurol Neurosci 27:151–161
Takahashi T, Kalka C, Masuda H, Chen D, Silver M, Kearney M, Magner M, Isner JM, Asahara T (1999) Ischemia- and cytokine-induced mobilization of bone marrow-derived endothelial progenitor cells for neovascularization. Nat Med 5:434–438
Takamiya M, Okigaki M, Jin D, Takai S, Nozawa Y, Adachi Y, Urao N, Tateishi K, Nomura T, Zen K, Ashihara E, Miyazaki M, Tatsumi T, Takahashi T, Matsubara H (2006) Granulocyte colony-stimulating factor-mobilized circulating c-Kit +/Flk−1+ progenitor cells regenerate endothelium and inhibit neointimal hyperplasia after vascular injury. Arterioscler Thromb Vasc Biol 26:751–757
Tamaki T, Akatsuka A, Ando K, Nakamura Y, Matsuzawa H, Hotta T, Roy RR, Edgerton VR (2002) Identification of myogenic-endothelial progenitor cells in the interstitial spaces of skeletal muscle. J Cell Biol 157:571–577
Tan Z, Zhou LJ, Li Y, Cui YH, Xiang QL, Lin GP, Wang TH (2012) E(2)-BSA activates caveolin−1 via PI(3)K/ERK1/2 and lysosomal degradation pathway and contributes to EPC proliferation. Int J Cardiol 158:46–53
Tepper OM, Galiano RD, Capla JM, Kalka C, Gagne PJ, Jacobowitz GR, Levine JP, Gurtner GC (2002) Human endothelial progenitor cells from type II diabetics exhibit impaired proliferation, adhesion, and incorporation into vascular structures. Circulation 106:2781–2786
Tepper OM, Capla JM, Galiano RD, Ceradini DJ, Callaghan MJ, Kleinman ME, Gurtner GC (2005) Adult vasculogenesis occurs through in situ recruitment, proliferation, and tubulization of circulating bone marrow-derived cells. Blood 105:1068–1077
Thored P, Wood J, Arvidsson A, Cammenga J, Kokaia Z, Lindvall O (2007) Long-term neuroblast migration along blood vessels in an area with transient angiogenesis and increased vascularization after stroke. Stroke 38:3032–3039
Timmermans F, Van Hauwermeiren F, De Smedt M, Raedt R, Plasschaert F, De Buyzere ML, Gillebert TC, Plum J, Vandekerckhove B (2007) Endothelial outgrowth cells are not derived from CD133+ cells or CD45+ hematopoietic precursors. Arterioscler Thromb Vasc Biol 27:1572–1579
Timmermans F, Plum J, Yoder MC, Ingram DA, Vandekerckhove B, Case J (2009) Endothelial progenitor cells: identity defined? J Cell Mol Med 13:87–102

Tsai NW, Hung SH, Huang CR, Chang HW, Chang WN, Lee LH, Wang HC, Lin YJ, Lin WC, Cheng BC, Chiang YF, Su YJ, Tsai TR, Lu CH (2014) The association between circulating endothelial progenitor cells and outcome in different subtypes of acute ischemic stroke. Clin Chim Acta 427:6–10

Ueno H, Koyama H, Mima Y, Fukumoto S, Tanaka S, Shoji T, Emoto M, Nishizawa Y, Inaba M (2011) Comparison of the effect of cilostazol with aspirin on circulating endothelial progenitor cells and small-dense LDL cholesterol in diabetic patients with cerebral ischemia: a randomized controlled pilot trial. J Atheroscler Thromb 18:883–890

Urbich C, Dimmeler S (2004) Endothelial progenitor cells: characterization and role in vascular biology. Circ Res 95:343–353

Urbich C, Aicher A, Heeschen C, Dernbach E, Hofmann WK, Zeiher AM, Dimmeler S (2005) Soluble factors released by endothelial progenitor cells promote migration of endothelial cells and cardiac resident progenitor cells. J Mol Cell Cardiol 39:733–742

Vale PR, Isner JM, Rosenfield K (2001) Therapeutic angiogenesis in critical limb and myocardial ischemia. J Interv Cardiol 14:511–528

van der Strate BW, Popa ER, Schipper M, Brouwer LA, Hendriks M, Harmsen MC, van Luyn MJ (2007) Circulating human CD34+ progenitor cells modulate neovascularization and inflammation in a nude mouse model. J Mol Cell Cardiol 42:1086–1097

Vasa M, Fichtlscherer S, Aicher A, Adler K, Urbich C, Martin H, Zeiher AM, Dimmeler S (2001) Number and migratory activity of circulating endothelial progenitor cells inversely correlate with risk factors for coronary artery disease. Circ Res 89:E1–E7

Vaughan EE, O'Brien T (2012) Isolation of circulating angiogenic cells. Methods Mol Biol 916:351–356

Virgintino D, Errede M, Rizzi M, Girolamo F, Strippoli M, Walchli T, Robertson D, Frei K, Roncali L (2013) The CXCL12/CXCR4/CXCR7 ligand-receptor system regulates neuro-glio-vascular interactions and vessel growth during human brain development. J Inherit Metab Dis 36:455–466

Walczak P, Zhang J, Gilad AA, Kedziorek DA, Ruiz-Cabello J, Young RG, Pittenger MF, van Zijl PC, Huang J, Bulte JW (2008) Dual-modality monitoring of targeted intraarterial delivery of mesenchymal stem cells after transient ischemia. Stroke 39:1569–1574

Wang F, Xue J, Wang D, Wang X, Lu S, Tan M (2010) Treatment of atherosclerosis by transplantation of bone endothelial progenitor cells over-expressed paraoxonase−1 gene by recombinant adeno-associated virus in rat. Biol Pharm Bull 33:1806–1813

Wasserman JK, Yang H, Schlichter LC (2008) Glial responses, neuron death and lesion resolution after intracerebral hemorrhage in young vs. aged rats. Eur J Neurosci 28:1316–1328

Wassmann S, Werner N, Czech T, Nickenig G (2006) Improvement of endothelial function by systemic transfusion of vascular progenitor cells. Circ Res 99:e74–e83

Wei H, Mao Q, Liu L, Xu Y, Chen J, Jiang R, Yin L, Fan Y, Chopp M, Dong J, Zhang J (2011a) Changes and function of circulating endothelial progenitor cells in patients with cerebral aneurysm. J Neurosci Res 89:1822–1828

Wei HJ, Wang D, Chen JL, Xu Y, Lei P, Jiang RC, Liu L, Dong JF, Zhang JN (2011b) Mobilization of circulating endothelial progenitor cells after endovascular therapy for ruptured cerebral aneurysms. Neurosci Lett 498:114–118

Wei N, Yu SP, Gu X, Taylor TM, Song D, Liu XF, Wei L (2013) Delayed intranasal delivery of hypoxic-preconditioned bone marrow mesenchymal stem cells enhanced cell homing and therapeutic benefits after ischemic stroke in mice. Cell Transplant 22:977–991

Weissleder R, Moore A, Mahmood U, Bhorade R, Benveniste H, Chiocca EA, Basilion JP (2000) In vivo magnetic resonance imaging of transgene expression. Nat Med 6:351–355

Werner N, Kosiol S, Schiegl T, Ahlers P, Walenta K, Link A, Bohm M, Nickenig G (2005) Circulating endothelial progenitor cells and cardiovascular outcomes. N Engl J Med 353:999–1007

Westenbrink BD, Oeseburg H, Kleijn L, van der Harst P, Belonje AM, Voors AA, Schoemaker RG, de Boer RA, van Veldhuisen DJ, van Gilst WH (2008) Erythropoietin stimulates normal endothelial progenitor cell-mediated endothelial turnover, but attributes to neovascularization only in the presence of local ischemia. Cardiovasc Drugs Ther 22:265–274

Wong AD, Ye M, Levy AF, Rothstein JD, Bergles DE, Searson PC (2013) The blood-brain barrier: an engineering perspective. Front Neuroeng 6:7.

Wu Y, Ip JE, Huang J, Zhang L, Matsushita K, Liew CC, Pratt RE, Dzau VJ (2006) Essential role of ICAM−1/CD18 in mediating EPC recruitment, angiogenesis, and repair to the infarcted myocardium. Circ Res 99:315–322

Xu Y, Tian Y, Wei HJ, Chen J, Dong JF, Zacharek A, Zhang JN (2011) Erythropoietin increases circulating endothelial progenitor cells and reduces the formation and progression of cerebral aneurysm in rats. Neuroscience 181:292–299

Yamaguchi J, Kusano KF, Masuo O, Kawamoto A, Silver M, Murasawa S, Bosch-Marce M, Masuda H, Losordo DW, Isner JM, Asahara T (2003) Stromal cell-derived factor−1 effects on ex vivo expanded endothelial progenitor cell recruitment for ischemic neovascularization. Circulation 107:1322–1328

Yang HN, Park JS, Woo DG, Jeon SY, Park KH (2012) Transfection of VEGF(165) genes into endothelial progenitor cells and in vivo imaging using quantum dots in an ischemia hind limb model. Biomaterials 33:8670–8684

Yi C, Xia W, Zheng Y, Zhang L, Shu M, Liang J, Han Y, Guo S (2006) Transplantation of endothelial progenitor cells transferred by vascular endothelial growth factor gene for vascular regeneration of ischemic flaps. J Surg Res 135:100–106

Yip HK, Chang LT, Chang WN, Lu CH, Liou CW, Lan MY, Liu JS, Youssef AA, Chang HW (2008) Level and value of circulating endothelial progenitor cells in patients after acute ischemic stroke. Stroke 39:69–74

Yip HK, Tsai TH, Lin HS, Chen SF, Sun CK, Leu S, Yuen CM, Tan TY, Lan MY, Liou CW, Lu CH, Chang WN (2011) Effect of erythropoietin on level of circulating endothelial progenitor cells and outcome in patients after acute ischemic stroke. Crit Care 15:R40

Yoder MC (2009) Defining human endothelial progenitor cells. J Thromb Haemost 7(Suppl 1): 49–52

Yoder MC (2012) Human endothelial progenitor cells. Cold Spring Harb Perspect Med 2(7):a006692

Yoder MC, Mead LE, Prater D, Krier TR, Mroueh KN, Li F, Krasich R, Temm CJ, Prchal JT, Ingram DA (2007) Redefining endothelial progenitor cells via clonal analysis and hematopoietic stem/progenitor cell principals. Blood 109:1801–1809

Yoo CH, Na HJ, Lee DS, Heo SC, An Y, Cha J, Choi C, Kim JH, Park JC, Cho YS (2013) Endothelial progenitor cells from human dental pulp-derived iPS cells as a therapeutic target for ischemic vascular diseases. Biomaterials 34:8149–8160

Yoon CH, Hur J, Park KW, Kim JH, Lee CS, Oh IY, Kim TY, Cho HJ, Kang HJ, Chae IH, Yang HK, Oh BH, Park YB, Kim HS (2005) Synergistic neovascularization by mixed transplantation of early endothelial progenitor cells and late outgrowth endothelial cells: the role of angiogenic cytokines and matrix metalloproteinases. Circulation 112:1618–1627

Yoon CH, Hur J, Oh IY, Park KW, Kim TY, Shin JH, Kim JH, Lee CS, Chung JK, Park YB, Kim HS (2006) Intercellular adhesion molecule−1 is upregulated in ischemic muscle, which mediates trafficking of endothelial progenitor cells. Arterioscler Thromb Vasc Biol 26:1066–1072

Yu SP, Wei Z, Wei L (2013) Preconditioning strategy in stem cell transplantation therapy. Transl Stroke Res 4:76–88

Zemani F, Silvestre JS, Fauvel-Lafeve F, Bruel A, Vilar J, Bieche I, Laurendeau I, Galy-Fauroux I, Fischer AM, Boisson-Vidal C (2008) Ex vivo priming of endothelial progenitor cells with SDF−1 before transplantation could increase their proangiogenic potential. Arterioscler Thromb Vasc Biol 28:644–650

Zengin E, Chalajour F, Gehling UM, Ito WD, Treede H, Lauke H, Weil J, Reichenspurner H, Kilic N, Ergun S (2006) Vascular wall resident progenitor cells: a source for postnatal vasculogenesis. Development 133:1543–1551

Zhan K, Bai L, Xu J (2013) Role of vascular endothelial progenitor cells in construction of new vascular loop. Microvasc Res 90:1–11

Zhang ZG, Zhang L, Jiang Q, Chopp M (2002) Bone marrow-derived endothelial progenitor cells participate in cerebral neovascularization after focal cerebral ischemia in the adult mouse. Circ Res 90:284–288

Zhang L, Li Y, Romanko M, Kramer BC, Gosiewska A, Chopp M, Hong K (2012) Different routes of administration of human umbilical tissue-derived cells improve functional recovery in the rat after focal cerebral ischemia. Brain Res 1489:104–112

Zhao YH, Yuan B, Chen J, Feng DH, Zhao B, Qin C, Chen YF (2013) Endothelial progenitor cells: therapeutic perspective for ischemic stroke. CNS Neurosci Ther 19:67–75

Zhou WJ, Zhu DL, Yang GY, Zhang Y, Wang HY, Ji KD, Lu YM, Gao PJ (2009) Circulating endothelial progenitor cells in Chinese patients with acute stroke. Hypertens Res 32:306–310

Zhu JH, Wang XX, Zhang FR, Shang YP, Tao QM, Chen JZ (2008) Safety and efficacy of autologous endothelial progenitor cells transplantation in children with idiopathic pulmonary arterial hypertension: open-label pilot study. Pediatr Transplant 12:650–655

Chapter 8
Adipose-Derived Stem Cells: Isolation and Culturing

Vaughan Feisst and Michelle B. Locke

Contents

Abstract The use of human adipose-derived stem cells (ASC) for research and clinical purposes has become increasingly common over the past years. The advantages of ASC over more traditional sources of adult stem cells such as bone marrow (BM) include it's ready availability in most people, the ease of significant volume fat removal by liposuction and subsequently the large number of adult mesenchymal stem cells (MSC) able to be isolated from each harvest. This chapter provides clear, focused methods for the isolation of the stromal vascular fraction (SVF) from lipoaspirate, which contains the ASC. Methods for subsequent culture, passage and cryopreservation of ASC are given, to enable other researchers to work with this exciting resource.

M. B. Locke (✉)
Faculty of Medicine and Health Sciences, South Auckland Clinical School, C/- Middlemore Hospital, The University of Auckland, Private Bag 93311, Otahuhu, Auckland 1640, New Zealand
e-mail: Michelle.Locke@middlemore.co.nz

V. Feisst
Dunbar Laboratory, School of Biological Sciences, The University of Auckland, Private Bag 92019, Auckland 1142, New Zealand
e-mail: v.feisst@auckland.ac.nz

L.-R. Zhao, J. H. Zhang (eds.), *Cellular Therapy for Stroke and CNS Injuries*, Springer Series in Translational Stroke Research, DOI 10.1007/978-3-319-11481-1_8

Abbreviations

°C	Degrees centigrade
ASC	Adipose-derived stem cells
BM	Bone marrow
CD	Cluster of differentiation
CO2	Carbon dioxide
DMEM	Dulbecco's modified eagle's medium
DMSO	Dimethyl sulfoxide
FACS	Fluorescence activate cell
FBS	Fetal bovine serum
g	Gravity force
μm	Micrometres
mL	Millilitres
mm	Millimetres
MSC	Mesenchymal stem cell
P	Passage
PBS	Phosphate buffered saline
RBC	Red blood cell
SVF	Stromal vascular fraction
UCB	Umbilical cord blood

8.1 Introduction

The use of human adipose-derived stem cells (ASC) for research and clinical purposes has exploded since 2001, when Zuk and colleagues published their sentinel paper on the mesenchymal differentiation of ASC (Zuk et al. 2001). Adipose tissue is derived from embryonic mesoderm, therefore ASC are a form of adult mesenchymal stem cell (MSC). Early research confirmed their differentiation into common mesenchymal tissues, such as bone, muscle and cartilage (Gimble and Guilak 2003; Wickham et al. 2003; Zuk et al. 2002). Recently, there is increasing evidence of their potential ability to transdifferentiate along some ectodermal cell lines, such as the creation of neural cell types (Erba et al. 2010; Yu et al. 2011).

While other sources of adult MSC exist, such as bone marrow (BM) and umbilical cord blood (UCB), adipose tissue represents a readily available resource. Over the past decades, the prevalence of obesity has been rising in the Western world. In the United States, more than one third (35.7%) of adults are obese (Centres for Disease Control and Prevention 2013). Data from the American Society for Aesthetic Plastic Surgery shows a corresponding increase in the number of patients undergoing elective liposuction (also known as lipoplasty or liposculpture) surgery (American Society for Aesthetic Plastic Surgery 2013). Over 15 years, from 1997 to 2012, the number of patients undergoing this procedure annually jumped from 177,000 to 313,000, an increase of 77% (American Society for Aesthetic Plastic

Surgery 2013). It is now the second most commonly performed cosmetic surgery procedure. Lipoaspirate, the waste product of liposuction, is commonly discarded following the surgery. However, it is an excellent source of ASC for research and potentially clinical purposes, with a stem cell yield which compares well to traditional BM harvest (Sekiya et al. 2002; Zuk et al. 2002). ASC have also been reported to be easier to culture and to grow faster than BM MSC (Mitchell et al. 2006). When a cell source is required for clinical use, liposuction is a straightforward procedure which can be performed on an outpatient basis under general or local anaesthetic with minimal morbidity. The reported complication rate is around 0.1 %, making it a safe, well tolerated method to obtain autologous cells for clinical treatment (Hanke and Coleman 1999; Housman et al. 2002).

In this chapter we focus on the technical method of isolating ASC from the raw tissue sample, commonly lipoaspirate following liposuction or excised fat.

8.2 Materials

8.2.1 Equipment

No special equipment is required for ASC isolation and culture. All equipment used can generally be found in any cell culture laboratory. Availability of the following standard equipment is required for cellular processing and culture:

- Class II laminar flow hood
- Warm water bath (37 °C)
- 50 mL tubes
- Cell culture flasks (ideally T25 or T75)
- Centrifuge
- Forceps and tissue scissors or scalpel if tissue block obtained
- Incubator (37 °C, 5 % CO_2)
- 100 μm cell filters
- 4 °C refrigerator
- −80 °C freezer
- Nalgene freezing container
- −196 °C (liquid nitrogen) cryogenic storage facility

8.2.2 Chemicals and Reagents

A list of chemicals and reagents which are required for cellular isolation and processing is provided below. The manufacturer's name and product code for the authors preferred products are provided in parentheses.

- Dulbecco's modified eagle's medium (DMEM) with nutrient mixture F12+Hepes (Life Technologies 11330057)
- Dimethyl sulfoxide (DMSO) (Sigma Aldritch D2650-5X5ML)
- Fetal bovine serum (FBS) (Life Technologies 10091148)
- GlutaMAX Supplement (Life Technologies 35050061)
- Penicillin-Streptomycin (Life Technologies 15140122)
- Collagenase, type I (Life Technologies 17100017)
- Trypsin or TrypLE™ (Life Technologies 12563029)
- Phosphate buffered saline (PBS) (Lonza 17-516F)

8.2.3 Medium

All culture mediums are created by mixing the listed ingredients in a laminar flow hood. Once prepared, medium should be used within four weeks. Medium should be gently warmed in a water bath to 37 °C before use.

1. Complete medium:
   - 500 ml DMEM F12 Hepes
   - 50 ml FBS
   - 5 ml GlutaMAX supplement
   - 5 ml Penicillin-Streptomycin

2. Cryopreservation medium:
   - 20 ml FBS
   - 5 ml DMSO

8.3 Methods

8.3.1 Source of Adipose Tissue

Our adipose tissue comes from lipoaspirate, the waste product following elective liposuction surgery. It is a mixture of suction-aspirated adipose tissue, blood and infiltration fluid. The infiltration fluid comprises a mixture of saline, adrenaline and often local anaesthetic. The most commonly used formula for infiltration fluid is the Hunstad formula, being 1000 mL of Ringers Lactate, 50 mL of 1 % lignocaine (lidocaine) and 1 mL (1 ampoule) of 1:1000 epinephrine, or Modified Hunstad, which has half the amount of lignocaine (Hunstad and Aitken 2006). Infiltration fluid is infused into each area to be treated prior to performing liposuction. This acts to reduce blood loss and provide analgesia. It is thought that approximately 20 % of the infiltrated fluid is removed with the liposuction. The fluid mixes with the aspirated fat and contributes to the lipoaspirate being in a semi-liquid state. This

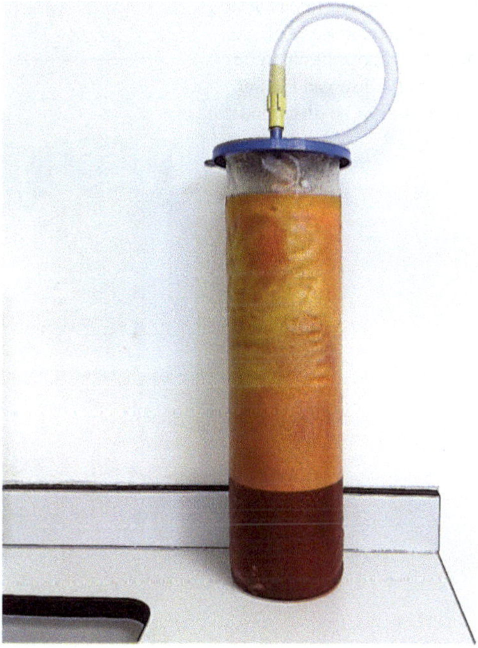

Fig. 8.1 Unprocessed lipoaspirate showing semi-liquid fat on the top (yellow) with blood and infiltration fluid in the lower layer (red)

lipoaspirate is commonly discarded as medical waste after the procedure. Appropriate institutional or regional ethical approval must be sought to allow the donation of this waste product for research purposes, and the patient's approval is also obtained prior to surgery.

Given that we solely use lipoaspirate, our source tissue for ASC is in a semi-liquid state on arrival at our laboratory (Fig. 8.1). However, it is possible to use excised fat for the same purpose. If a solid block of excised fat is to be used, it requires additional treatment to be manually broken down prior to processing as below.

8.3.2 Processing of Adipose Tissue

When lipoaspirate is transported to the lab, it travels in the original surgical waste container at room temperature. A block of excised tissue would routinely be transported to the laboratory wrapped in saline-soaked gauze to prevent desiccation and secured in a sterile, screw top container.

Our method for ASC isolation is based on the seminal paper by Zuk and colleagues from 2001 with minor modifications (Zuk et al. 2001). Following isolation, ASC are contained within the stromal vascular fraction (SVF) cell pellet in the bottom of the tube (see Fig. 8.3 and Fig. 8.4). Cryopreservation of the SVF results in red blood cell (RBC) lysis so we rarely perform a specific RBC lysis step, preferring instead to repeat the initial wash step to decrease the RBC load if necessary.

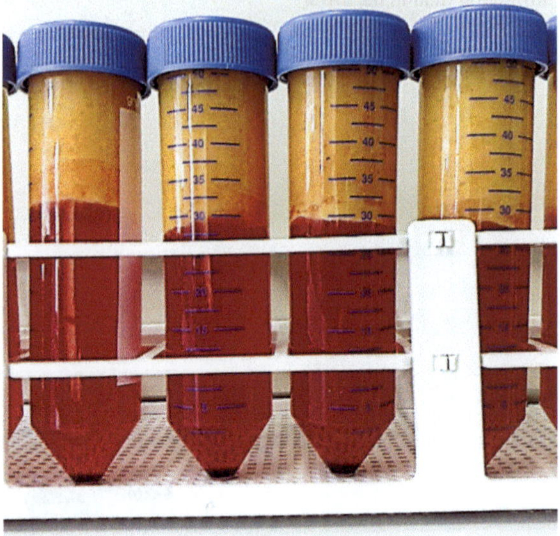

Fig. 8.2 Washed lipoaspirate showing the top layer of floating adipose tissue to be retained (yellow) and lower liquid layer to be discarded (red)

However, if you wish to count the cells in the SVF directly then it is likely that you will need to perform a RBC lysis.

1. Excised fat is placed in a sterile plastic tray with a small amount of Complete medium. Using sterile forceps and a pair of sterile scissors, or scalpel blade, it is cut into small pieces (ideally less than 5 mm by 5 mm) to facilitate digestion. These pieces are then placed into a 50 ml tube for processing as per lipoaspirate. *This step is not required if semi-liquid lipoaspirate is used.*
2. Top layer of lipoaspirate, containing adipose tissue (or excised and diced adipose tissue), is placed into 50 mL tubes, ensuring the tubes are no more than half full of lipoaspirate.
3. Tubes are filled to 50 mL with Complete medium, agitated to mix and centrifuged at 700 g for 10 min.
4. Floating adipose tissue is transferred into fresh 50 ml tubes and the liquid lower layer is discarded (see Fig. 8.2).
5. Repeat steps 2–4.
6. An equal volume of 0.075 % Collagenase type I in PBS is added to each tube of washed lipoaspirate for digestion.
7. Tubes are placed in a warm water bath for 1 h for digestion. An agitating water bath is ideal for digestion. If not available, a normal water bath may be used but

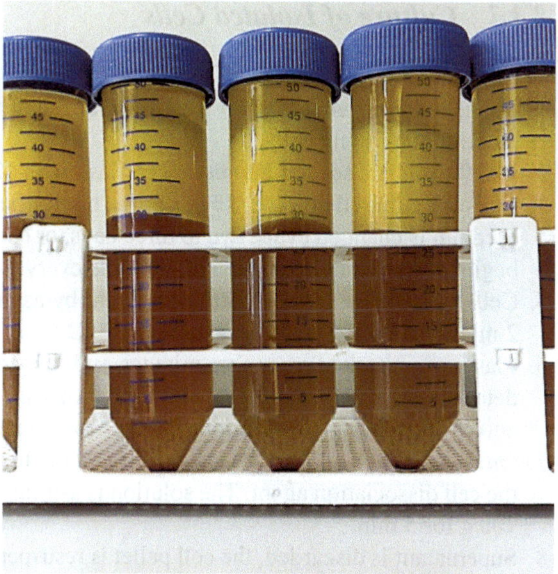

Fig. 8.3 Lipoaspirate after collagenase digestion and spin cycle, showing floating mature adipocytes in the top layer (yellow), media in the lower layer (orange) and a cell pellet at base of tube

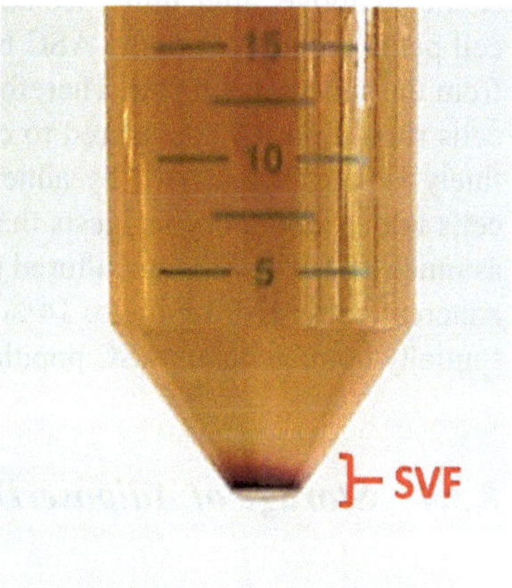

Fig. 8.4 Close up of SVF pellet after digestion and spin cycle

the tubes should be inverted several times or agitated by hand every five to ten minutes during the digestion process.

8. Tubes are centrifuged at 700 g for 10 min.
9. Supernatants containing floating adipocytes and media are discarded. Cell pellets are resuspended in Complete medium.
10. Resuspended cells are passed through a 100 μm cell filter into a clean 50 mL tube to remove connective tissue.

SVF is now ready for cryopreservation or culture.

8.3.3 Culture of Isolated Cells

1. A cell count is performed on fresh SVF or recovered cells using Trypan blue to identify non-viable cells.
2. 1×10^6 cells are seeded into each T75 flask with 12 mL of Complete medium.
3. Cells are cultured at 37 °C/5 % CO_2 in a humidified atmosphere.
4. Medium is changed every two to three days, or earlier if it discolours. Once cells begin to proliferate they generally double every 5–7 days.
5. Cells are passaged when ~80 % confluent by aspirating off the medium, adding 2 mL of TrypLE™, and incubating at 37 °C.
6. Flasks are checked after five minutes and tapped gently to encourage cellular detachment. If cells don't detach easily, flasks are placed back into the incubator and checked every three to four minutes for detachment.
7. Once cells have all detached, 8 mL of Complete medium is added to neutralise the cell dissociation agent. The solution is aspirated into tubes and centrifuged at 700 g for 5 min.
8. Supernatant is discarded, the cell pellet is resuspended in Complete medium and seeded 1:1 into fresh T75 flasks for ongoing cellular expansion.

As noted above, after initial isolation the ASC are contained within the SVF. This cell pellet includes not only ASC but also vascular cells and other cell populations from the subcutaneous fat. Therefore experiments performed using freshly isolated cells should not be considered to contain a pure population of ASC. ASC are routinely isolated from SVF by adherent cell culture. FACS assessment of cultured cells in our laboratory suggests that CD31+ and/or CD45+ vascular cells make up as much as 50 % of the uncultured population of SVF. This figure falls rapidly with adherent culture, dropping to 14 % at passage zero (P0) and 2 % by P2, with an essentially homogeneous ASC population present thereafter (data not shown).

8.3.4 Storage of Adipose-Derived Stem Cells

The SVF pellet or cultured ASC at any stage can be cryopreserved for long term storage. The authors use 10 % DMSO as their preferred cryopreservant.

1. A cell count is performed on fresh SVF or recovered cells using Trypan blue to identify non-viable cells.
2. Cells are centrifuged at 700 g for 5 min and supernatant is discarded.
3. Cells are resuspended in 0.5 mL of Complete medium per 1×10^6 cells.
4. 0.5 mL of Cyropresevation medium per 1×10^6 cells is added, and mixed with a pipette, making a final concentration of 10 % DMSO.
5. Tubes are cooled at − 1 °C per minute using a Nalgene freezing container (authors use Mr Frosty™ (ThermoFisher)) with 100 % isopropyl alcohol and placed into a − 80 °C freezer overnight.

6. The following morning, cryovials are removed and transferred to a liquid nitrogen storage facility for long term storage at −196 °C.

8.4 Authors Notes

- We strongly advise that cell culture media is changed on day one following cell passage and/or plating. This removes non adherent cells, which ensures that cells undergoing cell death do not affect the viability of healthy, adherent cells.
- When passaging our cells, we re-seed ASC at a 1:1 ratio. We are aware that other institutions seed at a lower density, commonly 1:4. However, we have found that cell culture can stall and proliferation cease if the cells are seeded at a lower density, presumably due to paracrine growth factors assisting in the maintenance of proliferation. A 1:1 ratio has proved to be the most effective in our facility.
- We use TrypLE™ rather than trypsin for cell passage as we have found it to be a more effective cell dissociation agent for adherent ASC.

Acknowledgements The authors gratefully acknowledge the patients who generously donate tissue for our research and the surgeons who facilitate the tissue donation. We thank Prof. Rod Dunbar, Director, Maurice Wilkins Centre for Molecular Biodiscovery, for his overarching support and assistance.

References

American Society for Aesthetic Plastic Surgery (2013) Cosmetic surgery national data bank statistics 2012. http://www.surgery.org/sites/default/files/ASAPS-2012-Stats.pdf. Accessed 3 Jan 2014

Centres for Disease Control and Prevention (2013) Overweight and obesity. http://www.cdc.gov/obesity/data/adult.html. Accessed 03 Jan 2014

Erba P, Terenghi G, Kingham PJ (2010) Neural differentiation and therapeutic potential of adipose tissue derived stem cells. Curr Stem Cell Res Ther 5(2):153–160

Gimble J, Guilak F (2003) Adipose-derived adult stem cells: isolation, characterization, and differentiation potential. Cytotherapy 5(5):362–369

Hanke CW, Coleman WP 3rd (1999) Morbidity and mortality related to liposuction. Questions and answers. Dermatol Clin 17(4):899–902

Housman TS, Lawrence N, Mellen BG, George MN, Filippo JS, Cerveny KA et al (2002) The safety of liposuction: results of a national survey. Dermatol Surg 28(11):971–978

Hunstad JP, Aitken ME (2006) Liposuction: techniques and guidelines. Clin Plast Surg 33(1):13–25, v.

Mitchell JB, McIntosh K, Zvonic S, Garrett S, Floyd ZE, Kloster A et al (2006) Immunophenotype of human adipose-derived cells: temporal changes in stromal-associated and stem cell-associated markers. Stem Cells 24(2):376–385

Sekiya I, Larson BL, Smith JR, Pochampally R, Cui J-G, Prockop DJ (2002) Expansion of human adult stem cells from bone marrow stroma: conditions that maximize the yields of early progenitors and evaluate their quality. Stem Cells 20(6):530–541

Wickham MQ, Erickson GR, Gimble JM, Vail P, Guilak F (2003) Multipotent stromal cells derived from the infrapatellar fat pad of the knee. Clin Orthop Relat Res 412:196–212
Yu JM, Bunnell BA, Kang S-K (2011) Neural differentiation of human adipose tissue-derived stem cells. Methods Mol Biol 702:219–231
Zuk PA, Zhu M, Mizuno H, Huang J, Futrell JW, Katz AJ et al (2001) Multilineage cells from human adipose tissue: implications for cell-based therapies. Tissue Eng 7(2):211–228
Zuk PA, Zhu M, Ashjian P, De Ugarte DA, Huang JI, Mizuno H et al (2002) Human adipose tissue is a source of multipotent stem cells. Mol Biol Cell 13(12):4279–4295

Chapter 9
Transplantation of Adipose-Derived Stem Cells in Stroke

Cheuk-Kwan Sun

Contents

C.-K. Sun (✉)
Department of Emergency Medicine, E-Da Hospital, I-Shou University, No. 1, Yi-Da Road, Jiao-Su Village, Yan-Chao District, Kaohsiung City 824, Taiwan, Republic of China
e-mail: lawrence.c.k.sun@gmail.com

L.-R. Zhao, J. H. Zhang (eds.), *Cellular Therapy for Stroke and CNS Injuries*, Springer Series in Translational Stroke Research, DOI 10.1007/978-3-319-11481-1_9

Abstract Not only is stroke second only to cardiac ischemia as a leading cause of death worldwide, but it also drastically impaired the quality of life of the survivors through its crippling neurological sequelae which account for the third leading cause of disability. Instead of merely a loss of functioning neurons from ischemia, stroke triggers a cascade of adverse events including inflammation, oxidative stress, and apoptosis that perpetuates the initial ischemic damage. Current therapeutic strategies, including the use of thrombolytic agents and other non-pharmaceutical approaches, have their limitations either because of the risk of complications or focusing only on the prevention of brain damage and rehabilitation. More importantly, none has been convincingly shown to improve neurological outcome in patients with stroke once the brain tissue is infarcted. Accumulating evidence has indicated that, instead of being only neuroprotective, stem cells actually possess neurorestorative function for promoting recovery of the injured brain tissue. Accordingly, cell transplant therapy with adipose-derived mesenchymal stem cells (ADSC) has recently emerged as a potentially feasible therapeutic option not only because of their abundance and relative ease of being harvested, but also because of the possibility of autologous implantation and their demonstrated multiple beneficial biological actions against stroke in experimental settings, namely paracrine effects, transdifferentiation, and immunomodulation, that could enhance brain plasticity such as neurogenesis, remyelination, synaptogenesis, and angiogenesis in the recovery process. The nature and source of ADSC as well as their demonstrated therapeutic potential against stroke, the clinical perspective in stroke treatment, and the potential risks are reviewed.

Abbreviations

ADSC	Adipose-derived mesenchymal stem cells
APC	Antigen-presenting cell
BDNF	Brain-derived neurotrophic factor
bFGF	Basic fibroblast growth factor
BMP2	Bone morphogenetic protein 2
CSPG	Chondroitin sulphate proteoglycans
CXCR4	Chemokine receptor type 4
DCX	Doublecortin
FACS	Fluorescence-activated cell sorting
FGF2	Fibroblast growth factor 2
G-CSF	granulocyte colony-stimulating factor
GDNF	Glial derived neurotrophic factor
GFAP	Glial fibrillary acidic protein
GM-CSF	Granulocyte-macrophage colony-stimulating factor
HGF	Hepatocyte growth factor
IDO	Indoleamine-2,3-dioxygenase
IGF-1	Insulin-like growth factor-1
IL	Interleukin
IL-1R	Interleukin 1 receptor

iPSC	Induced pluripotent stem cells
MACS	Magnetic activated cell sorting
MAP2	Microtubule-associated protein 2
MCAO	Middle cerebral artery occlusion
MHC-II	Major histocompatibility complex class II
NeuN	Neuronal nuclei
NF	Neurofilament
NGF	Nerve growth factor
NT-3	Neurotrophin-3
Olig-2	Oligodendrocyte
PAI-1	Plasminogen activator inhibitor-1
ROS	Reactive oxygen species
rtPA	Recombinant tissue plasminogen activator
SDF-1	Stromal cell-derived factor 1
SVF	Stromal vascular fraction
SYP	Synaptophysin
TGF-β1	Transforming growth factor beta 1
TLR-4	Toll-like receptor-4
TNF-alpha	Tumor necrosis factor-alpha
VCAM-1	Vascular cell adhesion molecule 1
VEGF	Vascular endothelial growth factor
vWF	Von Willebran factor

9.1 Introduction

9.1.1 The Layout of the Chapter

To cover the essential knowledge on this topic, the chapter is divided into several sections, including an outline of stroke as a life-threatening disease and the current medical strategies, the introduction of adipose-derived mesenchymal stem cells (ADSC) as a treatment tool for stroke and its therapeutic advantages, description of the general methodology of ADSC isolation, the mechanisms of action of ADSC against stroke, the current status of experimental and clinical studies as well as the perspectives of ADSC applications and the potential risks and limitations.

9.1.2 Stroke, Current Treatment Strategies, and Their Limitations

Stroke, which can be of ischemic or hemorrhagic origins due to a disruption of blood supply to the brain tissue, has been reported to be the second leading cause of death worldwide (Feigin et al. 2014). Not only is stroke a ruthless killer, but it also leaves behind devastating neurological deficits (i.e., those of motor, sensory, and cognitive) that account for their being the third leading cause of disability (Rosado-de-Castro et al. 2013). The cost for chronic care and rehabilitation as well as the remarkable impairment of the quality of life for the affected individuals impose an enormous socioeconomic burden on a society (Go et al. 2013).

The current mainstay of treatment for acute ischemic stroke, which accounts for over 80 % of all cases of stroke (Liu et al. 2013), includes the use of thrombolytic agents (e.g. recombinant tissue plasminogen activator, rtPA) that are supposed to be given within 4.5 h after the attack (Del Zoppo et al. 2009) with the hope of resuming patency of the supplying artery to restore the function of tissue surrounding the core region of infarction (i.e. ischemic penumbra) to minimize the ischemic insult rather than the irreversible infarction itself (Smith et al. 2011). However, the narrow therapeutic time window and the significant risk of symptomatic intracranial hemorrhage (i.e. up to 5.6 %) and death (Wardlaw et al. 2009; Seet and Rabinstein 2012) substantially hinder its use which is only suitable for only 2–4 % of patients with ischemic stroke (Molina 2011). The popularity of rtPA use is further hampered by its limited efficacy in disability prevention which is only six patients per 1000 ischemic strokes and its lack of beneficial impact on mortality rate (Hacke et al. 2004). As a result, various non-pharmaceutical strategies have been proposed including neuroprotective approaches such as hypothermia, ischemic/hypoxic conditioning, acupuncture, certain medical gases, and transcranial laser therapy as well as mechanical endovascular recanalization and recovery devices for treating the chronic phase of stroke (Chen et al. 2014b).

Since ischemic injury of the brain involves a cascade of events, a multi-faceted therapeutic approach is preferred (Chen et al. 2014b). The clinical possibility of cell therapy (i.e., "cell replacement therapy" or "cell transplant therapy") for central nervous system disorders, including stroke, gained much attention in the year 2000 (Bjorklund and Lindvall 2000; Zivin 2000) when the first clinical trial on neuronal cellular transplantation in patients with stroke was reported (Kondziolka et al. 2000). Although it was not stem cell that was transplanted, it opened up the avenue for exploring the therapeutic potential of stem cell transplantation for stroke (Cairns and Finklestein 2003).

9.1.3 Stem Cells: Their Natures, Sources, and Therapeutic Potentials

The two distinctive properties that distinguish stems cells from other somatic cells are "self-renewal" and "potency" (Kuhl and Kuhl 2013). Self-renewal refers to the process in which a stem cell undergoes mitotic cell division to produce at least one daughter cell with equal developmental potential as the mother cell, in other words, another stem cell. On the other hand, potency is the ability of a stem cell to differentiate into different mature specialized cell types (i.e. multi-lineage differentiation). Regarding the use of stems cells in the treatment of stroke, experimental studies using embryonic stem cells (Chang et al. 2013; Drury-Stewart et al. 2013), mesenchymal stem cells (Ikegame et al. 2011), hematopoietic stem cells (Tsuji et al. 2014), neural stem cells (Andres et al. 2011), induced pluripotent stem cells (iPSC) (Oki et al. 2012), and also multipotent adult progenitor cells (Mora-Lee et al. 2012) in animal models of stroke have been reported with unanimous positive therapeutic results. Although mesenchymal stem cells were first identified four decades ago as adherent cells with fibroblastoid morphology being able to differentiate into cells of mesodermal origin such as osteocytes, chondrocytes, and adipocytes (Friedenstein et al. 1974), they were later found to be also capable of differentiating into ectodermal and endodermal elements (Lakshmipathy and Verfaillie 2005; Ikegame et al. 2011). The fact further highlights their observed therapeutic versatility against a variety of diseases of different pathological origins as reflected in their abilities of vascular endothelial (Li et al. 2013a), neuronal (Gao et al. 2013), and musculoskeletal (Gardner et al. 2013) repairs.

9.2 Adipose-Derived Stem Cells: Therapeutic Advantages, Sources, and Isolation

9.2.1 Advantages of Therapeutic use of Adipose-Derived Mesenchymal Stem Cells Compared with Stem Cells of Other Origins

Compared with embryonic stem cells, autologous mesenchymal stem cells have the advantage of being self-derived without the concern of ethics and that of possible infection from unknown donors. The reported sources of mesenchymal stem cells include bone marrow (Skvortsova et al. 2008), adipose tissue (Gutierrez-Fernandez et al. 2013a), embryo (Liu et al. 2009), placenta (Kranz et al. 2010), dental pulp and periodontal ligament (Moshaverinia et al. 2014; Vasandan et al. 2014). Other sources, including palatine tonsil (Janjanin et al., 2008), dermis (Feisst et al. 2014), and skeletomuscular system (Aydin et al. 2014; Mason et al. 2014), have also been reported. In particular, the two most readily available sources of autologous ADSC, bone marrow and adipose tissue, have been widely investigated both experimentally and clinically regarding their therapeutic potentials in treating a myriad of diseases, especially ischemic and microvascular disorders (Calio et al. 2014; Liu et al. 2014).

Previous studies have shown that not only are adipose-derived stem cells (ADSC) (also known as "adipose-derived mesenchymal stem cells", "adipose tissue-derived multipotent stromal cells" or "adipose-derived mesenchymal stromal cells") relatively easy to obtain with less invasive procedures compared to bone marrow stem cells, but the former also exhibit better proliferative activity, differentiating capacity, immunomodulatory function, and trophic factor-releasing ability than the latter including greater production of vascular endothelial growth factor (VEGF), angiopoietin-1, and hepatocyte growth factor (HGF) (Ikegame et al., 2011) as well as interleukin 1 receptor (IL-1R), IL-6, IL-8, granulocyte colony-stimulating factor (G-CSF), granulocyte-macrophage colony-stimulating factor (GM-CSF), monocyte chemotactic protein 1, nerve growth factor (Banas et al., 2008; Ikegame et al. 2011), and transforming growth factor (TGF)-β1 (Melief et al. 2013b). Furthermore, the lack of major histocompatibility complex class II (MHC-II) expression in adipose-derived stem cells also enables their storage and allogeneic administration to individuals with acute ischemic stroke (Gutierrez-Fernandez et al. 2012). On the other hand, although the discovery of iPSC seems to offer a solution to the problem of the limited availability of stem cells through reprogramming of autologous somatic cells (Takahashi and Yamanaka, 2006), the requirement for the activation of potentially tumorigenic genes (e.g. c-myc) for its induction (Araki et al. 2011) has raised much clinical concerns. ADSC, therefore, appear to be an applicable clinical tool for routine clinical practice.

9.2.2 Source- and Donor-Dependent Variability in ADSC Quality

Previous animal experimental studies have used adipose tissue from different regions for autologous ADSC isolation, including inguinal (Chen et al. 2012) and peri-epididymal (Leu et al. 2010; Jiang et al. 2014) fat depots. Besides, other studies using human ADSC for animal studies also showed positive therapeutic results (Kang et al. 2003b; Kim et al. 2007; Yang et al. 2012; Liu et al. 2014). Therefore, it appears that the source of ADSC has no remarkable bearing on the treatment outcomes. As a result, although subcutaneous adipose tissue especially that from the abdomen, thigh, and buttock removed through fat-extracting body-shaping procedures such as liposuction and lipectomy (Chia and Theodorou 2012) used to be regarded as medical waste products, it now serves as a readily available clinical source of adipose-derived stem cells. On the issue of the optimal source of ADSC, a previous clinically-oriented study comparing the properties of ADSC isolated from human adipose tissues of different depots, including abdominal subcutaneous fat, omentum, pericardial adipose tissue, and thymic remnants, has demonstrated that ADSC isolated from different sources exhibited varied proliferation and differentiation capacities that should be taken into account to serve a specific therapeutic purpose (Russo et al. 2014). For instance, ADSC isolated from intrathoracic depots exhibited a longer average doubling time and a higher proportion of $CD34^+$ cells compared with those isolated from subcutaneous fat or the omentum. Moreover, subcutaneous and pericardial adipose tissue yielded ADSC with enhanced adipogenic differentiation potential, while ADSC from the omentum displayed high levels of osteogenic markers (Russo et al. 2014). Another study that underscores the importance of the origins of ADSC from a therapeutic point of view is its finding that adipose tissues of slightly different origins in close vicinity (i.e., epicardial fat, pericardial fat, and the right atrium) exhibited significantly different capacities of secreting trophic and inflammatory cytokines, different degrees of upregulation of inflammation- and fibrosis-related genes, as well as different therapeutic effects in a rat model of myocardial infarction to which mesenchymal stem cells from right atrium and epicardial fat were even found to be detrimental (Naftali-Shani et al. 2013). On the other hand, specific origins of ADSC may also be taken into account for special therapeutic needs. For instance, a clonogenic population of metabolically active stem cells has been reported to reside in adult human brown adipose tissue that may be activated to participate in energy homeostasis in vivo and, therefore, may be of therapeutic importance for obesity and related metabolic disorders (Silva et al. 2014).

The correlation between donor-dependent variability and the quality of ADSC has also been widely investigated. For instance, one study demonstrated a negative association of donor age with the proliferation and differentiation potential of ADSC (Choudhery et al. 2014). Donors of advanced age has also been reported to reduce the yield of ADSC expressing low-affinity nerve growth factor receptor (CD271) (Cuevas-Diaz Duran et al. 2013) and also produce ADSC of impaired angiogenic capacities (De Barros et al. 2013). On the other hand, infant-derived cells have been shown to be morphologically more elongated with long telomeres, and exhibit augmented angiogenic and osteogenic abilities compared with older cells (Wu et al. 2013). Nevertheless, other studies using adipose tissue from human adults showed that donor age, body-mass index, and harvest site do not influence cell yield and proliferation rate (Buschmann et al. 2013). Moreover, doubling time, telomere length, the osteogenic and chondrogenic differentiation capacity, as well as osteogenic paracrine activity were also found to be similar among ADSC from adult donors of different ages (Ding et al. 2013; Wu et al. 2013). Consistently, another experimental study investigating the impact of donor age on the function of adipose-derived stem cells also demonstrated that aged ADSC from rats still retained potential to support axon regeneration (Mantovani et al. 2012). One interesting finding is that ADSC from older donors were found to exhibit compromised adipogenic potential that actually favors their application in regeneration therapy (Ding et al. 2013). The overall promising proliferation and differentiation capabilities of ADSC regardless of the donor's age, therefore, open up an avenue to their clinical application, taken into account that the elderly will be the greatest beneficiaries of autologous stem cell treatment. On the other hand, the morphology, proliferation rate, and doubling time of ADSC have also been shown to vary with the nature of the coatings on which they were cultured (Marycz et al. 2013).

9.2.3 *Isolation, Culture, and Identification of Adipose-Derived Mesenchymal Stem Cells*

The isolation, culturing, and identification of ADSC are straightforward, including the procedures of mincing, digestion, filtering, centrifugation, culturing, and flow

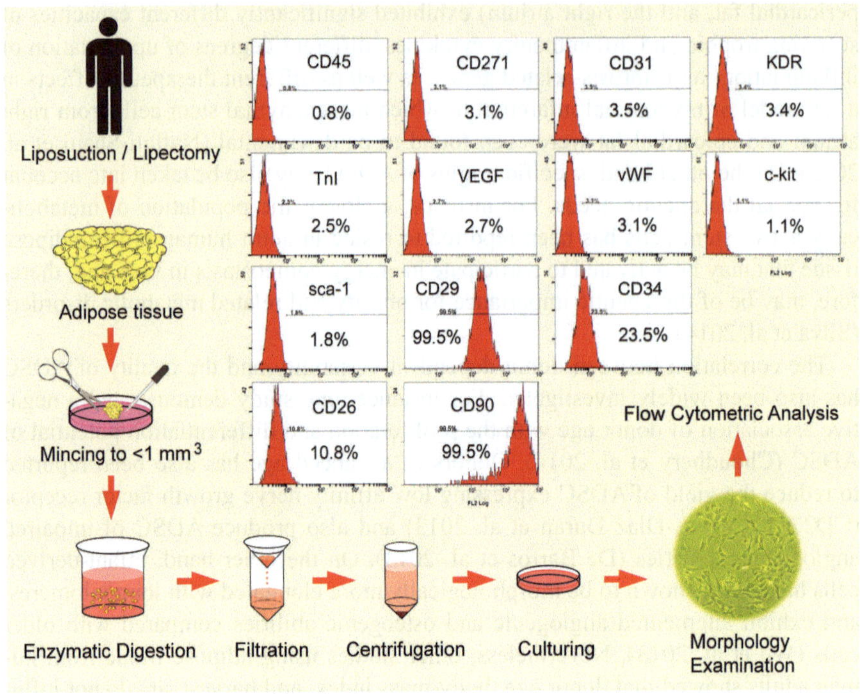

Fig. 9.1 Simplified procedures of harvesting, processing, culturing, and characterization of adipose-derived stem cells. After being removed from the human body through procedures such as liposuction and lipectomy, the adipose tissue is minced to small pieces of size less than 1 mm^3 to maximize the efficiency of enzymatic digestion for freeing the cells from connective tissue. After cell-harvesting through filtration and centrifugation, the cells are purified with the number of cells expanded through culturing. The cultured cells subsequently undergo flow cytometric analysis for the identification of surface markers characteristic of mesenchymal stem cells. (Note the typical spindle-shaped morphology of mesenchymal stem cells at right lower corner)

cytometric identification as described previously (Leu et al. 2010). Briefly, the harvested adipose tissue is minced into < 1 mm^3 size pieces using a pair of sharp, sterile surgical scissors to maximize the surface areas for enzyme digestion (Fig. 9.1). Then 200–300 μL of sterile saline is added to every 0.5 g of tissue to prevent dehydration. Sterile saline (37 °C) is added to the homogenized adipose tissue in a ratio of 3:1 (saline: adipose tissue), followed by the addition of stock collagenase solution to a final concentration of 0.5 Units/mL. The tubes with the contents are placed and secured on a Thermaline shaker and incubated with constant agitation for 60 ± 15 min at 37 °C. After 40 min of incubation, the content is triturated with a 25 mL pipette for 2–3 min. The cells obtained are placed back to the rocker for incubation. The contents of the flask were transferred to 50 mL tubes after digestion, followed by centrifugation at 600 g, for 5 min at room temperature. The fat layer and saline supernatant from the tube are poured out gently in one smooth motion or removed using vacuum suction. The cell pellet thus obtained is resuspended in

40 mL saline and then centrifuged again at 600 g for 5 min at room temperature. After being resuspended again in 5 mL saline, the cell suspension is filtered through a 100 μm filter into a 50 mL conical tube to which 2 mL of saline is added to rinse the remaining cells through the filter. The flow-through is pipetted to a 40 μm filter into a new 50 mL conical tube. The tubes are centrifuged for a third time at 600 g for 5 min at room temperature. The cells, which are a mixture of lymphocytes, macrophages, fibroblasts, endothelial cells, and other cell populations, are resuspended in saline. An aliquot of cell suspension can then be removed for cell culturing in DMEM-low glucose medium contain 10% FBS for two weeks. Flow cytometric analysis is subsequently used for the identification of cellular characteristics after cell labeling with appropriate antibodies. The flow cytometric characteristics and typical morphology of ADSC are also shown in Fig. 9.1. A previous study comparing the phenotypes of different mesenchymal stem cells isolated from human term placental chorionic villi, umbilical cord, adult bone marrow and adipose tissue demonstrated that, although the phenotypes were mostly similar among stem cells of different origins, vascular cell adhesion molecule 1 (VCAM-1) (i.e. CD106) was highly expressed on chorionic villi-derived mesenchymal stem cells, whereas it was moderately expressed on bone marrow-derived mesenchymal stem cells and absent on ADSC (Yang et al. 2013). Another study also showed consistent results (Zhu et al. 2012).

9.2.4 *Automated Devices for Adipose-Derived Stem Cell Isolation*

Compared with other tissues from which stem cells are isolated, adipose tissue has been shown to have at least two log greater concentrations of available stem and progenitor cells. This knowledge enables the direct utilization of these useful cellular elements without prior ex vivo expansion (Hicok and Hedrick 2011). Indeed, the Celution system, which is a closed, commercially available automated platform for adipose tissue processing for the isolation of adipose-derived stem and progenitor cells, has been described in 2011 (Hicok and Hedrick 2011). The system has been reported to take only 2.5 h for processing and successfully applied clinically (Marino et al. 2013). The "stromal vascular fraction" (SVF) thus obtained comprises both live and dead cells. Therefore, one noteworthy concern is that the cell debris may contribute to subsequent inflammatory responses that would potentially alter cell differentiation (Ye and Gimble 2011). Accordingly, several approaches have been proposed for retrieving the viable cells from SVF, including fluorescence-activated cell sorting (FACS), magnetic activated cell sorting (MACS), and dielectrophoresis (Wu and Morrow 2012). The former two involve the use of antibodies, while the latter retrieves life cells based on the presence of charge on their surface.

9.3 ADSC as a Therapeutic Option Against Stroke: Principles and Mechanisms

9.3.1 Therapeutic Actions of ADSC Implicated in Pathophysiological Changes of Stroke

The therapeutic role of ADSC against stroke could best be understood by reviewing the essential pathological changes and the physiological recovery mechanisms involved. Ischemia-induced inflammatory responses in stroke involve not just the neurons, but also other components of the neurovascular unit (del Zoppo 2009). This finding underscores the importance of immunomodulation in the management of stroke instead of merely restoring tissue perfusion (Iadecola and Anrather 2011). In addition, investigation of brain recovery from ischemic stroke has revealed the plasticity of the repairing process that involves axonal outgrowth and myelination (Ueno et al. 2012). Moreover, beside necrosis, apoptosis initiated after the stroke attack also results in irreversible loss of cellular elements in the central nervous system (Ouyang and Giffard 2013). Furthermore, although resuming patency of the obstructed vessel through fibrinolysis or angioplasty theoretically salvages the region at risk of ischemic infarction, the resulting ischemia-reperfusion injury actually triggers a cascade of inflammatory events (Iadecola and Anrather 2011; Liu et al. 2014). The major contributor to injuries following reperfusion is the reactive oxygen species (ROS) generated both from inflammatory cells and damaged mitochondria (Manzanero et al. 2013).

Pathologically, similar to the microscopic changes observed in animal models of stroke, evidence of reactive gliosis has been reported in human subjects after ischemic stroke including increased numbers of glial fibrillary acidic protein (GFAP)-positive reactive astrocytes and ED1-positive activated microglia as well as enhanced expression of chondroitin sulphate proteoglycans (CSPG) in the cortical penumbra regions (Huang et al. 2014). Hence, stroke involves a series of pathological changes that require a number of corresponding measures for the subsequent repairing. This is reflected in the results of a previous study that demonstrated the activation of hundreds of genes responsible for not only tissue repair, but also nervous system development and cell proliferation both in the penumbra and core of infarct as early as 24 h after ischemic stroke in rats (Ramos-Cejudo et al. 2012), highlighting the complexity of the repairing process. Since it is proposed that stem cells, which are known to participate in physiological tissue repair in various organs, may have a significant role to play in the recovery process after stroke (Gutierrez-Fernandez et al. 2012), numerous previous studies have been conducted to investigate the therapeutic potential of ADSC using the known recovery mechanisms of stroke as referring parameters.

9.3.2 Observed Therapeutic Effects of ADSC Against Stroke

To date, most results of the therapeutic use of ADSC against stroke came from animal studies for which middle cerebral artery occlusion (MCAO) is the commonly used model. The parameters for assessment were based on the established pathological changes after stroke at molecular, cellular, and functional levels. For instance, the findings of increased levels of chemokine receptor type 4 (CXCR4), stromal cell-derived factor 1 (SDF-1), IL-8/Gro, Doublecortin (DCX) (i.e., marker of migrating neuroblasts), von Willebran factor (vWF), and endothelial cell markers as well as enhanced microvessel proliferation after ADSC treatment in a rat ischemic stroke model in one study (Leu et al. 2010), together with consistent observation of augmented expressions of basic fibroblast growth factor (bFGF) and VEGF with enhanced angiogenesis in the brain in another animal investigation (Wang et al. 2008), highlight the roles of ADSC in nerve repair and revascularization in the ischemic brain. Reinforcing evidence was provided by another study that demonstrated elevated levels of VEGF, synaptophysin (SYP), oligodendrocyte (Olig-2) and neurofilament (NF) in rats after ADSC treatment compared to those in untreated animals 14 days after MCAO (Gutierrez-Fernandez et al. 2013b). The reduction in expression of GFAP in the previous studies also signifies an amelioration of reactive gliosis after ADSC treatment (Leu et al. 2010; Gutierrez-Fernandez et al. 2013b; Jiang et al. 2014). Besides, the suppressed mRNA expressions of Bax and caspase 3 as well as the increased expression of Bcl-2 in animals with stroke after ADSC treatment compared to those in the untreated group suggest an anti-apoptotic function of ADSC (Leu et al. 2010; Jiang et al. 2014). On the other hand, intravenous infusion of human ADSC has also been reported to attenuate neurological deficits (Kim et al. 2007; Yang et al. 2012), brain edema, atrophy, glial proliferation, inflammation, and apoptosis (Kim et al. 2007) in a rat model of hemorrhagic stroke.

However, the effect of ADSC treatment on infarct volume after experimental ischemic stroke is equivocal. Although one study demonstrated a significant reduction (Leu et al. 2010), other studies demonstrated no notable change in infarct volume (Gutierrez-Fernandez et al. 2013b; Jiang et al. 2014) despite the same number of cells being administered each time (2×10^6) and the unanimous findings of significantly improved neurological function, reduced cell death, and enhanced cellular proliferation in all studies (Leu et al. 2010; Gutierrez-Fernandez et al. 2013b; Jiang et al. 2014). The discrepancies in infarct volume among the studies may partly be explained by the differences in the choice of ligation procedure (i.e. permanent (Gutierrez-Fernandez et al. 2013b) vs. transient (Leu et al. 2010; Jiang et al. 2014)), the timing and frequency of ADSC administration (once at 30 min after stroke (Gutierrez-Fernandez et al. 2013b) vs. 3 times at 0, 12 and 24 h after stroke (Leu et al. 2010) vs. once at 3 days after stroke induction (Jiang et al. 2014)), the time of sacrificing animals for histological analysis after induction (14 days (Gutierrez-Fernandez et al. 2013b) vs. 21 days (Leu et al. 2010) vs. 28 days (Jiang et al. 2014)), and the route of ADSC administration (systemic intravenous

(Leu et al. 2010; Gutierrez-Fernandez et al., 2013b) vs. intra-carotid arterial (Jiang et al. 2014)). Therefore, although the timing of sacrificing animals and the route of cell injection do not seem to be the significant causes of discrepancies in the size of infarct, it appears that an early timing and increased frequency of ADSC administration (Leu et al. 2010) offered significant benefit in the reduction of infarct volume in a rodent experimental setting of ischemic stroke. Accordingly, a meta-analysis demonstrated that the efficacy in structural restoration drops by 1.5 % for each day delay in treatment and that a significant dose-response relationship exists between the number of stem cells administered and the improvement of structural outcome after ischemic stroke (Lees et al. 2012). Another interesting comparison based on animal experimentation between the therapeutic effects of autologous and allogeneic cells on structural and functional outcomes after ischemic stroke revealed that the former is more effective in preserving structural integrity, while the latter is more beneficial for functional outcome (Lees et al. 2012).

9.3.3 Mechanisms Underlying Therapeutic Actions of ADSC from Experimental Studies

Taken together, experimental investigations, both in vitro and in vivo, have provided significant insight into some of the mechanisms involved in repair of the nervous system after stroke. The mechanisms underlying the above-mentioned therapeutic benefits of ADSC against stroke can be summarized into (i) paracrine effects, (ii) transdifferentiation, and (iii) immunomodulation (Fig. 9.2).

9.3.3.1 Paracrine Effects

Although several previous experimental studies have demonstrated the presence of ADSC in the brain up to several weeks after being administered, the scarcity of stem cells in brain tissue could not account for the observed therapeutic outcomes (Leu et al. 2010; Jiang et al. 2014). The finding of stem cells not yet embedded into the brain tissue (Gutierrez-Fernandez et al. 2011; Ikegame et al. 2011) also precludes the possibility of their direct participation as fully functional neurons, implying their help through other mechanisms in the recovery process (Gutierrez-Fernandez et al. 2011). In concert with that finding, ADSC has been reported to produce a number of trophic factors including VEGF, angiopoietin-1, and HGF (Ikegame et al. 2011), insulin-like growth factor-1 (IGF-1) (Wei et al. 2009), TGF-β1 (Melief et al. 2013b), bone morphogenetic protein 2 (BMP2) and fibroblast growth factor 2 (FGF2) (Moriyama et al. 2012) as well as nervous system-related molecules including nerve growth factor (Banas et al. 2008; Ikegame et al. 2011), brain-derived neurotrophic factor (BDNF) (Iadecola and Anrather 2011; Liu et al. 2014), GFAP, nestin, and microtubule-associated protein 2 (MAP2) (Yang et al. 2011). Therefore, based on the actions of these trophic factors, it is rational to attribute the observed enhancement of angiogenesis and neurogenesis as well as the abatement

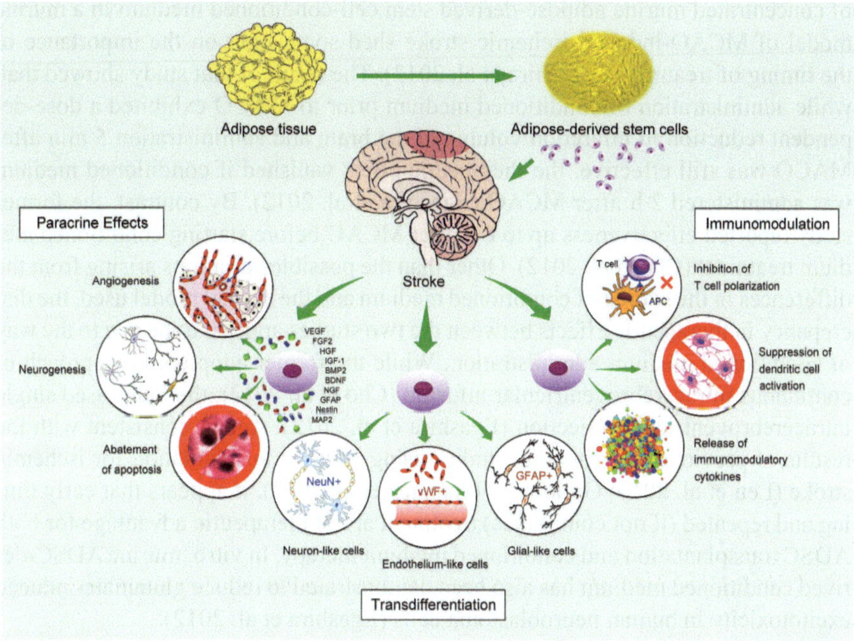

Fig. 9.2 Summary of reported mechanisms underlying adipose-derived stem cell treatment for stroke. The three major mechanisms by which adipose-derived stem cells exert therapeutic functions include paracrine effects, transdifferentiation, and immunomodulation. The paracrine effects stem from the release of a variety of trophic factors from stem cells that elicit a number of biological responses such as angiogenesis, neurogenesis, and abatement of apoptosis. Transdifferentiation of stem cells involves the transformation of implanted stem cells into specific cellular elements with distinct functions and cell markers (e.g., neuron-like, endothelium-like, or glial-like cells). Immunomodulation includes stem cell-mediated modification of the immunological system, such as inhibition of T cell polarization for alleviating immune responses, suppression of transformation of monocytes to antigen-presenting immunogenic cells (e.g., dendritic cells) for inducing tolerance, and the release of various immunomodulatory cytokines for suppressing inflammatory reactions. *APC*: Antigen-presenting cell; *VEGF*: Vascular endothelial growth factor; *FGF2*: Fibroblast growth factor 2; *HGF*: Hepatocyte growth factor; *IGF-1*: Insulin-like growth factor-1; *BMP2*: Bone morphogenetic protein 2; *BDNF*: Brain-derived neurotrophic factor; *NGF*: Nerve growth factor; GFAP: Glial fibrillary acidic protein; *MAP2*: Microtubule-associated protein 2; *NeuN*: Neuronal nuclei; *vWF*: von Willebran factor

of apoptosis to the paracrine effects of the administered ADSC (Leu et al. 2010; Gutierrez-Fernandez et al. 2012).

Consistently, another intriguing finding is the discovery of therapeutic effects against stroke using cell-free ADSC culture medium (Cho et al. 2012; Egashira et al. 2012). One study applying human adipose-derived stem cell-conditioned medium to the lateral ventricle of a rat model of ischemic stroke 8 h after MCAO continuously for 7 days demonstrated not only a reduction of infarction volume and preservation of motor function, but also enhanced endothelial cell proliferation, reduced neural cell apoptosis, and suppressed astrogliosis in the penumbra regions (Cho et al. 2012). Another similar study using intracerebroventricular administration of concentrated murine adipose-derived stem cell-conditioned medium in a murine model of MCAO-induced ischemic stroke shed some light on the importance of the timing of treatment (Egashira et al. 2012). The result of that study showed that, while administration of conditioned medium prior to MACO exhibited a dose-dependent reduction in infarction volume of the brain and administration 5 min after MACO was still effective, the therapeutic effect vanished if conditioned medium was administered 2 h after MCAO (Egashira et al. 2012). By contrast, the former study reported effectiveness up to 8 h after MCAO before starting conditioned medium treatment (Cho et al. 2012). Other than the possible variations arising from the differences in the source of conditioned medium and the animal model used, the discrepancy in therapeutic effects between the two studies appears to be due to the way of conditioned medium administration. While the former adopted the approach of continuous intracerebroventricular infusion (Cho et al. 2012), the latter used single intracerebroventricular injection (Egashira et al. 2012). Again, consistent with the results of previous experimental studies using ADSC transplantation for ischemic stroke (Leu et al. 2010; Gutierrez-Fernandez et al. 2012), it appears that early timing and repeated (if not continuous) treatment are of therapeutic advantage for both ADSC transplantation and conditioned medium therapy. In vitro, murine ADSC-derived conditioned medium has also been demonstrated to reduce glutamate-induced excitotoxicity in human neuroblastoma cells (Egashira et al. 2012).

9.3.3.2 Transdifferentiation

The role of direct cell participation regarding the use of ADSC for the treatment of ischemic stroke remains controversial. Previous studies using bone marrow-derived mesenchymal stem cells demonstrated that physical presence of the infused stem cells depends on the route of administration. Implantation of stem cells in the injured brain was evident when the cells were given through the carotid artery (Gutierrez-Fernandez et al. 2011; Jiang et al. 2014) but not through the intravenous route (Gutierrez-Fernandez et al. 2011). Neurological deficits, however, were improved regardless of the presence of implanted stem cells in the brain (Gutierrez-Fernandez et al. 2011; Jiang et al. 2014), raising the question regarding the therapeutic significance of stem cell implantation in stroke. Indeed, it has been shown that only a small fraction (around 0.02 %) of intravenously administered bone marrow-derived hematopoietic stem cells migrate to the ischemic brain, and most of the transplanted cells express microglial but not neural protein markers (Schwarting et al. 2008). For ADSC, while a study failed to identify evidence of migration or implantation of cells into the damaged brain after their intravenous injection in an animal model of stroke despite significant functional recovery (Gutierrez Fernandez et al. 2013b), other experimental studies (Kim et al. 2007; Leu et al. 2010; Yang et al. 2012) have demonstrated presence of the transplanted ADSC several weeks after intravenous administration with the expression of von Willebran factor, a marker of endothelial cell (Kim et al. 2007; Leu et al. 2010). Another study using ADSC to treat a rat model of hemorrhagic stroke through right lateral cerebral ventricular injection demonstrated the differentiation of the infused ADSC into neuron-like (NeuN+) and glial-like cells (GFAP+) in region surrounding the hematoma (Chen et al. 2012). Despite the relatively small number of ADSC to explain the overall functional recovery in the reported studies, their presence signifies "transdifferentitation" as a possible mechanism underlying the positive therapeutic impact (Gutierrez-Fernandez et al. 2013a). Indeed, the capacity of neural differentiation for ADSC has been extensively investigated (Cardozo et al. 2010; Kompisch et al. 2010; Liao et al. 2010; Qian et al. 2010; Abdanipour et al. 2011; Yu et al. 2011; Ahmadi et al. 2012). It has also been reported that, compared with bone marrow-derived mesenchymal stem cells, ADSC have superior neurogenic potential (Kang et al. 2004). Consistently, previous studies using ADSC after induced neural differentiation for treating experimental ischemic stroke were also found to be effective in improving functional recovery (Kang et al. 2003b; Yang et al. 2011). On the other hand, another finding of interest is the requirement for direct physical contact between human ADSC and murine neural stem cells in vitro for induction of neuronal differentiation of the latter, further emphasizing the existence of a mechanism that involves cell-cell interaction other than that of transdifferentiation and paracrine effects in promoting neurogenesis (Kang et al. 2003a).

9.3.3.3 Immunomodulation

Taking into account the immunological nature of stroke-elicited damage and the subsequent repairing process (Iadecola and Anrather 2011), it is not surprising to find that ADSC exert their therapeutic actions at least partly through immunomodulation. Indeed, ADSC have been reported to produce a variety of immunomodulatory cytokines, including IL-1R, IL-6, IL-8, IL-18, toll-like receptor (TLR)-4, TGF-β1, plasminogen activator inhibitor-1 (PAI-1), G-CSF, GM-CSF, and monocyte chemotactic protein 1 (Banas et al. 2008; Leu et al. 2010; Ikegame et al. 2011; Melief et al. 2013b). Moreover, ADSC have been shown to suppress the differentiation of monocytes towards antigen-presenting immunogenic cells and promote differentiation towards an anti-inflammatory IL-10-producing cell type through the production of IL-6 (Melief et al. 2013a). Consistently, coculturing ADSC with allogeneic dendritic cells revealed that ADSC could negatively modulate immunity and induce immune tolerance through downregulating costimulatory molecules (i.e., CD80, CD83, CD86, and secretion of IL-12 and tumor necrosis factor (TNF)-alpha), while induce dendritic cell tolerance through upregulating indoleamine-2,3-dioxygenase (IDO). Cocultured dendritic cells were also found to inhibit CD4+ T cell activation and naive T cells toward Th1 helper cell polarization (Peng et al. 2012). Again, another credit given to ADSC as compared with bone marrow-derived mesenchymal stem cells in the aspect of immunomodulation in stroke treatment is the finding of a higher immunomodulatory capacity in the former than that in the latter (Melief et al. 2013b).

9.4 Clinical Use of ADSC Against Stroke: Present Status, Perspectives, and Limitations

9.4.1 Clinical Application of ADSC: Probabilities and Possibilities

Given the promising experimental outcomes of applying stem cells to the treatment of stroke and the in-depth understanding of the underlying mechanisms, a number of clinical trials are either reported or still on-going in recent years despite the majority of them are small, nonrandomized, and uncontrolled. The cells administered included bone marrow mononuclear cells (Correa et al. 2005; Li et al. 2013b), bone marrow-derived mesenchymal stem cells, (Bang et al. 2005; Suarez-Monteagudo et al. 2009; Lee et al. 2010; Bringas et al. 2011; Honmou et al. 2011), human teratocarcinoma-derived neurons (Kondziolka et al. 2000), peripheral blood hematopoietic progenitor/stem cells (Chen et al. 2014a), umbilical cord-derived mesenchymal stem cells (Han et al. 2011; Jiang et al. 2013), as well as human (Rabinovich et al. 2005) and porcine fetal cells (Savitz et al. 2005). Except for premature termination of the study adopting porcine fetal cells because of overt complications (Savitz et al. 2005), the results of other published trials support the safety and effectiveness of stem cell/progenitor cells as a therapeutic tool in the clinical setting of ischemic and hemorrhagic stroke as reflected in the overall significantly improved neurological functions of the treated patients up to 5 years of follow-up (Lee et al. 2010). On the other hand, results on the use of ADSC in clinical trial have not been reported. To date, there is only one study still recruiting patients to explore the safety and effectiveness of applying autologous ADSC in patients after stroke on the National Institutes of Health clinical trial registry database (www.clinicaltrials.gov). Therefore, albeit optimistic, the exact therapeutic impact of ADSC on disease progression and functional recovery in the clinical setting of stroke remains to be elucidated for the years to come.

9.4.2 ADSC Against Stroke: Concerns and Speculations

Despite the promising outcomes of applying ADSC to the treatment of stroke in experimental settings, there have been serious concerns about possible tumorigenesis in the clinical scenario because of the multilineage differentiation potential of ADSC (Lee et al. 2012). A study investigating the fate of human ADSC from different human donors after being subcutaneously injected into immunodeficient SCID mice showed that the cells survived for at least 17 months with subsequent differentiation into fibroblasts of the subdermic connective tissue and into mature adipocytes of fat tissue, exclusively at the site of injection without evidence of migration or fusion with host cells (Lopez-Iglesias et al. 2011), underscoring the safety of ADSC transplantation. Moreover, the use of terminally differentiated ADSC may be a possible option for minimizing the risk especially when the protocols for in vitro transdifferentiation of ADSC into neuronal lineage have been well-documented (Cardozo et al. 2010; Kompisch et al. 2010; Liao et al. 2010; Qian et al. 2010; Abdanipour et al. 2011; Yu et al. 2011; Ahmadi et al. 2012). Indeed, the use of induced ADSC has been endorsed as a promising therapeutic option in stroke treatment (Yang et al. 2011; Shen et al. 2013). Furthermore, it has been shown that iPSC can be generated from human ADSC without transducing c-myc so that the proliferative and differentiation capacity of ADSC can be enhanced without increasing the risk of oncogenesis (Aoki et al. 2010).

On the other hand, taken into account the therapeutic advantage of early stem cell administration at the acute stage of stroke, the use of automated devices for adipose-derived stem cell isolation for direct injection without ex vivo expansion and purification may be a feasible option for daily clinical practice because of the high concentration of useful stem and progenitor cells from adipose tissue compared with other sources (Hicok and Hedrick 2011). At the other end of the spectrum, the use of gene-transfer techniques for producing stem cells over-expressing different neurotrophic factors, such as BDNF, glial derived neurotrophic factor (GDNF), or neurotrophin-3 (NT-3), has been reported to be effective options in the treatment of ischemic stroke in animal models (Chen et al. 2013). Finally, considering the wide therapeutic applicability and easy harvesting of ADSC, the establishment of autologous or allogeneic cell banks for ADSC storage to facilitate urgent or scheduled use is no longer a far-fetched idea (West et al. 2014).

Conclusions

Taken into consideration the possibility of autologous transplantation without significant reduction in therapeutic potency with the donor's age, the absence of serious ethical issues and concerns regarding disease transmission from allogeneic sources, the abundance, the relative ease of acquisition and culturing, the superior immunomodulatory function compared with stem cells from other sources, as well as the promising therapeutic efficacy in the treatment of stroke in the experimental settings, it is conceivable that ADSC will have an important role to play in the clinical setting of stroke treatment. Results from large-scaled, randomized, and well-controlled clinical trials are eagerly awaited to turn the possibility into reality.

Conflict of Interest The author declares no conflict of interests. No part of the manuscript has been previously published in any language and all illustrations are original.

References

Abdanipour A, Tiraihi T, Delshad A (2011) Trans-differentiation of the adipose tissue-derived stem cells into neuron-like cells expressing neurotrophins by selegiline. Iran Biomed J 15:113–121

Ahmadi N, Razavi S, Kazemi M, Oryan S (2012) Stability of neural differentiation in human adipose derived stem cells by two induction protocols. Tissue Cell 44:87–94

Andres RH, Horie N, Slikker W, Keren-Gill H, Zhan K, Sun G, Manley NC, Pereira MP, Sheikh LA, McMillan EL, Schaar BT, Svendsen CN, Bliss TM, Steinberg GK (2011) Human neural stem cells enhance structural plasticity and axonal transport in the ischaemic brain. Brain 134:1777–1789

Aoki T, Ohnishi H, Oda Y, Tadokoro M, Sasao M, Kato H, Hattori K, Ohgushi H (2010) Generation of induced pluripotent stem cells from human adipose-derived stem cells without c-MYC. Tissue Eng Part A 16:2197–2206

Araki R, Hoki Y, Uda M, Nakamura M, Jincho Y, Tamura C, Sunayama M, Ando S, Sugiura M, Yoshida MA, Kasama Y, Abe M (2011) Crucial role of c-Myc in the generation of induced pluripotent stem cells. Stem Cells 29:1362–1370

Aydin A, Duruksu G, Erman G, Subasi C, Aksoy A, Unal ZS, Karaoz E (2014) Neurogenic differentiation capacity of subacromial bursal tissue-derived stem cells. J Orthop Res 32:151–158

Banas A, Teratani T, Yamamoto Y, Tokuhara M, Takeshita F, Osaki M, Kawamata M, Kato T, Okochi H, Ochiya T (2008) IFATS collection: in vivo therapeutic potential of human adipose tissue mesenchymal stem cells after transplantation into mice with liver injury. Stem Cells 26:2705–2712

Bang OY, Lee JS, Lee PH, Lee G (2005) Autologous mesenchymal stem cell transplantation in stroke patients. Ann Neurol 57:874–882

Bjorklund A, Lindvall O (2000) Cell replacement therapies for central nervous system disorders. Nat Neurosci 3:537–544

Bringas ML, Suarez C, Sanchez C, Alvarez LM, Valdes P, Salazar S, Chongo D, Jahanshahi M (2011) Cognitive changes after stem cell transplantation in a patient with subcortical stroke. BMJ Case Rep 2011:bcr0320113944

Buschmann J, Gao S, Harter L, Hemmi S, Welti M, Werner CM, Calcagni M, Cinelli P, Wanner GA (2013) Yield and proliferation rate of adipose-derived stromal cells as a function of age, body mass index and harvest site-increasing the yield by use of adherent and supernatant fractions? Cytotherapy 15:1098–1105

Cairns K, Finklestein SP (2003) Growth factors and stem cells as treatments for stroke recovery. Phys Med Rehabil Clin N Am 14:S135–142

Calio ML, Marinho DS, Ko GM, Rodrigues R, Carbonel AF, Oyama LM, Ormanji M, Guirao TP, Calio PL, Reis LA, de Jesus Simoes M, do Nascimento TL, Teixeira Ferreira A, Bertoncini CR (2014) Transplantation of bone marrow mesenchymal stem cells decreases oxidative stress, apoptosis, and hippocampal damage in brain of a spontaneous stroke model. Free Radic Biol Med 70:141–154

Cardozo A, Ielpi M, Gomez D, Argibay P (2010) Differential expression of Shh and BMP signaling in the potential conversion of human adipose tissue stem cells into neuron-like cells in vitro. Gene Expr 14:307–319

Chang DJ, Oh SH, Lee N, Choi C, Jeon I, Kim HS, Shin DA, Lee SE, Kim D, Song J (2013) Contralaterally transplanted human embryonic stem cell-derived neural precursor cells (ENStem-A) migrate and improve brain functions in stroke-damaged rats. Exp Mol Med 45:e53

Chen C, Wang Y, Yang GY (2013) Stem cell-mediated gene delivering for the treatment of cerebral ischemia: progress and prospectives. Curr Drug Targets 14:81–89

Chen DC, Lin SZ, Fan JR, Lin CH, Lee W, Lin CC, Liu YJ, Tsai CH, Chen JC, Cho DY, Lee CC, Shyu WC (2014a) Intracerebral implantation of autologous peripheral blood stem cells in stroke patients: a randomized phase II study. Cell Transplant *(in press)*

Chen F, Qi Z, Luo Y, Hinchliffe T, Ding G, Xia Y, Ji X (2014b) Non-pharmaceutical therapies for stroke: mechanisms and clinical implications. Prog Neurobiol 115:246–269

Chen J, Tang YX, Liu YM, Chen J, Hu XQ, Liu N, Wang SX, Zhang Y, Zeng WG, Ni HJ, Zhao B, Chen YF, Tang ZP (2012) Transplantation of adipose-derived stem cells is associated with neural differentiation and functional improvement in a rat model of intracerebral hemorrhage. CNS Neurosci Ther 18:847–854

Chia CT, Theodorou SJ (2012) 1000 consecutive cases of laser-assisted liposuction and suction-assisted lipectomy managed with local anesthesia. Aesthetic Plast Surg 36:795–802

Cho YJ, Song HS, Bhang S, Lee S, Kang BG, Lee JC, An J, Cha CI, Nam DH, Kim BS, Joo KM (2012) Therapeutic effects of human adipose stem cell-conditioned medium on stroke. J Neurosci Res 90:1794–1802

Choudhery MS, Badowski M, Muise A, Pierce J, Harris DT (2014) Donor age negatively impacts adipose tissue-derived mesenchymal stem cell expansion and differentiation. J Transl Med 12:8

Correa P, Felix R, Mendonca ML, Freitas G, Azevedo J, Dohmann H, Alves S, Mesquita C (2005) Dual-head coincidence gamma camera FDG-PET before and after autologous bone marrow mononuclear cell implantation in ischaemic stroke. Eur J Nucl Med Mol Imaging 32:999

Cuevas-Diaz Duran R, Gonzalez-Garza MT, Cardenas-Lopez A, Chavez-Castilla L, Cruz-Vega DE, Moreno-Cuevas JE (2013) Age-related yield of adipose-derived stem cells bearing the low-affinity nerve growth factor receptor. Stem Cells Int 2013:372164

De Barros S, Dehez S, Arnaud E, Barreau C, Cazavet A, Perez G, Galinier A, Casteilla L, Planat-Benard V (2013) Aging-related decrease of human ASC angiogenic potential is reversed by hypoxia preconditioning through ROS production. Mol Ther 21:399–408

del Zoppo GJ (2009) Inflammation and the neurovascular unit in the setting of focal cerebral ischemia. Neuroscience 158:972–982

Del Zoppo GJ, Saver JL, Jauch EC, Adams HP, Jr. (2009) Expansion of the time window for treatment of acute ischemic stroke with intravenous tissue plasminogen activator: a science advisory from the American Heart Association/American Stroke Association. Stroke 40:2945–2948

Ding DC, Chou HL, Hung WT, Liu HW, Chu TY (2013) Human adipose-derived stem cells cultured in keratinocyte serum free medium: Donor's age does not affect the proliferation and differentiation capacities. J Biomed Sci 20:59

Drury-Stewart D, Song M, Mohamad O, Guo Y, Gu X, Chen D, Wei L (2013) Highly efficient differentiation of neural precursors from human embryonic stem cells and benefits of transplantation after ischemic stroke in mice. Stem Cell Res Ther 4:93

Egashira Y, Sugitani S, Suzuki Y, Mishiro K, Tsuruma K, Shimazawa M, Yoshimura S, Iwama T, Hara H (2012) The conditioned medium of murine and human adipose-derived stem cells exerts neuroprotective effects against experimental stroke model. Brain Res 1461:87–95

Feigin VL, Forouzanfar MH, Krishnamurthi R, Mensah GA, Connor M, Bennett DA, Moran AE, Sacco RL, Anderson L, Truelsen T, O'Donnell M, Venketasubramanian N, Barker-Collo S, Lawes CM, Wang W, Shinohara Y, Witt E, Ezzati M, Naghavi M, Murray C (2014) Global and regional burden of stroke during 1990–2010: findings from the Global Burden of Disease Study 2010. Lancet 383:245–254

Feisst V, Brooks AE, Chen CJ, Dunbar PR (2014) Characterization of mesenchymal progenitor cell populations directly derived from human dermis. Stem Cells Dev 23(6):631–642

Friedenstein AJ, Deriglasova UF, Kulagina NN, Panasuk AF, Rudakowa SF, Luria EA, Ruadkow IA (1974) Precursors for fibroblasts in different populations of hematopoietic cells as detected by the in vitro colony assay method. Exp Hematol 2:83–92

Gao S, Zhao P, Lin C, Sun Y, Wang Y, Zhou Z, Yang D, Wang X, Xu H, Zhou F, Cao L, Zhou W, Ning K, Chen X, Xu J (2013) Differentiation of human-adipose derived stem cells into neuron-like cells which are compatible with photocurable three-dimensional scaffolds. Tissue Eng Part A 20(7–8):1271–1284

Gardner OF, Archer CW, Alini M, Stoddart MJ (2013) Chondrogenesis of mesenchymal stem cells for cartilage tissue engineering. Histol Histopathol 28:23–42

Go AS, Mozaffarian D, Roger VL, Benjamin EJ, Berry JD, Borden WB, Bravata DM, Dai S, Ford ES, Fox CS, Franco S, Fullerton HJ, Gillespie C, Hailpern SM, Heit JA, Howard VJ, Huffman MD, Kissela BM, Kittner SJ, Lackland DT, Lichtman JH, Lisabeth LD, Magid D, Marcus GM, Marelli A, Matchar DB, McGuire DK, Mohler ER, Moy CS, Mussolino ME, Nichol G, Paynter NP, Schreiner PJ, Sorlie PD, Stein J, Turan TN, Virani SS, Wong ND, Woo D, Turner MB (2013) Heart disease and stroke statistics–2013 update: a report from the American Heart Association. Circulation 127:e6–e245

Gutierrez-Fernandez M, Rodriguez-Frutos B, Alvarez-Grech J, Vallejo-Cremades MT, Exposito-Alcaide M, Merino J, Roda JM, Diez-Tejedor E (2011) Functional recovery after hematic

administration of allogenic mesenchymal stem cells in acute ischemic stroke in rats. Neuroscience 175:394–405
Gutierrez-Fernandez M, Fuentes B, Rodriguez-Frutos B, Ramos-Cejudo J, Vallejo-Cremades MT, Diez-Tejedor E (2012) Trophic factors and cell therapy to stimulate brain repair after ischaemic stroke. J Cell Mol Med 16:2280–2290
Gutierrez-Fernandez M, Rodriguez-Frutos B, Otero-Ortega L, Ramos-Cejudo J, Fuentes B, Diez-Tejedor E (2013a) Adipose tissue-derived stem cells in stroke treatment: from bench to bedside. Discov Med 16:37–43
Gutierrez-Fernandez M, Rodriguez-Frutos B, Ramos-Cejudo J, Teresa Vallejo-Cremades M, Fuentes B, Cerdan S, Diez-Tejedor E (2013b) Effects of intravenous administration of allogenic bone marrow- and adipose tissue-derived mesenchymal stem cells on functional recovery and brain repair markers in experimental ischemic stroke. Stem Cell Res Ther 4:11
Hacke W, Donnan G, Fieschi C, Kaste M, von Kummer R, Broderick JP, Brott T, Frankel M, Grotta JC, Haley EC Jr Kwiatkowski T, Levine SR, Lewandowski C, Lu M, Lyden P, Marler JR, Patel S, Tilley BC, Albers G, Bluhmki E, Wilhelm M, Hamilton S, Investigators AT, Investigators ET, Investigators Nr-PSG (2004) Association of outcome with early stroke treatment: pooled analysis of ATLANTIS, ECASS, and NINDS rt-PA stroke trials. Lancet 363:768–774
Han H, Chang SK, Chang JJ, Hwang SH, Han SH, Chun BH (2011) Intrathecal injection of human umbilical cord blood-derived mesenchymal stem cells for the treatment of basilar artery dissection: a case report. J Med Case Rep 5:562
Hicok KC, Hedrick MH (2011) Automated isolation and processing of adipose-derived stem and regenerative cells. Methods Mol Biol 702:87–105
Honmou O, Houkin K, Matsunaga T, Niitsu Y, Ishiai S, Onodera R, Waxman SG, Kocsis JD (2011) Intravenous administration of auto serum-expanded autologous mesenchymal stem cells in stroke. Brain 134:1790–1807
Huang L, Wu ZB, Zhuge Q, Zheng W, Shao B, Wang B, Sun F, Jin K (2014) Glial scar formation occurs in the human brain after ischemic stroke. Int J Med Sci 11:344–348
Iadecola C, Anrather J (2011) The immunology of stroke: from mechanisms to translation. Nat Med 17:796–808
Ikegame Y, Yamashita K, Hayashi S, Mizuno H, Tawada M, You F, Yamada K, Tanaka Y, Egashira Y, Nakashima S, Yoshimura S, Iwama T (2011) Comparison of mesenchymal stem cells from adipose tissue and bone marrow for ischemic stroke therapy. Cytotherapy 13:675–685
Janjanin S, Djouad F, Shanti RM, Baksh D, Gollapudi K, Prgomet D, Rackwitz L, Joshi AS, Tuan RS (2008) Human palatine tonsil: a new potential tissue source of multipotent mesenchymal progenitor cells. Arthritis Res Ther 10:R83
Jiang Y, Zhu W, Zhu J, Wu L, Xu G, Liu X (2013) Feasibility of delivering mesenchymal stem cells via catheter to the proximal end of the lesion artery in patients with stroke in the territory of the middle cerebral artery. Cell Transplant 22:2291–2298
Jiang W, Liang G, Li X, Li Z, Gao X, Feng S, Wang X, Liu M, Liu Y (2014) Intracarotid transplantation of autologous adipose-derived mesenchymal stem cells significantly improves neurological deficits in rats after MCAo. J Mater Sci Mater Med 25:1357–1366
Kang SK, Jun ES, Bae YC, Jung JS (2003a) Interactions between human adipose stromal cells and mouse neural stem cells in vitro. Brain Res Dev Brain Res 145:141–149
Kang SK, Lee DH, Bae YC, Kim HK, Baik SY, Jung JS (2003b) Improvement of neurological deficits by intracerebral transplantation of human adipose tissue-derived stromal cells after cerebral ischemia in rats. Exp Neurol 183:355–366
Kang SK, Putnam LA, Ylostalo J, Popescu IR, Dufour J, Belousov A, Bunnell BA (2004) Neurogenesis of Rhesus adipose stromal cells. J Cell Sci 117:4289–4299
Kim JM, Lee ST, Chu K, Jung KH, Song EC, Kim SJ, Sinn DI, Kim JH, Park DK, Kang KM, Hyung Hong N, Park HK, Won CH, Kim KH, Kim M, Kun Lee S, Roh JK (2007) Systemic transplantation of human adipose stem cells attenuated cerebral inflammation and degeneration in a hemorrhagic stroke model. Brain Res 1183:43–50
Kompisch KM, Lange C, Steinemann D, Skawran B, Schlegelberger B, Muller R, Schumacher U (2010) Neurogenic transdifferentiation of human adipose-derived stem cells? a critical

protocol reevaluation with special emphasis on cell proliferation and cell cycle alterations. Histochem Cell Biol 134:453–468

Kondziolka D, Wechsler L, Goldstein S, Meltzer C, Thulborn KR, Gebel J, Jannetta P, DeCesare S, Elder EM, McGrogan M, Reitman MA, Bynum L (2000) Transplantation of cultured human neuronal cells for patients with stroke. Neurology 55:565–569

Kranz A, Wagner DC, Kamprad M, Scholz M, Schmidt UR, Nitzsche F, Aberman Z, Emmrich F, Riegelsberger UM, Boltze J (2010) Transplantation of placenta-derived mesenchymal stromal cells upon experimental stroke in rats. Brain Res 1315:128–136

Kuhl SJ, Kuhl M (2013) On the role of Wnt/beta-catenin signaling in stem cells. Biochim Biophys Acta 1830:2297–2306

Lakshmipathy U, Verfaillie C (2005) Stem cell plasticity. Blood Rev 19:29–38

Lee JS, Hong JM, Moon GJ, Lee PH, Ahn YH, Bang OY (2010) A long-term follow-up study of intravenous autologous mesenchymal stem cell transplantation in patients with ischemic stroke. Stem Cells 28:1099–1106

Lee JS, Bae GY, Lee MO, Cha HJ (2012) Oncogenic challenges in stem cells and the link to cancer initiation. Arch Pharm Res 35:235–244

Lees JS, Sena ES, Egan KJ, Antonic A, Koblar SA, Howells DW, Macleod MR (2012) Stem cell-based therapy for experimental stroke: a systematic review and meta-analysis. Int J Stroke 7:582–588

Leu S, Lin YC, Yuen CM, Yen CH, Kao YH, Sun CK, Yip HK (2010) Adipose-derived mesenchymal stem cells markedly attenuate brain infarct size and improve neurological function in rats. J Transl Med 8:63

Li M, Li S, Yu L, Wu J, She T, Gan Y, Hu Z, Liao W, Xia H (2013a) Bone mesenchymal stem cells contributed to the neointimal formation after arterial injury. PLoS One 8:e82743

Li ZM, Zhang ZT, Guo CJ, Geng FY, Qiang F, Wang LX (2013b) Autologous bone marrow mononuclear cell implantation for intracerebral hemorrhage-a prospective clinical observation. Clin Neurol Neurosurg 115:72–76

Liao D, Gong P, Li X, Tan Z, Yuan Q (2010) Co-culture with Schwann cells is an effective way for adipose-derived stem cells neural transdifferentiation. Arch Med Sci 6:145–151

Liu X, Ye R, Yan T, Yu SP, Wei L, Xu G, Fan X, Jiang Y, Stetler RA, Liu G, Chen J (2013) Cell based therapies for ischemic stroke: from basic science to bedside. Prog Neurobiol 115:92–115

Liu XL, Zhang W, Tang SJ (2014) Intracranial transplantation of human adipose-derived stem cells promotes the expression of neurotrophic factors and nerve repair in rats of cerebral ischemia-reperfusion injury. Int J Clin Exp Pathol 7:174–183

Liu YP, Seckin H, Izci Y, Du ZW, Yan YP, Baskaya MK (2009) Neuroprotective effects of mesenchymal stem cells derived from human embryonic stem cells in transient focal cerebral ischemia in rats. J Cereb Blood Flow Metab 29:780–791

Lopez-Iglesias P, Blazquez-Martinez A, Fernandez-Delgado J, Regadera J, Nistal M, Miguel MP (2011) Short and long term fate of human AMSC subcutaneously injected in mice. World J Stem Cells 3:53–62

Mantovani C, Raimondo S, Haneef MS, Geuna S, Terenghi G, Shawcross SG, Wiberg M (2012) Morphological, molecular and functional differences of adult bone marrow- and adipose-derived stem cells isolated from rats of different ages. Exp Cell Res 318:2034–2048

Manzanero S, Santro T, Arumugam TV (2013) Neuronal oxidative stress in acute ischemic stroke: sources and contribution to cell injury. Neurochem Int 62:712–718

Marino G, Moraci M, Armenia E, Orabona C, Sergio R, De Sena G, Capuozzo V, Barbarisi M, Rosso F, Giordano G, Iovino F, Barbarisi A (2013) Therapy with autologous adipose-derived regenerative cells for the care of chronic ulcer of lower limbs in patients with peripheral arterial disease. J Surg Res 185:36–44

Marycz K, Krzak-Ros J, Donesz-Sikorska A, Smieszek A (2013) The morphology, proliferation rate, and population doubling time factor of adipose-derived mesenchymal stem cells cultured on to non-aqueous SiO, TiO, and hybrid sol-gel-derived oxide coatings. J Biomed Mater Res A doi:10.1002/jbm.a.35072

Mason S, Tarle SA, Osibin W, Kinfu Y, Kaigler D (2014) Standardization and safety of alveolar bone-derived stem cell isolation. J Dent Res 93:55–61

Melief SM, Geutskens SB, Fibbe WE, Roelofs H (2013a) Multipotent stromal cells skew monocytes towards an anti-inflammatory interleukin-10-producing phenotype by production of interleukin-6. Haematologica 98:888–895

Melief SM, Zwaginga JJ, Fibbe WE, Roelofs H (2013b) Adipose tissue-derived multipotent stromal cells have a higher immunomodulatory capacity than their bone marrow-derived counterparts. Stem Cells Transl Med 2:455–463

Molina CA (2011) Reperfusion therapies for acute ischemic stroke: current pharmacological and mechanical approaches. Stroke 42:S16–19

Mora-Lee S, Sirerol-Piquer MS, Gutierrez-Perez M, Gomez-Pinedo U, Roobrouck VD, Lopez T, Casado-Nieto M, Abizanda G, Rabena MT, Verfaille C, Prosper F, Garcia-Verdugo JM (2012) Therapeutic effects of hMAPC and hMSC transplantation after stroke in mice. PLoS One 7:e43683

Moriyama M, Moriyama H, Ueda A, Nishibata Y, Okura H, Ichinose A, Matsuyama A, Hayakawa T (2012) Human adipose tissue-derived multilineage progenitor cells exposed to oxidative stress induce neurite outgrowth in PC12 cells through p38 MAPK signaling. BMC Cell Biol 13:21

Moshaverinia A, Xu X, Chen C, Ansari S, Zadeh HH, Snead ML, Shi S (2014) Application of stem cells derived from the periodontal ligament or gingival tissue sources for tendon tissue regeneration. Biomaterials 35:2642–2650

Naftali-Shani N, Itzhaki-Alfia A, Landa-Rouben N, Kain D, Holbova R, Adutler-Lieber S, Molotski N, Asher E, Grupper A, Millet E, Tessone A, Winkler E, Kastrup J, Feinberg MS, Zipori D, Pevsner-Fischer M, Raanani E, Leor J (2013) The origin of human mesenchymal stromal cells dictates their reparative properties. J Am Heart Assoc 2:e000253

Oki K, Tatarishvili J, Wood J, Koch P, Wattananit S, Mine Y, Monni E, Tornero D, Ahlenius H, Ladewig J, Brustle O, Lindvall O, Kokaia Z (2012) Human-induced pluripotent stem cells form functional neurons and improve recovery after grafting in stroke-damaged brain. Stem Cells 30:1120–1133

Ouyang YB, Giffard RG (2013) microRNAs affect BCL-2 family proteins in the setting of cerebral ischemia. Neurochem Int (*in press*)

Peng W, Gao T, Yang ZL, Zhang SC, Ren ML, Wang ZG, Zhang B (2012) Adipose-derived stem cells induced dendritic cells undergo tolerance and inhibit Th1 polarization. Cell Immunol 278:152–157

Qian DX, Zhang HT, Ma X, Jiang XD, Xu RX (2010) Comparison of the efficiencies of three neural induction protocols in human adipose stromal cells. Neurochem Res 35:572–579

Rabinovich SS, Seledtsov VI, Banul NV, Poveshchenko OV, Senyukov VV, Astrakov SV, Samarin DM, Taraban VY (2005) Cell therapy of brain stroke. Bull Exp Biol Med 139:126–128

Ramos-Cejudo J, Gutierrez-Fernandez M, Rodriguez-Frutos B, Exposito Alcaide M, Sanchez-Cabo F, Dopazo A, Diez-Tejedor E (2012) Spatial and temporal gene expression differences in core and periinfarct areas in experimental stroke: a microarray analysis. PLoS One 7:e52121

Rosado-de-Castro PH, Pimentel-Coelho PM, da Fonseca LM, de Freitas GR, Mendez-Otero R (2013) The rise of cell therapy trials for stroke: review of published and registered studies. Stem Cells Dev 22:2095–2111

Russo V, Yu C, Belliveau P, Hamilton A, Flynn LE (2014) Comparison of human adipose-derived stem cells isolated from subcutaneous, omental, and intrathoracic adipose tissue depots for regenerative applications. Stem Cells Transl Med 3:206–217

Savitz SI, Dinsmore J, Wu J, Henderson GV, Stieg P, Caplan LR (2005) Neurotransplantation of fetal porcine cells in patients with basal ganglia infarcts: a preliminary safety and feasibility study. Cerebrovasc Dis 20:101–107

Schwarting S, Litwak S, Hao W, Bahr M, Weise J, Neumann H (2008) Hematopoietic stem cells reduce postischemic inflammation and ameliorate ischemic brain injury. Stroke 39:2867–2875

Seet RC, Rabinstein AA (2012) Symptomatic intracranial hemorrhage following intravenous thrombolysis for acute ischemic stroke: a critical review of case definitions. Cerebrovasc Dis 34:106–114

Shen CC, Yang YC, Chiao MT, Chan SC, Liu BS (2013) Low-level laser stimulation on adipose-tissue-derived stem cell treatments for focal cerebral ischemia in rats. Evid Based Complement Alternat Med 2013:594906

Silva FJ, Holt DJ, Vargas V, Yockman J, Boudina S, Atkinson D, Grainger DW, Revelo MP, Sherman W, Bull DA, Patel AN (2014) Metabolically active human brown adipose tissue derived stem cells. Stem Cells 32:572–581

Skvortsova VI, Gubskiy LV, Tairova RT, Povarova OV, Cheglakov IB, Holodenko RV, Holodenko IV, Yarygin KN, Yarygin VN (2008) Use of bone marrow mesenchymal (stromal) stem cells in experimental ischemic stroke in rats. Bull Exp Biol Med 145:122–128

Smith WS, English JD, Johnston SC (2011) Chapter 370. Cerebrovascular diseases. In: Longo D, Fauci A, Kasper D, Hauser S, Jameson J, Loscalzo J (eds) Harrison's principles of internal medicine, 18th edn. McGraw Hill, New York

Suarez-Monteagudo C, Hernandez-Ramirez P, Alvarez-Gonzalez L, Garcia-Maeso I, de la Cueta-ra-Bernal K, Castillo-Diaz L, Bringas-Vega ML, Martinez-Aching G, Morales-Chacon LM, Baez-Martin MM, Sanchez-Catasus C, Carballo-Barreda M, Rodriguez-Rojas R, Gomez-Fernandez L, Alberti-Amador E, Macias-Abraham C, Balea ED, Rosales LC, Del Valle Perez L, Ferrer BB, Gonzalez RM, Bergado JA (2009) Autologous bone marrow stem cell neurotransplantation in stroke patients. an open study. Restor Neurol Neurosci 27:151–161

Takahashi K, Yamanaka S (2006) Induction of pluripotent stem cells from mouse embryonic and adult fibroblast cultures by defined factors. Cell 126:663–676

Tsuji M, Taguchi A, Ohshima M, Kasahara Y, Sato Y, Tsuda H, Otani K, Yamahara K, Ihara M, Harada-Shiba M, Ikeda T, Matsuyama T (2014) Effects of intravenous administration of umbilical cord blood CD34 cells in a mouse model of neonatal stroke. Neuroscience 263C:148–158

Ueno Y, Chopp M, Zhang L, Buller B, Liu Z, Lehman NL, Liu XS, Zhang Y, Roberts C, Zhang ZG (2012) Axonal outgrowth and dendritic plasticity in the cortical peri-infarct area after experimental stroke. Stroke 43:2221–2228

Vasandan AB, Shankar SR, Prasad P, Sowmya Jahnavi V, Bhonde RR, Jyothi Prasanna S (2014) Functional differences in mesenchymal stromal cells from human dental pulp and periodontal ligament. J Cell Mol Med 18:344–354

Wang JH, Liu N, Du HW, Weng JS, Chen RH, Xiao YC, Zhang YX (2008) [Effects of adipose-derived stem cell transplantation on the angiogenesis and the expression of bFGF and VEGF in the brain post focal cerebral ischemia in rats]. Xi Bao Yu Fen Zi Mian Yi Xue Za Zhi 24:958–961

Wardlaw JM, Murray V, Berge E, Del Zoppo GJ (2009) Thrombolysis for acute ischaemic stroke. Cochrane Database Syst Rev 7:CD000213

Wei X, Zhao L, Zhong J, Gu H, Feng D, Johnstone BH, March KL, Farlow MR, Du Y (2009) Adipose stromal cells-secreted neuroprotective media against neuronal apoptosis. Neurosci Lett 462:76–79

West CC, Murray IR, Gonzalez ZN, Hindle P, Hay DC, Stewart KJ, Peault B (2014) Ethical, legal and practical issues of establishing an adipose stem cell bank for research. J Plast Reconstr Aesthet Surg 67(6):745–751

Wu AY, Morrow DM (2012) Clinical use of dieletrophoresis separation for live adipose derived stem cells. J Transl Med 10:99

Wu W, Niklason L, Steinbacher DM (2013) The effect of age on human adipose-derived stem cells. Plast Reconstr Surg 131:27–37

Yang KL, Lee JT, Pang CY, Lee TY, Chen SP, Liew HK, Chen SY, Chen TY, Lin PY (2012) Human adipose-derived stem cells for the treatment of intracerebral hemorrhage in rats via femoral intravenous injection. Cell Mol Biol Lett 17:376–392

Yang YC, Liu BS, Shen CC, Lin CH, Chiao MT, Cheng HC (2011) Transplantation of adipose tissue derived stem cells for treatment of focal cerebral ischemia. Curr Neurovasc Res 8:1–13

Yang ZX, Han ZB, Ji YR, Wang YW, Liang L, Chi Y, Yang SG, Li LN, Luo WF, Li JP, Chen DD, Du WJ, Cao XC, Zhuo GS, Wang T, Han ZC (2013) CD106 identifies a subpopulation of mesenchymal stem cells with unique immunomodulatory properties. PLoS One 8:e59354

Ye J, Gimble JM (2011) Regulation of stem cell differentiation in adipose tissue by chronic inflammation. Clin Exp Pharmacol Physiol 38:872–878

Yu JM, Bunnell BA, Kang SK (2011) Neural differentiation of human adipose tissue-derived stem cells. Methods Mol Biol 702:219–231

Zhu X, Du J, Liu G (2012) The comparison of multilineage differentiation of bone marrow and adipose-derived mesenchymal stem cells. Clin Lab 58:897–903

Zivin JA (2000) Cell transplant therapy for stroke: hope or hype. Neurology 55:467

Part II
Stem Cell Research in CNS Injuries

Chapter 10
Endogenous Neurogenesis After Traumatic Brain Injury

Michelle H. Theus and Daniel J. Liebl

Contents

Abstract Adult neurogenesis in the central nervous system (CNS) is a distinctive process that leads to the renewal of neuronal populations in brain regions such as the olfactory bulb and hippocampal dentate gyrus. The existence of self-renewing, migratory neural stem/progenitor cells (NSPCs) in the adult brain has led to discoveries about their homeostatic role in neurogenesis and injury-induced changes following CNS trauma. Expansion and ectopic migration of quiescent endogenous NSPCs is thought to stabilize the injured milieu with the potential of providing cellular replacement of damaged or lost neurons. A better understanding of how resident NSPCs are robustly activated as well as limited will provide a way forward for maximizing the potential of these cells to reconstitute the cellular architecture in an attempt to regain function after injury. Here, we will focus specifically on traumatic brain injury and its effects on the neurogenic compartments in the adult brain and the subsequent responses.

D. J. Liebl (✉)
The Miami Project to Cure Paralysis and Department of Neurological Surgery, University of Miami, 1095 NW 14th Terrace, Miami, FL 33136, USA
e-mail: dliebl@med.miami.edu

The Miami Project to Cure Paralysis, Department of Neurological Surgery, University of Miami Miller School of Medicine, 1095 NW 14th Terrace, R-48, Miami, FL 33136, USA

M. H. Theus
The Department of Biomedical Sciences and Pathobiology, Virginia-Maryland Regional College of Veterinary Medicine, 215 Duck Pond Drive, Blacksburg, VA 24061, USA

L.-R. Zhao, J. H. Zhang (eds.), *Cellular Therapy for Stroke and CNS Injuries*,
Springer Series in Translational Stroke Research, DOI 10.1007/978-3-319-11481-1_10

Abbreviations

bFGF	Basic fibroblast growth factor
BDNF	Brain-derived trophic factor
BMPs	Bone morphogenic proteins
CC	Corpus callosum
CCI	Controlled cortical impact
CNS	Central nervous system
DAI	Diffuse axonal injury
DCX	Doublecortin
DG	Dentate gyrus
EAE	Experimental autoimmune encephalomyelitis
EGF	Epidermalgrowthfactor
Epo	Erythropoietin
FG	FluoroGold
FGF-2	Fibroblast growth factor-2
GCL	Granule cell layer
GCV	Ganciclovir
GFAP	Glial fibrillary acidic protein
HSV-TK	Herpes simplex virus thymidine kinase
IGF-1	Insulin-like growth factor-1
LFP	Lateral fluid percussion
NGF	Nerve growth factor
NMDA	N-methyl-D-aspartate
NSPCs	Neural stem/progenitor cells
OPCs	Oligodendrocyte precursor cells
PDGF	Platelet-derived growth factor
RMS	Rostral migratory stream
SGZ	Subgranular zone
SHH	Sonic hedgehog
SVZ	Subventricular zone
TBI	Traumatic brain injury
TGFα	Transforming Growth Factor-α
Thbs4	Thrombospondin 4
VEGF	Vascular endothelial growth factor
YFP	Yellow fluorescent protein

10.1 Adult Neurogenesis and CNS Injury

Adult neurogenesis in the central nervous system (CNS) is a distinctive process that leads to the renewal of neuronal populations in brain regions such as the olfactory bulb and hippocampal dentate gyrus (Brus et al. 2013). Knowledge of the existence of self-renewing, migratory neural stem cells has led to discoveries about their homeostatic role in olfaction as well as in learning and memory processes in the adult brain (Zhang et al. 2008). Given this continued neuronal cell replacement capacity throughout adulthood, experimental studies have now expanded to include the response of these neurogenic regions to traumatic CNS injury. Recent studies have begun to assess the potential of these cells to generate new neurons capable of functionally counterbalancing neuronal loss or providing support of the micro-environment in and surrounding the damaged milieu following traumatic brain injury (TBI) (Sun et al. 2005; Kernie and Parent 2010). Endogenous NSPCs are also present at non-neurogenic sites including the ependyma of the central canal in the adult spinal cord, which will be discussed more in-depth in later chapters. Briefly, unlike the subventricular zone (SVZ) and dentate gyrus compartments, neural stem cells residing in the adult spinal cord do not contribute to active neurogenesis. In non-pathologic conditions, these cells are glial progenitors; they generate essentially astrocytes and oligodendrocytes but not neurons. Only when these cells are transplanted into known neurogenic sites such as the dentate gyrus are they able to produce new neurons (Horner et al. 2000; Shihabuddin et al. 2000). These studies demonstrate that the extent of endogenous neurogenesis is a function of the local microenvironment and may result in regional differences with regard to injury-induced responses generated by resident neural stem cells.

Injury to the CNS leads to expansion and ectopic migration of quiescent endogenous neural stem/progenitor cells (NSPCs). In particular, proliferation rates in the SVZ and subgranular zone (SGZ) of the dentate gurus are increased in response to traumatic, seizure-inducing, ischemic and demyelinating brain injury (Arvidsson et al. 2002; Parent et al. 2002; Picard-Riera et al. 2002b; Rice et al. 2003). Ependymal cell proliferation was also observed in response to several types of spinal cord trauma including compression, contusion and transection injury (Matthews et al. 1979; Vaquero et al. 1981; Beattie et al. 1997). Directed differentiation of these cells, however, is depended on local cues in or around the area of injury. For example, most NSPCs differentiate into astrocytes in the injured cortex or spinal cord but repopulate neuronal populations in the selectively damaged hippocampus following traumatic insult. Interestingly, oligodendrocyte differentiation has been shown to predominate in response to certain demyelinating conditions (Picard-Riera et al. 2004). Thus, it seems that local cues originating either from the lesion or from the environment in which cells migrate are important to direct their differentiation into the appropriate cell fate. Investigators are just beginning to elucidate the molecular and cellular mechanism(s) regulating these injury-induced responses. A better understanding of how resident NSPCs are robustly activated as well as limited in the context of CNS trauma will provide a way forward for maximizing the potential of these cells to reconstitute the cellular architecture in an attempt to regain function after injury. Here, we will focus specifically on traumatic brain injury and its effects on the neurogenic compartments in the adult brain and the subsequent responses.

The multi-potential nature of endogenous NSPCs makes them a key component in innovative stem cell strategies to improve TBI outcome. Although, there have been substantial improvements in the acute care for traumatic CNS injuries, there are currently no long-term clinical therapies to effectively stimulate regeneration of brain tissue that will lead to functional restoration (McArthur et al. 2004). Therapeutic strategies involving NSPCs have recently become the focus of pre-clinical stud-

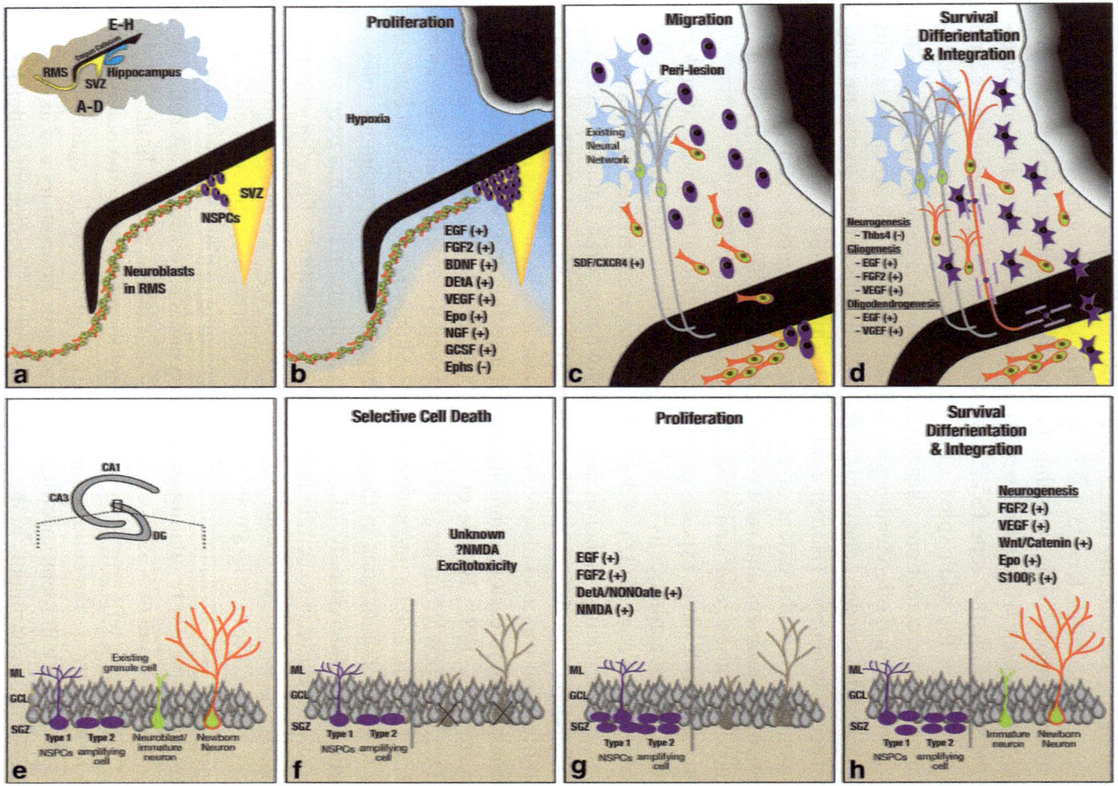

Fig. 10.1 Mechanisms regulating cellular changes in the two neurogenic compartments following traumatic brain injury (TBI). Similarities and differences exist between TBI-induced responses in the SVZ (A-D) and dentate gyrus (E-H). **a** Under naïve conditions in the SVZ-RMS NSPCs (*purple*) line the lateral ventricles and produce neuroblasts (*orange/green*) that migrate through the RMS to the olfactory bulb. **b** Following TBI, NSPCs are stimulated to proliferate and survive post-injury.Expression of multiple growth promoting molecules are enhanced and contribute to the proliferative effects. Ephrins and Eph receptors,

ies, as current efforts using a variety of neuroprotective agents have failed to show efficacy in TBI patients. Research efforts have, therefore, shifted to the potential of enhancing the endogenous response of stem cells in the brain as a possible means to significantly impact the acute and long-term management of clinical TBI. One major advantage to augmenting the endogenous NSPC response is that it alleviates the need for transplanting an exogenous cell source. The advantageous and disadvantageous of stem cell transplant will be discussed in later chapters. The endogenous stem cell response is considered by many to be too weak to provide sufficient benefits in the post-repair phases of injury; however, recent findings indicate that these resident cells play a much greater role in tissue stability than originally thought.

Traumatic brain injury has been shown to stimulate dramatic and identifiable changes in the SVZ and dentate gyrus (see Fig. 10.1). Mechanical trauma to the CNS results in the disruption of the cellular microenvironment leading to massive necrotic loss of neuronal and glia populations at the area of injury. The progressive cascade of secondary events, including changes in the vascular and extracellular matrix lead to ischemia, inflammation, excitotoxicity, free radial damage and diffuse axonal injury (DAI) which contribute to neuronal cell death. Neuronal loss can be observed in both a focal and diffuse pattern. Focal neuronal loss is typically seen within the hemorrhagic lesion of the grey matter following models of contusion injury such as the controlled cortical impact. Among diffuse injury sites, the hippocampus is known to be especially vulnerable following brain trauma. Apoptotic loss of critical neuronal populations responsible for learning and memory can occur up to a year post-injury in humans (Williams et al. 2001; Taylor et al. 2006). Interestingly, animal models of DAI have shown that despite axonal trauma in neighboring neurons, many of these cells undergo reorganization and repair suggesting they are amenable to change within the damaged milieu. Given the progress we have made in understanding endogenous neurogenesis; active repair strategies that stimulate this innate response have tremendous potential to slow progressive cell loss by protecting vulnerable cell populations, promoting plasticity and regulating environment-specific cell fate restriction for neuronal or glial replacement.

---

negative regulators, are down regulated during population expansion. Low oxygen levels (*blue*) precede proliferative changes in the SVZ-RMS and may activate key pathways necessary for proliferation and migration effects. **c** NSPCs and neuroblasts actively migrate to the peri-lesion site after TBI. While the mechanism(s) regulating this event are under investigation, the SDF/CXCR4 has been implicated. **d** Differentiation of SVZ-derived cells occurs in the damaged cortex. Several key factors that regulate gliogenesis in these cells also promote oligodendrogenesis. The survival and integration of newborn neurons remains limited under these conditions. **e** Neurogenesis in the dentate gyrus. **f** Unlike the SVZ, neuroblasts and immature neurons are depleted due to necrosis and apoptosis, the main inducer of which is still unknown. **g** Proliferation of NSPCs counterbalances immature neuronal loss in the DG and is positively regulated by several known factors. **h** Restoration of mature granule neuron numbers is mediated by NSPC expansion and can be enhanced by delivery of growth factors. *SVZ* subventricular zone, *RMS* rostral migratory stream, *ML* molecular layer, *GCL* granule cell layer, *SGZ* sub-granule zone, *EGF* epidermal growth factor, *FGF* fibroblast growth factor, *DEtA/NONOate* nitric oxide donor, *VEGF* vascular endothelial growth factor, *BDNF* brain-derived growth factor, *NGF* nerve growth factor, *Epo* erythropoietin, *SDF* stromal-derived growth factor, *GCSF* granulocyte colony stimulating factor, *Thbs4* thrombospondin-4, *NMDA* N-Methyl-D-aspartic acid

10.2 Effects of TBI on the Neurogenic Compartments

The number of NSPCs in both the SVZ and hippocampus are significantly increased after TBI, a process that is more robust in juvenile animals (Singer et al. 2011; Kleindienst et al. 2013; Taylor et al. 2013). NSPCs from the SVZ appear to have the potential to migrate to areas of focal cortical damage, while hippocampal neurogenesis occurs in the setting of diffuse injury. Overall, the number of new neurons generated after TBI remains small in comparison with astrocyte and oligodendroctye differentiation. The following sections will examine injury-induced responses in these two prominent neurogenic compartments in the adult brain.

10.2.1 Neurogenesis in the SVZ

The SVZ represents the remnant of an enlarged perinatal ventricular germinal zone which narrows during development and persists throughout adulthood as a neurogenic compartment consisting of multiple cell lineages that derive from an original self-renewing stem cell. (Tramontin et al. 2003). This relatively quiescent or slowly proliferating cell expresses glial fibrillary acidic protein (GFAP; type B cells) and differentiates to become a rapidly dividing transit-amplifying progenitor (type C cells) which generate the typically seen chain and migrating neuroblasts in the rostral migratory stream (RMS) (type A cells) (Alvarez-Buylla and Garcia-Verdugo 2002). Interestingly, during development the SVZ generates both neuronal and glial cell types of the CNS; however, only neurogenesis (birth of new neurons) persists in adulthood. The normal role for adult NSPCs in the anterior telencephalic SVZ is to supply the RMS with neuronal precursors that continually replace glomerular neurons in the olfactory bulb (Luskin 1994; Pencea et al. 2001). Continuous production of neuroblasts that migrate and integrate into the olfactory bulb circuitry is controlled by the local environment within the SVZ niche, which is separated from the ventricle cavity by ependymal cells (Lois and Alvarez-Buylla 1993). It is generally thought that enhancing the neurogenic activities of endogenous NSPCs may provide additive therapeutic benefits after brain injury.

Consistent with models of ischemia, experimental TBI induces an increase in SVZ proliferation, and a secondary redirected stream of precursors into the surrounding tissues (Salman et al. 2004). Several rodent models of TBI have been extensively used to demonstrate increases in bilateral proliferation in the adult SVZ up to 14 days after injury; namely the lateral fluid percussion (LFP) injury (Chirumamilla et al. 2002; Yoshimura et al. 2003; Urrea et al. 2007; Bye et al. 2011) and unilateral controlled cortical impact (CCI) models (Chirumamilla et al. 2002; Lu et al. 2003; Ramaswamy et al. 2005; Urrea et al. 2007; Theus et al. 2010; Radomski et al. 2013). This effect has not been tested in blast-induced or repetitive TBI injury models to date. These proliferative changes have been shown to persist at least one year following LFP injury in rats and is linked to progressive ventriculomegaly (Chen et al. 2003), whether these cells continue to migrate to areas of active neural repair at this time has yet to be determined. TBI-induced proliferation in the SVZ also appears to correlate with injury severity. Recent studies show that both mild and moderate TBI alters the SVZ-RMS niche with a greater proliferative response seen following moderate CCI injury. Interestingly, this translates into defects associated with ipsilateral but not contralateral olfactory bulb neurogenesis and impaired olfaction up to 30 days post-injury (Radomski et al. 2013). TBI-induced proliferation in the SVZ niche does not result in a corresponding increase in olfactory bulb neurogenesis. This is most likely due to the mass exodus of cells migrating from the SVZ-RMS system to the lesion and peri-lesion sites which counterbalances the increases in proliferation. Regional changes in numerous soluble neurotrophic factors such as nerve growth factor (NGF), brain-derived trophic factor (BDNF) and fibroblast growth factor-2 (FGF-2) have been implicated in stimulating proliferation in the SVZ following TBI (Oyesiku et al. 1999; Truettner et al. 1999; Yoshimura et al. 2003) (see Table 10.1). The differential interactions among these and other neurotrophins may be important in determining the final number and phenotype of the SVZ-responding cells. The morphogens Wnt, sonic hedgehog (Shh) and bone morphogenic proteins (BMPs) may also play a major role in the NSPC response to brain injury. For example, Wnts are endogenously expressed in the adult SVZ (Wexler et al. 2009) and β-catenin activity has been shown to be up regulated in proliferating progenitor cells after TBI (White et al. 2010). Whether these and other diffusible molecules, seen to be enhanced in the SVZ after TBI, are directly related to the expansion and migration effect has yet to be determined.

In addition to soluble mitogenic factors that may participate in stimulating the SVZ response to TBI, studies have demonstrated changes in the EphB3 receptor, a local inhibitory molecule that maintains homeostasis in the SVZ (Theus et al. 2010). In contrast to soluble mediators of NSPC functions, ephrins and Eph receptors are membrane-bound signaling partners of both A and B-classes that have recently been shown to tightly regulate proliferation and migration in the SVZ (Conover et al. 2000; Holmberg et al. 2005; Furne et al. 2009). The mechanisms controlling their expression in this region are not fully understood and have not been, until recently, studied in the context of brain injury. Examination of the temporal change in EphB3 expression showed a dramatic decrease which correlated with CCI-induced proliferation, cell survival and lowered oxygen conditions or hypoxia in the SVZ (Theus et al. 2010; Baumann et al. 2013). In addition to TBI-induced proliferation, cell death or turnover is reduced in the SVZ, an underappreciated and under studied cellular response in the neurogenic compartment. These studies further identified a novel mechanism controlling cell survival of the NSPCs involving EphB3 and EphA4 as dependence receptors in a cell autonomous fashion (Furne et al. 2009; Theus et al. 2010; Nelersa et al. 2012). Overall, the removal of such inhibitory control on adult neurogenesis may be an early event that allows simultaneous expansion and migration of the NSPC population in response to TBI-induced neurotrophic signals produced after trauma.

Migration of SVZ-derived neural stem and/or progenitor cells, analyzed using lentiviral, microsphere and dye labeling, has been demonstrated to occur following

Table10.1 Positive and negative regulators of adult neurogenesis in the traumatic injured brain

GF-2	(+) Proliferation, astrogenesis	(+) Proliferation, Neurogenesis	Improved Neurological score	Sun D et al., 2009 Thau-Zuchamn et al., 2012 Yoshimura et al., 2003 Laskowski et al., 2005
EtA/NONOate	(+) Proliferation	(+) Proliferation	Improved Neurological score	Lu et al., 2003; Dunyere et al., 2003
EGF	(+) Proliferation, astrogenesis, oligodendrogenesis	(+) Neurogenesis	Improved Neurological score	Thau-Zuchamn et al., 2012;Lee et al., 2010
DNF	(+) Proliferation	ND	No behavioral improvement	Truettner et al., 1999; Blaha et al.,2000
GF	(+) Proliferation	ND	ND	Oyesiku et al., 1999
/nt/b-catenin	ND	(+) Neurogenesis	ND	Zhang et al., 2013
po	(+) Cell Survival	(+) Neurogenesis/Angiogenesis	Improved Sensorimotor/cognitive function	Mahmood et al., 2007; Xiong et al., 2011; Schober et al., 2013
100B	ND	(+) Neurogenesis	Improved Cognitive recovery	Kleindienst et al., 2005
DF-CXCR4	(+) Migration	ND	ND	Younghye et al., 2013
CSF	(+) Proliferation	ND	Improved Motor function	Yang et al., 2010c
Membrane-Bound Molecules	**SVZ**	**SGZ**	**Effect**	**References**
phB3/ephrinB3	(-) Proliferation, (+) Survival	ND	Improved motor function	Theus et al., 2010; unpublished data

ND; Not Determined
-) Increased
) Decreased
GF, epidermal growth factor; FGF, fibroblast growth factor; DEtA/NONOate, nitric oxide donor; VEGF, vascular endothelial growth factor; BDNF, brain-derived growth factor; NGF, nerve growth factor; Epo, erythropoietin; SDF, stromal-derived growth ictor; GCSF, granulocyte colony stimulating factor

TBI (Salman et al. 2004; Ramaswamy et al. 2005; Saha et al. 2013). Although, the mechanism(s) regulating their recruitment are unknown, these cells migrate to the striatum, corpus callosum (CC), and peri-lesion cortex, but not the hippocampus, and differentiate mainly in glial cells, which may contribute to glial scar formation. Many of these migrating cells are positive for doublecortin (DCX), a marker of immature neuroblasts, suggesting that local neurogenesis could be driven by injury-specific extracellular cues. In the peri-lesion cortex, expression of the mature neuronal marker NeuN was seen in cells tagged prior to injury by intraventricular infusion of a lipophilic dye, suggesting that these cells were derived from the SVZ compartment. Whether neuroblasts that migrate from the SVZ can effectively replace cortical neuronal loss caused by TBI remains to be determined. Similarly, it is conceivable that migrating NSPCs could be directed toward an oligodendrocyte lineage to aid in remyelination of white matter tracks damaged by TBI. The potential for directing the phenotypic differentiation of SVZ-derived cells, that have been clearly shown to migrate to areas of injury, may become an intriguing therapeutic goal. On the other hand, the SVZ has been shown to be extremely heterogenic and the fact that progenitor populations existing in different subregions (Doetsch et al. 1997; Sundholm-Peters et al. 2004; Azim et al. 2012) may respond differently to trauma is underappreciated. Therefore, a better understanding of the cellular and molecular cues driving the directed migration and differentiation of SVZ-derived cells to areas of tissue damage is needed in the context of TBI.

Significant spontaneous functional recovery has been shown to occur following TBI (Anderson et al. 2000; Demeurisse 2000; Sinha et al. 2013); injury-induced neurogenesis has been thought to be one potential contributor to this effect (Chirumamilla et al. 2002; Chen et al. 2003; Richardson et al. 2007). Several studies have shown that a variety of pharmacologic agents increase neurogenesis and lead to improved outcomes following TBI. The drugs, however, have effects on the brain that are independent of their ability to promote neurogenesis, and it therefore becomes difficult to attribute improvements in behavior to their effects on the neurogenic compartments. Although it is well-established that SVZ progenitors are activated by injury, the relevance of this response and the mechanism(s) responsible still remain elusive. In order to demonstrate that functional recovery requires NSPC activation, more cell-specific assays are needed. Several experimental approaches have been used to test the contribution of NSPCs to spontaneous recovery after CNS injury. Specifically, the generation of genetically engineered mice that can regulate neurogenesis more directly, using an inducible system to drive the expression of the herpes simplex virus thymidine kinase (HSV-TK), have aided in testing the functional significance of NSPCs by ablating them in a temporally controlled manner (Singer et al. 2009; Sun et al. 2013). Other methods such as administration of Ara-C or cranial irradiation have also been used; however, lack of specificity and potentially toxic side effects that lead to unwanted immune activation may affect other mediators of recovery or damage not related to neurogenesis (Doetsch et al. 1999; Naylor et al. 2008; Hellstrom et al. 2009; Lau et al. 2009).

Using a genetic approach, studies have demonstrated a supportive role for NSPCs in the stabilization of the cortical microenvironment after TBI. Administration of ganciclovir (GCV) was used to ablate actively dividing NSPCs expressing HSV-TK under the control of the nestin[δ] promoter, which drives expression selectively in neural progenitors (Yu et al. 2008). Ablation of the NSPC population in the adult brain attenuated spontaneous functional recovery, exacerbated neuronal injury, promoted astrocyte hypertrophy and reduced the local glial response (unpublished findings). These findings indicate that NSPCs play a role in TBI-induced astrogenesis by contributing to glial numbers and by limiting local resident glial reactivity to support recovery of the injured tissue. This is consistent with previous reports showing that SVZ-derived astrocytes, which have different properties compared with cortical astrocytes namely high levels of thrombospondin 4 (Thbs4), represented the majority of migrating SVZ-derived NSPCs that target the acute injured cortex and not doublecortin (DCX +) neuroblasts at 2 weeks post-photothrombotic injury (Benner et al. 2013). Interestingly, Thbs4 knockout mice show a phenotypic switch of the fate labeled SVZ-derived cells present at the injury site, where DCX + cells predominated over GFAP expressing cells. Thbs4 regulates this response by directly activating Notch1 receptor. Notch signaling in neural stem cells strongly promotes gliogenesis and reduces neuronal fate. This altered cellular response resulted in abnormal glial scar formation and increased microvascular hemorrhage. It is unclear, however, whether the DCX + migrating neuroblasts survived and fully differentiated into appropriate functionally mature neurons in the damaged cortex during sub-acute and chronic phases of injury. Eventual neuronal replacement from these cells might not be successful without first stabilizing the injured environment.

It is evident from these studies that the migrating SVZ-derived NSPCs do indeed play an important role in the acute response to TBI and other types of brain injuries through astrogliosis and trophic support. It also supports the idea that highly migratory SVZ-derived NSPCs are amenable to phenotypic switching, although the potential of these cells to generate oligodendrocytes following TBI needs to be determined. During post-natal development, NSPCs generate oligodendrocytes that integrate into white matter (Levison and Goldman, 1993) which ceases in adulthood. However, reactivation following demyelinating conditions such as EAE or lysolecithin-induced demyelination has been shown to mobilize the cells to the CC and striatum where they differentiated into astrocytes and oligodendrocytes (Picard-Riera et al. 2002a; Nait-Oumesmar et al. 2008). Recent studies indicate that a subpopulation of epidermal-growth-factor (EGF)-activated type-B cells are oligodendrocyte precursor cells (OPCs) and contribute to myelin repair in the CC following a demyelinating lesion and traumatic axonal injury (Gonzalez-Perez and Alvarez-Buylla 2011; Sullivan et al. 2013). Therefore, different subpopulations of SVZ-derived cells (i.e. neural stem, transit-amplifying, OPC progenitor and neuroblast) may contribute to injury stability and repair by migrating to multiple regions of damage following trauma, a mechanism(s) that depends on the local microenvironmental cues. The potential for directing the phenotypic differentiation of endogenous NSPCs may become an intriguing therapeutic goal, once the molecular signals driving these phenomena are better understood.

To determine whether enhancing neurogenesis can improve functional outcomes following TBI, studies have examined several growth factors that regulate the expansion and differentiation of SVZ-derived cells in injured tissues. In particular, pharmacologic studies using exogenously administered growth factors have correlated increases in proliferation, survival and migration of SVZ-derived cells with greater functional recovery after TBI (Sun et al. 2009, 2010; Thau-Zuchman et al. 2010; Yang et al. 2010). Among these factors, basic fibroblast growth factor (bFGF), epidermal growth factor (EGF), platelet-derived growth factor (PDGF), BDNF, NGF, erythropoietin (Epo) and vascular endothelial growth factor (VEGF) have been widely accepted as important mediators of neurogenesis and potent mitogenic factors for NSPCs in both *in vitro* and *in vivo* settings (Alagappan et al. 2009, 2013; Mani et al. 2010; Alvarez-Palazuelos et al. 2011; Bath et al. 2012; Moore et al. 2013; Woodbury and Ikezu 2013). For instance, peripheral administration of Epo promoted NSPC proliferation, neurogenesis, oligodendrogenesis, and neurovascular remodeling following TBI in rats that led to enhanced spatial learning performance and sensory motor function (Ning et al. 2011; Xiong et al. 2011). These studies suggest that growth factor treatment could augment NSPC activity directly resulting in identifiable improvements in functional recovery. A number of studies show that diffusible trophic factors may be suitable for expanding the population of NSPCs in the early phase of repair following TBI, while others have been shown them to direct cell fate choice on select NSPC sub-populations (Alvarez-Palazuelos et al. 2011). Based on these observations, strategies to expand the entire neurogenic compartment prophylactically or immediately following injury would be ideal in order to provide greater numbers of NSPCs that are able to migrate to areas of injury and stabilize the microenvironment. Once stabilized, such treatments may be modified with delivery of additional factors better able to control neuronal or oligodendrocyte cell fate determination. Although, neuronal replacement strategies using this endogenous repair mechanism could further enhance recovery, it has yet to be determined whether these olfactory-bound NSPCs are capable of differentiating into excitatory neurons with appropriate cortical projections in traumatic injured tissues. Likewise, it is unclear whether providing activity-dependent dendritic input is necessary for long-term survival of newly generated cortical neurons.

10.2.2 Neurogenesis in the Hippocampus

While neurogenesis in the SVZ contributes to olfactory function, the generation of new neurons within the DG of the adult hippocampus is critical for spatial learning, object recognition and memory. Although selectively more vulnerable to TBI, the neurogenic region of the hippocampus may have a unique ability to locally replace damaged neurons, as this injury site is normally an area of active endogenous neurogenesis. Studies in both rodents and humans demonstrate that cells with astrocyte morphology reside in the SGZ of the DG and continually generate neuroblasts, which migrate into the granule cell layer (GCL), differentiate into mature excitatory neurons and extend axons into the CA3 region (Eriksson et al. 1998; Roy et al. 2000; Seri et al. 2001), whereas SVZ neuroblasts mainly contribute to the inhibitory interneuron population in the OB. Although some of these new neurons die by apoptosis over a few weeks, many demonstrate functional integration and evoked electrical potential within the hippocampal circuitry (Hastings and Gould 1999; Markakis and Gage 1999; van Praag et al. 2002). Unlike the SVZ, SGZ-derived NSPCs are relatively quiescent and have a much slower proliferative rate (Seri et al. 2001) and subsequent neuroblast production undergo short-distance migration into the GCL. Long distance migration does not seem to occur in pathological situations like it does in the SVZ (Picard-Riera et al. 2004; Brus et al. 2013). Additionally, temporal profiling of these cells indicate that differentially regulated, cell-autonomous factors change over time which could make these cells more or less responsive to niche cues during normal aging (Gilley et al. 2011). Many studies have highlighted the presence of increased neurogenesis in the injured DG. These studies, performed in multiple models of injury and disease, demonstrate that new SGZ-derived neurons are able to replace lost neurons resulting in restored function of the hippocampus (Nakatomi et al. 2002; Jin et al. 2003; Lu et al. 2003; Naylor et al. 2008; Ning et al. 2011).

One of the functional hallmarks of TBI is the pronounced deficit in learning and memory, which is usually associated with the histological presence of cell death in the DG, CA1 and CA3 layers (Smith et al. 1991, 1994, 1995; Hamm et al. 1993; Reeves et al. 1997; Rola et al. 2006). Because increased neurogenesis in this region has been linked to enhanced learning and memory function, injury-induced neurogenesis could function to replace damaged neurons, and thus contribute to neuronal circuit repair and restoration of neurological function in patients with TBI. Determining how this normally latent pool of NSPCs can be activated or even enhanced offers considerable potential for the development of targeted neurogenic-based therapeutics to ameliorate acute and chronic TBI-induced cognitive decline. Recent BrdU-labeling and cell lineage tracing studies using transgenic mice carrying an inducible yellow fluorescent protein (YFP) gene under the nestinδ promoter, indicate that TBI evokes a sustained increase in neurogenesis in the adult hippocampus lasting up to 6 months post-TBI and exerts differential effects acutely in hippocampal DG-derived progenitors depending on their state of lineage progression (Kernie et al. 2001; Yu et al. 2008; Singer et al. 2009). For example, through unknown mechanisms the early, nestin-expressing progenitors are stimulated to divide after CCI and LFP injuries (Rice et al. 2003; Sun et al. 2005; Yu et al. 2008), whereas the late DCX-expressing neuroblasts or immature neurons are extremely vulnerable to brain injury and undergo acute selective cell death (Yu et al. 2008; Zhou et al. 2012). Cell death is observed in the ipsilateral SGZ, GCL and occasionally in the hilus following CCI-injury. Cell death begins as early as 6 h, peaks at 24 h and reaches normal levels at 14 days post-injury (Rola et al. 2006). It appears that the time course of cellular loss of GCL neurons and SGZ neuroblasts in the DG mimics that of the cortical injured site. Conversely, few dying cells are seen in the contralateral DG. Surprisingly, while the late neuroblast progenitors are selectively vulnerable to cell death in the SGZ, cell survival is actually enhanced in the SVZ (Theus et al. 2010). The differences in their vulnerability may be related to the proximity of the neurogenic compartment to the injury, specific niche cues or intrinsic factor expression regulating cell autonomous activities.

It is currently unknown why immature neurons in the DG are so vulnerable to TBI. Recent studies of newborn neurons in the dentate gyrus show that they are electrophysiologically distinct from mature granular neurons in the dentate gyrus (Tovar and Westbrook 1999; Lopez de Armentia and Sah 2003; Ye et al. 2005; Ge et al. 2006). Newborn immature neurons are also more easily excited than mature neurons due to their unique ion channel expression (Ben-Ari 2002). Calcium influx-mediated excitotoxicity has been considered a major cause of neuronal death when neurons are challenged post-TBI with the excitatory amino acid, glutamate (Globus et al. 1995; Saatman et al. 1996). Furthermore, direct application of N-methyl-D-aspartate (NMDA) to hippocampal organotypic slices demonstrated acute excitotoxic injury in the CA1, CA3 and DG regions followed by stimulation of proliferation in the DG, suggesting that TBI-induced vulnerability and subsequent neurogenesis in the DG may, in part, be a consequence of excitotoxic injury (Bunk et al. 2010). This is consistent with *in vivo* studies using NMDA receptor antagonists which suppressed injury-neurogenesis in the DG (Bernabeu and Sharp 2000; Arvidsson et al. 2001). Excitotoxicity seemingly overlaps with other types of cell death, such as apoptosis and necrosis, but depends on the intensity threshold of the initiating stimulus. Recent studies indicate that most dying cells in the DG at 24 hours post-moderate TBI did not undergoing apoptosis but instead co-expressed RIP-1, a marker for a subtype of necrotic cell death (Chan et al. 2003). Interestingly, increasing the severity of injury increases apoptotic markers in the GCL (Colicos and Dash 1996; Clark et al. 2001).

Although acute cell loss is observed in the DG after TBI, subsequent recovery or re-population of the injured neurogenic compartment appears to occur in response to cellular loss. Recent findings indicate that the acute cell death, seen within the first week post-injury, of the late DCX-expressing committed neuroblasts can be counterbalanced by the active proliferation of the quiescent type-1 early nestin-expressing progenitors following trauma (Yu et al. 2008). Indeed, similar to the SVZ response, induction of proliferation is observed in both the contralateral and ipsilateral DG after TBI (Dash et al. 2001; Kernie et al. 2001; Lu et al. 2003; Yoshimura et al. 2003; Emery et al. 2005). Following both CCI and LFP injury models, proliferation in the DG begins as early as 2 days post-injury, peaks during the first week following injury and returns to baseline by 35 days (Rice et al. 2003; Sun et al. 2005). Interestingly, overall cell and new neuron production in the contralateral DG of TBI-injured mice is not significantly different from uninjured controls even though proliferation and DCX-positive cells are increased without any appreciable loss in cell numbers. However, in the ipsilateral hippocampus a significant increase in the number of newly generated cells, expressing neuronal markers, are predominant after longer maturation periods.

The differentiation status of these newly born cells has been examined at several time points after injury (Sun et al. 2005, 2009; Rola et al. 2006; Blaiss et al. 2011). Fate labeling studies indicate that they contribute to TBI-induced neurogenesis in the hippocampus. Previous studies have shown a four-fold increase in the number of BrdU-labeled cells appearing in the SGL at 5 days after injury; however, by 10-weeks only 46 % of them remained in the injured DG compared to 65% of the originally labeled cells in un-injured mice. This translated into a 2-fold preserved increase in newly born cells in the DG of TBI-injured mice compared to un-injured controls. Double-labeling of the cells indicated that 80% of the BrdU-labeled cells in the DG expressed NeuN and calbindin, indicative of a mature neuronal dentate phenotype; 10% were GFAP-positive for astrocytes and the remaining were undergoing cell death (Sun et al. 2007). Similar trends were observed in the control mice, where overall TBI-injured mice had significantly more newly generated neurons and astrocytes as a result of the increased proliferation in the acute phases of injury. No differences were observed in the cell fate. Injection of the retrograde neuronal tracer FluoroGold (FG) into the CA3 hippocampus resulted in greater numbers of FG labeled BrdU-positive cells in the DG at 10 weeks post-injury suggesting connectivity with the CA3 target region. Genetic fate labeling of NSPCs using YFP driven by the nestinδ promoter also showed stable DG neurogenesis at 2 months post-TBI, where YFP-labeled cells persist as NeuN-positive and GFAP-expressing progenitors in the DG (Blaiss et al. 2011). No YFP-labeled cells were observed as reactive astrocytes in the hilus or DG. Importantly, ablation of TBI-induced neurogenesis following ganciclovir treatment in the nestinδ–HSV–TK transgenic mice exacerbated the cognitive effects of TBI indicating that injury-induced neurogenesis in the DG plays an important role in restoring learning and memory function.

TBI-induced proliferation appears to have a similar temporal trend between the DG and SVZ. However, the DG has a more pronounced neurogenic response to injury as compared to the combined neurogenic and gliogenic response of the SVZ following cortical injury (Kernie et al. 2001). Unlike the SVZ response, NSPCs residing in the injured hippocampus remain localized and active in their neurogenic niche and appear to retain their differentiation and migration patterns during repair. Interestingly, *in vitro* studies show a similar cell autonomous role between SVZ- and DG-derived NSPCs, where they have similar growth, migration and differentiation potentials in culture. Transplantation of DG-derived NSPCs to the SVZ compartments demonstrates that environmental influences are paramount, since these newly transplanted DG-derived NSPCs migrated to the OB and differentiated into interneurons similar to SVZ-derived NSPCs (Suhonen et al. 1996). Thus, it is thought that differential gene expression within the specialized DG niche, such as those regulating ECM adhesion, cell-cell contacts, chemoattractive and repulsive cues, regulate DG homeostasis and could play important roles in DG-derived NSPCs following TBI. The possibility of enhancing these regulatory cues in the molecular layer of the DG could significantly impact NSPC responses and subsequent cognitive recovery by restoring long-term potentiation and its effects on learning and memory (van Praag et al. 1999; Wang et al. 2005).

Several studies have attempted to provide neurotrophic support to the hippocampus in acute CNS injury to boost neurogenesis and cognitive recovery. Under non-injured conditions, insulin-like growth factor-1 (IGF-1) is reported to increase hippocampal proliferation and neurogenesis (Aberg et al. 2000). Chronic IGF treatment following focal lesion promoted functional recovery related not to an increase in newly born cells but to increased dendritic arborization and cellular plasticity.

TBI-induced neurogenesis is also improved following intraventricular infusion of bFGF under non-pathological situation (Kuhn et al. 1997; Tropepe et al. 1999) and following TBI (Sun et al. 2009). EPO has also been shown to amplify DG neurogenesis and spatial learning following TBI, an effect that was ablated following Ara-C treatment which limited the NSPC response (Zhang et al. 2012). This study demonstrated that EPO acts directing on NSPCs in the DG to improve neurogenic-mediated recovery. Other factors regulating DG neurogenesis, include noggin, BMPs, shh, Notch, transforming Growth Factor-α (TGFα), Eph/ephrins, and VEGF (Alvarez-Buylla and Lim 2004) have been less well characterized in TBI. Upregulation of VEGF is observed after CCI (Skold et al. 2005) and has been implicated in neurogenesis and migration in a rodent model of focal cerebral ischemia and TBI (Wang et al. 2007; Lee and Agoston 2010; Thau-Zuchman et al. 2012). A better understanding of these important developmental signals and their actions following trauma will improve strategies for their delivery in the neurogenic niche (Lim et al. 2007).

10.3 Summary

Although much of the studies highlighted above have been conducted in rodents, new data from human brain samples indicates an increase in the number of proliferating NSPCs and increased markers for NSPC expression in the peri-lesional cortex after TBI. This suggests that TBI may also induce neurogenesis in the human brain (Zheng et al. 2013). In light of recent studies that support a positive role for endogenous NSPCs in stabilizing the injury microenvironment and promoting repair, challenges still remain in maximizing the benefits of this self-repair response. Directing the activated endogenous precursor cells down the desired lineage, in order to produce mature, functionally integrated neurons in non-neurogenic sites is one example. An enormous body of work in the past decade has identified a number of factors that regulate neurogenesis, including calorie restriction, voluntary exercise, neural activity, several growth factors, and pharmaceutical agents such as antidepressants (Griffin et al. 2009; Ma et al. 2009; Bechara and Kelly 2013). Other epigenetic factors may also exist to positively influence neurogenesis in both hippocampal and olfactory structures which could boost their regenerative potential in the post-trauma brain. Amongst all the optimism surrounding the potential of injury-induced neurogenesis, there remain significant concerns, including post-traumatic epilepsy (Diaz-Arrastia et al. 2009; Kernie and Parent 2010) or possible tumor formation. Therefore, strategies this enhance neurogenesis following TBI must stem from a strong mechanistic understanding of the complex cellular response within the dynamic injury environment to aid in the development of clinically relevant therapeutics.

References

Aberg MA, Aberg ND, Hedbacker H, Oscarsson J, Eriksson PS (2000) Peripheral infusion of IGF-I selectively induces neurogenesis in the adult rat hippocampus. J Neurosci 20:2896–2903

Alagappan D, Lazzarino DA, Felling RJ, Balan M, Kotenko SV, Levison SW (2009) Brain injury expands the numbers of neural stem cells and progenitors in the SVZ by enhancing their responsiveness to EGF. ASN Neuro 1(2):pii: e00009

Alagappan D, Balan M, Jiang Y, Cohen RB, Kotenko SV, Levison SW (2013) Egr-1 is a critical regulator of EGF-receptor-mediated expansion of subventricular zone neural stem cells and progenitors during recovery from hypoxia-hypoglycemia. ASN Neuro 5:183–193

Alvarez-Buylla A, Garcia-Verdugo JM (2002) Neurogenesis in adult subventricular zone. J Neurosci 22:629–634

Alvarez-Buylla A, Lim DA (2004) For the long run: maintaining germinal niches in the adult brain. Neuron 41:683–686

Alvarez-Palazuelos LE, Robles-Cervantes MS, Castillo-Velazquez G, Rivas-Souza M, Gonzalez-Castaneda RE, Luquin S, Gonzalez-Perez O (2011) Regulation of neural stem cell in the human SVZ by trophic and morphogenic factors. Curr Signal Transduct Ther 6:320–326

Anderson VA, Catroppa C, Rosenfeld J, Haritou F, Morse SA (2000) Recovery of memory function following traumatic brain injury in pre-school children. Brain Inj 14:679–692

Arvidsson A, Kokaia Z, Lindvall O (2001) N-methyl-D-aspartate receptor-mediated increase of neurogenesis in adult rat dentate gyrus following stroke. Eur J Neurosci 14:10–18

Arvidsson A, Collin T, Kirik D, Kokaia Z, Lindvall O (2002) Neuronal replacement from endogenous precursors in the adult brain after stroke. Nat Med 8:963–970

Azim K, Raineteau O, Butt AM (2012) Intraventricular injection of FGF-2 promotes generation of oligodendrocyte-lineage cells in the postnatal and adult forebrain. Glia 60:1977–1990

Bath KG, Akins MR, Lee FS (2012) BDNF control of adult SVZ neurogenesis. Dev Psychobiol 54:578–589

Baumann G, Travieso L, Liebl DJ, Theus MH (2013) Pronounced hypoxia in the subventricular zone following traumatic brain injury and the neural stem/progenitor cell response. Exp Biol Med (Maywood) 238:830–841

Beattie MS, Bresnahan JC, Komon J, Tovar CA, Van Meter M, Anderson DK, Faden AI, Hsu CY, Noble LJ, Salzman S, Young W (1997) Endogenous repair after spinal cord contusion injuries in the rat. Exp Neurol 148:453–463

Bechara RG, Kelly AM (2013) Exercise improves object recognition memory and induces BDNF expression and cell proliferation in cognitively enriched rats. Behav Brain Res 245:96–100

Ben-Ari Y (2002) Excitatory actions of gaba during development: the nature of the nurture. Nat Rev Neurosci 3:728–739

Benner EJ, Luciano D, Jo R, Abdi K, Paez-Gonzalez P, Sheng H, Warner DS, Liu C, Eroglu C, Kuo CT (2013) Protective astrogenesis from the SVZ niche after injury is controlled by Notch modulator Thbs4. Nature 497:369–373

Bernabeu R, Sharp FR (2000) NMDA and AMPA/kainate glutamate receptors modulate dentate neurogenesis and CA3 synapsin-I in normal and ischemic hippocampus. J Cereb Blood Flow Metab 20:1669–1680

Blaiss CA, Yu TS, Zhang G, Chen J, Dimchev G, Parada LF, Powell CM, Kernie SG (2011) Temporally specified genetic ablation of neurogenesis impairs cognitive recovery after traumatic brain injury. J Neurosci 31:4906–4916

Brus M, Keller M, Levy F (2013) Temporal features of adult neurogenesis: differences and similarities across mammalian species. Front Neurosci 7:135

Bunk EC, Konig HG, Bonner HP, Kirby BP, Prehn JH (2010) NMDA-induced injury of mouse organotypic hippocampal slice cultures triggers delayed neuroblast proliferation in the dentate gyrus: an in vitro model for the study of neural precursor cell proliferation. Brain Res 1359:22–32

Bye N, Carron S, Han X, Agyapomaa D, Ng SY, Yan E, Rosenfeld JV, Morganti-Kossmann MC (2011) Neurogenesis and glial proliferation are stimulated following diffuse traumatic brain injury in adult rats. J Neurosci Res 89:986–1000

Chan FK, Shisler J, Bixby JG, Felices M, Zheng L, Appel M, Orenstein J, Moss B, Lenardo MJ (2003) A role for tumor necrosis factor receptor-2 and receptor-interacting protein in programmed necrosis and antiviral responses. J Biol Chem 278:51613–51621

Chen XH, Iwata A, Nonaka M, Browne KD, Smith DH (2003) Neurogenesis and glial proliferation persist for at least one year in the subventricular zone following brain trauma in rats. J Neurotrauma 20:623–631

Chirumamilla S, Sun D, Bullock MR, Colello RJ (2002) Traumatic brain injury induced cell proliferation in the adult mammalian central nervous system. J Neurotrauma 19:693–703

Clark RSB, Chen M, Kochanek PM, Watkins SC, Jin KL, Draviam R, Nathaniel PD, Pinto R, Marion DW, Graham SH (2001) Detection of single- and double-strand DNA breaks after traumatic brain injury in rats: comparison of in situ labeling techniques using DNA polymerase I, the Klenow fragment of DNA polymerase I, and terminal deoxynucleotidyl transferase. J Neurotrauma 18:675–689

Colicos MA, Dash PK (1996) Apoptotic morphology of dentate gyrus granule cells following experimental cortical impact injury in rats: possible role in spatial memory deficits. Brain Res 739:120–131

Conover JC, Doetsch F, Garcia-Verdugo JM, Gale NW, Yancopoulos GD, Alvarez-Buylla A (2000) Disruption of Eph/ephrin signaling affects migration and proliferation in the adult subventricular zone. Nat Neurosci 3:1091–1097

Dash PK, Mach SA, Moore AN (2001) Enhanced neurogenesis in the rodent hippocampus following traumatic brain injury. J Neurosci Res 63:313–319

Demeurisse G (2000) Mechanisms of functional restoration after brain injury. Acta Neurol Belg 100:77–83

Diaz-Arrastia R, Agostini MA, Madden CJ, Van Ness PC (2009) Posttraumatic epilepsy: the endophenotypes of a human model of epileptogenesis. Epilepsia 50(Suppl 2):14–20

Doetsch F, Garcia-Verdugo JM, Alvarez-Buylla A (1997) Cellular composition and three-dimensional organization of the subventricular germinal zone in the adult mammalian brain. J Neurosci 17:5046–5061

Doetsch F, Garcia-Verdugo JM, Alvarez-Buylla A (1999) Regeneration of a germinal layer in the adult mammalian brain. Proc Natl Acad Sci U S A 96:11619–11624

Emery DL, Fulp CT, Saatman KE, Schutz C, Neugebauer E, McIntosh TK (2005) Newly born granule cells in the dentate gyrus rapidly extend axons into the hippocampal CA3 region following experimental brain injury. J Neurotrauma 22:978–988

Eriksson PS, Perfilieva E, Bjork-Eriksson T, Alborn AM, Nordborg C, Peterson DA, Gage FH (1998) Neurogenesis in the adult human hippocampus. Nat Med 4:1313–1317

Furne C, Ricard J, Cabrera JR, Pays L, Bethea JR, Mehlen P, Liebl DJ (2009) EphrinB3 is an anti-apoptotic ligand that inhibits the dependence receptor functions of EphA4 receptors during adult neurogenesis. Biochim Biophys Acta 1793:231–238

Ge S, Goh EL, Sailor KA, Kitabatake Y, Ming GL, Song H (2006) GABA regulates synaptic integration of newly generated neurons in the adult brain. Nature 439:589–593

Gilley JA, Yang CP, Kernie SG (2011) Developmental profiling of postnatal dentate gyrus progenitors provides evidence for dynamic cell-autonomous regulation. Hippocampus 21:33–47

Globus MY, Alonso O, Dietrich WD, Busto R, Ginsberg MD (1995) Glutamate release and free radical production following brain injury: effects of posttraumatic hypothermia. J Neurochem 65:1704–1711

Gonzalez-Perez O, Alvarez-Buylla A (2011) Oligodendrogenesis in the subventricular zone and the role of epidermal growth factor. Brain Res Rev 67:147–156

Griffin EW, Bechara RG, Birch AM, Kelly AM (2009) Exercise enhances hippocampal-dependent learning in the rat: evidence for a BDNF-related mechanism. Hippocampus 19:973–980

Hamm RJ, Lyeth BG, Jenkins LW, O'Dell DM, Pike BR (1993) Selective cognitive impairment following traumatic brain injury in rats. Behav Brain Res 59:169–173

Hastings NB, Gould E (1999) Rapid extension of axons into the CA3 region by adult-generated granule cells. J Comp Neurol 413:146–154
Hellstrom NA, Bjork-Eriksson T, Blomgren K, Kuhn HG (2009) Differential recovery of neural stem cells in the subventricular zone and dentate gyrus after ionizing radiation. Stem Cells 27:634–641
Holmberg J, Armulik A, Senti KA, Edoff K, Spalding K, Momma S, Cassidy R, Flanagan JG, Frisen J (2005) Ephrin-A2 reverse signaling negatively regulates neural progenitor proliferation and neurogenesis. Genes Dev 19:462–471
Horner PJ, Power AE, Kempermann G, Kuhn HG, Palmer TD, Winkler J, Thal LJ, Gage FH (2000) Proliferation and differentiation of progenitor cells throughout the intact adult rat spinal cord. J Neurosci 20:2218–2228
Jin K, Sun Y, Xie L, Batteur S, Mao XO, Smelick C, Logvinova A, Greenberg DA (2003) Neurogenesis and aging: FGF-2 and HB-EGF restore neurogenesis in hippocampus and subventricular zone of aged mice. Aging Cell 2:175–183
Kernie SG, Parent JM (2010) Forebrain neurogenesis after focal Ischemic and traumatic brain injury. Neurobiol Dis 37:267–274
Kernie SG, Erwin TM, Parada LF (2001) Brain remodeling due to neuronal and astrocytic proliferation after controlled cortical injury in mice. J Neurosci Res 66:317–326
Kleindienst A, Grunbeck F, Buslei R, Emtmann I, Buchfelder M (2013) Intraperitoneal treatment with S100B enhances hippocampal neurogenesis in juvenile mice and after experimental brain injury. Acta Neurochir (Wien) 155:1351–1360
Kuhn HG, Winkler J, Kempermann G, Thal LJ, Gage FH (1997) Epidermal growth factor and fibroblast growth factor-2 have different effects on neural progenitors in the adult rat brain. J Neurosci 17:5820–5829
Lau BW, Yau SY, Lee TM, Ching YP, Tang SW, So KF (2009) Intracerebroventricular infusion of cytosine-arabinoside causes prepulse inhibition disruption. Neuroreport 20:371–377
Lee C, Agoston DV (2010) Vascular endothelial growth factor is involved in mediating increased de novo hippocampal neurogenesis in response to traumatic brain injury. J Neurotrauma 27:541–553
Levison SW, Goldman JE (1993) Both oligodendrocytes and astrocytes develop from progenitors in the subventricular zone of postnatal rat forebrain. Neuron 10:201–212
Lim DA, Huang YC, Alvarez-Buylla A (2007) The adult neural stem cell niche: lessons for future neural cell replacement strategies. Neurosurg Clin N Am 18:81–92, ix
Lois C, Alvarez-Buylla A (1993) Proliferating subventricular zone cells in the adult mammalian forebrain can differentiate into neurons and glia. Proc Natl Acad Sci U S A 90:2074–2077
Lopez de Armentia M, Sah P (2003) Development and subunit composition of synaptic NMDA receptors in the amygdala: NR2B synapses in the adult central amygdala. J Neurosci 23:6876–6883
Lu D, Mahmood A, Zhang R, Copp M (2003) Upregulation of neurogenesis and reduction in functional deficits following administration of DEtA/NONOate, a nitric oxide donor, after traumatic brain injury in rats. J Neurosurg 99:351–361
Luskin M (1994) Neuronal cell lineage in the vertebrate central nervous system. FASEB J 8:722–730
Ma DK, Kim WR, Ming GL, Song H (2009) Activity-dependent extrinsic regulation of adult olfactory bulb and hippocampal neurogenesis. Ann N Y Acad Sci 1170:664–673
Mani N, Khaibullina A, Krum JM, Rosenstein JM (2010) Vascular endothelial growth factor enhances migration of astroglial cells in subventricular zone neurosphere cultures. J Neurosci Res 88:248–257
Markakis EA, Gage FH (1999) Adult-generated neurons in the dentate gyrus send axonal projections to field CA3 and are surrounded by synaptic vesicles. J Comp Neurol 406:449–460
Matthews MA, Onge MF St, Faciane CL (1979) An electron microscopic analysis of abnormal ependymal cell proliferation and envelopment of sprouting axons following spinal cord transection in the rat. Acta Neuropathol 45:27–36

McArthur DL, Chute DJ, Villablanca JP (2004) Moderate and severe traumatic brain injury: epidemiologic, imaging and neuropathologic perspectives. Brain Pathol 14:185–194

Moore L, Bain JM, Loh JM, Levison SW (2013) PDGF-responsive progenitors persist in the subventricular zone across the lifespan. ASN Neuro 6(2): pii: e00137

Nait-Oumesmar B, Picard-Riera N, Kerninon C, Baron-Van Evercooren A (2008) The role of SVZ-derived neural precursors in demyelinating diseases: from animal models to multiple sclerosis. J Neurol Sci 265:26–31

Nakatomi H, Kuriu T, Okabe S, Yamamoto S, Hatano O, Kawahara N, Tamura A, Kirino T, Nakafuku M (2002) Regeneration of hippocampal pyramidal neurons after ischemic brain injury by recruitment of endogenous neural progenitors. Cell 110:429–441

Naylor AS, Bull C, Nilsson MK, Zhu C, Bjork-Eriksson T, Eriksson PS, Blomgren K, Kuhn HG (2008) Voluntary running rescues adult hippocampal neurogenesis after irradiation of the young mouse brain. Proc Natl Acad Sci U S A 105:14632–14637

Nelersa CM, Barreras H, Runko E, Ricard J, Shi Y, Glass SJ, Bixby JL, Lemmon VP, Liebl DJ (2012) High-content analysis of proapoptotic EphA4 dependence receptor functions using small-molecule libraries. J Biomol Screen 17:785–795

Ning R, Xiong Y, Mahmood A, Zhang Y, Meng Y, Qu C, Chopp M (2011) Erythropoietin promotes neurovascular remodeling and long-term functional recovery in rats following traumatic brain injury. Brain Res 1384:140–150

Oyesiku NM, Evans CO, Houston S, Darrell RS, Smith JS, Fulop ZL, Dixon CE, Stein DG (1999) Regional changes in the expression of neurotrophic factors and their receptors following acute traumatic brain injury in the adult rat brain. Brain Res 833:161–172

Parent JM, Valentin VV, Lowenstein DH (2002) Prolonged seizures increase proliferating neuroblasts in the adult rat subventricular zone-olfactory bulb pathway. J Neurosci 22:3174–3188

Pencea V, Bingaman KD, Freedman LJ, Luskin MB (2001) Neurogenesis in the subventricular zone and rostral migratory stream of the neonatal and adult primate forebrain. Exp Neurol 172:1–16

Picard-Riera N, Decker L, Delarasse C, Goude K, Nait-Oumesmar B, Liblau R, Pham-Dinh D, Evercooren AB (2002a) Experimental autoimmune encephalomyelitis mobilizes neural progenitors from the subventricular zone to undergo oligodendrogenesis in adult mice. Proc Natl Acad Sci U S A 99:13211–13216

Picard-Riera N, Decker L, Delarasse C, Goude K, Nait-Oumesmar B, Liblau R, Pham-Dinh D, Baron-Van Evercooren A (2002b) Experimental autoimmune encephalomyelitis mobilizes neural progenitors from the subventricular zone to undergo oligodendrogenesis in adult mice. Proc Natl Acad Sci U S A 99:13211–13216

Picard-Riera N, Nait-Oumesmar B, Baron-Van Evercooren A (2004) Endogenous adult neural stem cells: limits and potential to repair the injured central nervous system. J Neurosci Res 76:223–231

Radomski KL, Zhou Q, Yi KJ, Doughty ML (2013) Cortical contusion injury disrupts olfactory bulb neurogenesis in adult mice. BMC Neurosci 14:142

Ramaswamy S, Goings GE, Soderstrom KE, Szele FG, Kozlowski DA (2005) Cellular proliferation and migration following a controlled cortical impact in the mouse. Brain Res 1053:38–53

Reeves TM, Lyeth BG, Phillips LL, Hamm RJ, Povlishock JT (1997) The effects of traumatic brain injury on inhibition in the hippocampus and dentate gyrus. Brain Res 757:119–132

Rice AC, Khaldi A, Harvey HB, Salman NJ, White F, Fillmore H, Bullock MR (2003) Proliferation and neuronal differentiation of mitotically active cells following traumatic brain injury. Exp Neurol 183:406–417

Richardson RM, Sun D, Bullock MR (2007) Neurogenesis after traumatic brain injury. Neurosurg Clin N Am 18:169–181, xi

Rola R, Mizumatsu S, Otsuka S, Morhardt DR, Noble-Haeusslein LJ, Fishman K, Potts MB, Fike JR (2006) Alterations in hippocampal neurogenesis following traumatic brain injury in mice. Exp Neurol 202:189–199

Roy NS, Wang S, Jiang L, Kang J, Benraiss A, Harrison-Restelli C, Fraser RA, Couldwell WT, Kawaguchi A, Okano H, Nedergaard M, Goldman SA (2000) In vitro neurogenesis by progenitor cells isolated from the adult human hippocampus. Nat Med 6:271–277

Saatman KE, Murai H, Bartus RT, Smith DH, Hayward NJ, Perri BR, McIntosh TK (1996) Calpain inhibitor AK295 attenuates motor and cognitive deficits following experimental brain injury in the rat. Proc Natl Acad Sci U S A 93:3428–3433

Saha B, Peron S, Murray K, Jaber M, Gaillard A (2013) Cortical lesion stimulates adult subventricular zone neural progenitor cell proliferation and migration to the site of injury. Stem Cell Res 11:965–977

Salman H, Ghosh P, Kernie SG (2004) Subventricular zone neural stem cells remodel the brain following traumatic injury in adult mice. J Neurotrauma 21:283–292

Seri B, Garcia-Verdugo JM, McEwen BS, Alvarez-Buylla A (2001) Astrocytes give rise to new neurons in the adult mammalian hippocampus. J Neurosci 21:7153–7160

Shihabuddin LS, Horner PJ, Ray J, Gage FH (2000) Adult spinal cord stem cells generate neurons after transplantation in the adult dentate gyrus. J Neurosci 20:8727–8735

Singer BH, Jutkiewicz EM, Fuller CL, Lichtenwalner RJ, Zhang H, Velander AJ, Li X, Gnegy ME, Burant CF, Parent JM (2009) Conditional ablation and recovery of forebrain neurogenesis in the mouse. J Comp Neurol 514:567–582

Singer BH, Gamelli AE, Fuller CL, Temme SJ, Parent JM, Murphy GG (2011) Compensatory network changes in the dentate gyrus restore long-term potentiation following ablation of neurogenesis in young-adult mice. Proc Natl Acad Sci U S A 108:5437–5442

Sinha S, Gunawat P, Nehra A, Sharma BS (2013) Cognitive, functional, and psychosocial outcome after severe traumatic brain injury: a cross-sectional study at a tertiary care trauma center. Neurol India 61:501–506

Skold MK, von Gertten C, Sandberg-Nordqvist AC, Mathiesen T, Holmin S (2005) VEGF and VEGF receptor expression after experimental brain contusion in rat. J Neurotrauma 22: 353–367

Smith DH, Okiyama K, Thomas MJ, Claussen B, McIntosh TK (1991) Evaluation of memory dysfunction following experimental brain injury using the Morris water maze. J Neurotrauma 8:259–269

Smith DH, Lowenstein DH, Gennarelli TA, McIntosh TK (1994) Persistent memory dysfunction is associated with bilateral hippocampal damage following experimental brain injury. Neurosci Lett 168:151–154

Smith DH, Soares HD, Pierce JS, Perlman KG, Saatman KE, Meaney DF, Dixon CE, McIntosh TK (1995) A model of parasagittal controlled cortical impact in the mouse: cognitive and histopathologic effects. J Neurotrauma 12:169–178

Suhonen JO, Peterson DA, Ray J, Gage FH (1996) Differentiation of adult hippocampus-derived progenitors into olfactory neurons in vivo. Nature 383:624–627

Sullivan GM, Mierzwa AJ, Kijpaisalratana N, Tang H, Wang Y, Song SK, Selwyn R, Armstrong RC (2013) Oligodendrocyte lineage and subventricular zone response to traumatic axonal injury in the corpus callosum. J Neuropathol Exp Neurol 72:1106–1125

Sun D, Colello RJ, Daugherty WP, Kwon TH, McGinn MJ, Harvey HB, Bullock MR (2005) Cell proliferation and neuronal differentiation in the dentate gyrus in juvenile and adult rats following traumatic brain injury. J Neurotrauma 22:95–105

Sun D, McGinn MJ, Zhou Z, Harvey HB, Bullock MR, Colello RJ (2007) Anatomical integration of newly generated dentate granule neurons following traumatic brain injury in adult rats and its association to cognitive recovery. Exp Neurol 204:264–272

Sun D, Bullock MR, McGinn MJ, Zhou Z, Altememi N, Hagood S, Hamm R, Colello RJ (2009) Basic fibroblast growth factor-enhanced neurogenesis contributes to cognitive recovery in rats following traumatic brain injury. Exp Neurol 216:56–65

Sun D, Bullock MR, Altememi N, Zhou Z, Hagood S, Rolfe A, McGinn MJ, Hamm R, Colello RJ (2010) The effect of epidermal growth factor in the injured brain after trauma in rats. J Neurotrauma 27:923–938

Sun C, Sun H, Wu S, Lee CC, Akamatsu Y, Wang RK, Kernie SG, Liu J (2013) Conditional ablation of neuroprogenitor cells in adult mice impedes recovery of poststroke cognitive function and reduces synaptic connectivity in the perforant pathway. J Neurosci 33:17314–17325

Sundholm-Peters NL, Yang HK, Goings GE, Walker AS, Szele FG (2004) Radial glia-like cells at the base of the lateral ventricles in adult mice. J Neurocytol 33:153–164

Taylor AN, Rahman SU, Tio DL, Sanders MJ, Bando JK, Truong AH, Prolo P (2006) Lasting neuroendocrine-immune effects of traumatic brain injury in rats. J Neurotrauma 23:1802–1813

Taylor SR, Smith C, Harris BT, Costine BA, Duhaime AC (2013) Maturation-dependent response of neurogenesis after traumatic brain injury in children. J Neurosurg Pediatr 12:545–554

Thau-Zuchman O, Shohami E, Alexandrovich AG, Leker RR (2010) Vascular endothelial growth factor increases neurogenesis after traumatic brain injury. J Cereb Blood Flow Metab 30: 1008–1016

Thau-Zuchman O, Shohami E, Alexandrovich AG, Leker RR (2012) Combination of vascular endothelial and fibroblast growth factor 2 for induction of neurogenesis and angiogenesis after traumatic brain injury. J Mol Neurosci 47:166–172

Theus MH, Ricard J, Bethea JR, Liebl DJ (2010) EphB3 limits the expansion of neural progenitor cells in the subventricular zone by regulating p53 during homeostasis and following traumatic brain injury. Stem Cells 28:1231–1242

Tovar KR, Westbrook GL (1999) The incorporation of NMDA receptors with a distinct subunit composition at nascent hippocampal synapses in vitro. J Neurosci 19:4180–4188

Tramontin AD, Garcia-Verdugo JM, Lim DA, Alvarez-Buylla A (2003) Postnatal development of radial glia and the ventricular zone (VZ): a continuum of the neural stem cell compartment. Cereb Cortex 13:580–587

Tropepe V, Sibilia M, Ciruna BG, Rossant J, Wagner EF, van der Kooy D (1999) Distinct neural stem cells proliferate in response to EGF and FGF in the developing mouse telencephalon. Dev Biol 208:166–188

Truettner J, Schmidt-Kastner R, Busto R, Alonso OF, Loor JY, Dietrich WD, Ginsberg MD (1999) Expression of brain-derived neurotrophic factor, nerve growth factor, and heat shock protein HSP70 following fluid percussion brain injury in rats. J Neurotrauma 16:471–486

Urrea C, Castellanos DA, Sagen J, Tsoulfas P, Bramlett HM, Dietrich WD (2007) Widespread cellular proliferation and focal neurogenesis after traumatic brain injury in the rat. Restor Neurol Neurosci 25:65–76

van Praag H, Christie BR, Sejnowski TJ, Gage FH (1999) Running enhances neurogenesis, learning, and long-term potentiation in mice. Proc Natl Acad Sci U S A 96:13427–13431

van Praag H, Schinder AF, Christie BR, Toni N, Palmer TD, Gage FH (2002) Functional neurogenesis in the adult hippocampus. Nature 415:1030–1034

Vaquero J, Ramiro MJ, Oya S, Cabezudo JM (1981) Ependymal reaction after experimental spinal cord injury. Acta Neurochir (Wien) 55:295–302

Wang S, Kee N, Preston E, Wojtowicz JM (2005) Electrophysiological correlates of neural plasticity compensating for ischemia-induced damage in the hippocampus. Exp Brain Res 165:250–260

Wang YM, Jin KL, Mao XO, Xie L, Banwait S, Marti HH, Greenberg DA (2007) VEGF-overexpressing transgenic mice show enhanced post-ischemic neurogenesis and neuromigration. J Neurosci Res 85:740–747

Wexler EM, Paucer A, Kornblum HI, Palmer TD, Geschwind DH (2009) Endogenous Wnt signaling maintains neural progenitor cell potency. Stem Cells 27:1130–1141

White BD, Nathe RJ, Maris DO, Nguyen NK, Goodson JM, Moon RT, Horner PJ (2010) Beta-catenin signaling increases in proliferating NG2+ progenitors and astrocytes during post-traumatic gliogenesis in the adult brain. Stem Cells 28:297–307

Williams S, Raghupathi R, MacKinnon MA, McIntosh TK, Saatman KE, Graham DI (2001) In situ DNA fragmentation occurs in white matter up to 12 months after head injury in man. Acta Neuropathol 102:581–590

Woodbury ME, Ikezu T (2013) Fibroblast Growth Factor-2 Signaling in Neurogenesis and Neurodegeneration. J Neuroimmune Pharmacol 9(2):92–101

Xiong Y, Zhang Y, Mahmood A, Meng Y, Qu C, Chopp M (2011) Erythropoietin mediates neurobehavioral recovery and neurovascular remodeling following traumatic brain injury in rats by increasing expression of vascular endothelial growth factor. Transl Stroke Res 2:619–632

Yang DY, Chen YJ, Wang MF, Pan HC, Chen SY, Cheng FC (2010) Granulocyte colony-stimulating factor enhances cellular proliferation and motor function recovery on rats subjected to traumatic brain injury. Neurol Res 32:1041–1049

Ye GL, Yi S, Gamkrelidze G, Pasternak JF, Trommer BL (2005) AMPA and NMDA receptor-mediated currents in developing dentate gyrus granule cells. Brain Res Dev Brain Res 155:26–32

Yoshimura S, Teramoto T, Whalen MJ, Irizarry MC, Takagi Y, Qiu J, Harada J, Waeber C, Breakefield XO, Moskowitz MA (2003) FGF-2 regulates neurogenesis and degeneration in the dentate gyrus after traumatic brain injury in mice. J Clin Invest 112:1202–1210

Yu TS, Zhang G, Liebl DJ, Kernie SG (2008) Traumatic brain injury-induced hippocampal neurogenesis requires activation of early nestin-expressing progenitors. J Neurosci 28:12901–12912

Zhang CL, Zou Y, He W, Gage FH, Evans RM (2008) A role for adult TLX-positive neural stem cells in learning and behaviour. Nature 451:1004–1007

Zhang Y, Chopp M, Mahmood A, Meng Y, Qu C, Xiong Y (2012) Impact of inhibition of erythropoietin treatment-mediated neurogenesis in the dentate gyrus of the hippocampus on restoration of spatial learning after traumatic brain injury. Exp Neurol 235:336–344

Zheng W, ZhuGe Q, Zhong M, Chen G, Shao B, Wang H, Mao X, Xie L, Jin K (2013) Neurogenesis in adult human brain after traumatic brain injury. J Neurotrauma 30:1872–1880

Zhou H, Chen L, Gao X, Luo B, Chen J (2012) Moderate traumatic brain injury triggers rapid necrotic death of immature neurons in the hippocampus. J Neuropathol Exp Neurol 71:348–359

Chapter 11
The Contribution of Mesenchymal Stromal Cells in Traumatic Brain Injury

Elisa R. Zanier, Francesca Pischiutta, Emanuela Parotto, Maddalena Caruso, Ornella Parolini and Maria-Grazia De Simoni

Contents

M.-G. De Simoni (✉) · E. R. Zanier (✉) · F. Pischiutta
Department of Neuroscience, IRCCS-Istituto di Ricerche Farmacologiche Mario Negri, 20156 Milan, Italy
e-mail: desimoni@marionegri.it

E. R. Zanier
e-mail: elisa.zanier@marionegri.it

E. Parotto
Institute of Anesthesia and Intensive Care, University of Padova, Padova, Italy

M. Caruso · O. Parolini
Centro di Ricerca E. Menni, Fondazione Poliambulanza Istituto Ospedaliero, Brescia, Italy

L.-R. Zhao, J. H. Zhang (eds.), *Cellular Therapy for Stroke and CNS Injuries*,
Springer Series in Translational Stroke Research, DOI 10.1007/978-3-319-11481-1_11

Abstract Traumatic brain injury (TBI) is the leading cause of mortality and disability among young people in high-income countries. No single-agent treatment has been successfully translated to the clinical setting, hence there is still the need to focus on strategies that simultaneously act on multiple injury mechanisms. Mesenchymal stem/stromal cells (MSCs) are ideal candidates since they act on multiple mechanisms of protection and repair, improving structural and functional outcome after experimental TBI. The magnitude of protection varies extremely in different studies. Besides conceptual issues and methodological differences between injury models and laboratories, heterogeneity of MSC populations also affects the outcomes. This chapter focuses on the biology of MSCs, on mechanisms of brain protection and repair and on open questions that need to be addressed in order to increase effectiveness, reduce variability and safely move from preclinical studies to clinical application.

Abbreviations

AKI	Acute kidney injury
BBB	Blood-brain barrier
BDNF	Brain-derived neurotrophic factor
BM	Bone marrow
BMMCs	BM-derived mononuclear cells
BrdU	5-bromo-2-deoxyuridine
CBF	Cerebral blood flow
CCl_4	Carbon tetrachloride
CCI	Controlled cortical impact
CNS	Central nervous system
CTs	Clinical trials
DCX	Doublecortin
GCS	Glasgow coma scale
GDNF	Glial cell-derived neurotrophic factor
GFAP	Glial fibrillary acidic protein
GMP	Good manufacturing practices
ia	Intra-arterial
ic	Intracerebral
icv	intracerebroventricular
IDO	Indoleamine 2, 3-dioxygenase
IL	Interleukin
INF	Interferon
ISTC	International Society for Cellular Therapy
iv	Intravenous
MAPC	Multipotent adult progenitor cells
MCP-1	Monocyte chemotactic protein-1
MIP-2	Macrophage inflammatory protein-2
MMP	Matrix metalloproteinase
MRI	Magnetic resonance imaging

MSC	Mesenchymal stem/stromal cell
MSCs	Mesenchymal stem/stromal cells
NGF	Nerve growth factor
NO	Nitric oxide
NSCs	Neural stem cells
NT-3	Neurotrophin-3
PGE-2	Prostaglandin E2
ROS	Reactive oxygen species
Sgz	Subgranular zone
Svz	Subventricular zone
TBI	Traumatic brain injury
TGF	Transforming growth factor
TIMP-3	Tissue inhibitor of matrix metalloproteinase-3
tPA	Tissue plasminogen activator
TNF-α	Tumor necrosis factor-α
Trk	Tyrosine kinase receptor
TSG-6	Tumor necrosis factor-stimulated gene-6
VCAM-1	Vascular cell adhesion molecule-1

11.1 Biology of MSCs

The last 50 years in the field of stem cell biology have been stimulated by increasing interest in "mesenchymal stem/stromal cells" (MSCs) (Caplan and Dennis 2006; Bianco et al. 2008; de Girolamo et al. 2013). The early description by Friedenstein and coworkers in the 1960s and 1970s of a minor population of cells in rodent bone marrow (BM) which adhered rapidly to plastic, with a fibroblast-like appearance, able to form clonal colonies in vitro and to differentiate towards the osteogenic lineage both in vitro and in vivo (Friedenstein 1990) paved the way to extensive research. This led to the discovery that cell populations with similar features could be isolated from many other species and, importantly, from almost all human tissues (da Silva Meirelles et al. 2006). At present, cells, with MSC characteristics have been derived from various human sites, including adipose tissue (Zuk et al. 2001; Rodriguez et al. 2005; Xu et al. 2005; Bourin et al. 2013), skeletal muscle (Jankowski et al. 2002), liver (Campagnoli et al. 2001), synovial membrane (De Bari et al. 2001), umbilical cord blood (Erices et al. 2000; Rogers and Casper 2004), periosteum (Fukumoto et al. 2003), dental pulp (Shi and Gronthos 2003), peripheral blood (Villaron et al. 2004), placental tissue (Bailo et al. 2004; Parolini et al. 2008), amniotic fluid (In 't Anker et al. 2003), tendon (Salingcarnboriboon et al. 2003) and menstrual blood (Meng et al. 2007; Musina et al. 2008; Rossignoli et al. 2013).

The ontogeny of human MSCs, their specific anatomical location and their native identity in the resident tissue are still poorly understood and at times controversial. Pericytes resident in the wall of blood microvessels and adventitial cells located around the larger arteries and veins have been proposed as the most likely candidates for the MSC primary precursors in post-natal tissues (reviewed in Murray et al. 2013). Our knowledge of the properties of these cells in vitro is more advanced and their in vitro characterization constitutes the base for their operational definition, albeit with variability within MSC populations and some remaining dark areas.

According to the *consensus* set out by the International Society for Cellular Therapy (ISTC) (Dominici et al. 2006), the minimum criteria required to define human MSCs are that in vitro they must: (1) be plastic-adherent; (2) express the cell surface antigens CD105, CD73 and CD90; (3) not express the cell surface antigens CD45, CD34, CD14, CD11b, CD79α, CD19, or HLA-DR and (4) differentiate into osteoblasts, adipocytes and chondroblasts under standard in vitro differentiating conditions. In this widely adopted proposal, the authors also recommend designating these cells more appropriately as "mesenchymal stromal cells" given that not all of them are stem cells (Horwitz et al. 2005; Dominici et al. 2006). These criteria are applicable specifically to human MSCs and cannot be entirely extended to cells isolated from other species, which may differ both in the marker expression and in some of their general characteristics and potential (Peister et al. 2004; Chamberlain et al. 2007). The biological attributes listed in the ISTC *consensus* paper are not exclusive to MSCs but are shared by other cell types (Haniffa et al. 2009). MSCs from various sources, while satisfying the minimal ISTC criteria, may differ in several aspects, ranging from their phenotype to their functions.

Despite significant progress in recent years in understanding the MSCs properties, several aspects of their biology are not yet fully understood and further comparative analysis of MSCs from different sources are required (Wagner et al. 2005; Kern et al. 2006; Barry et al. 2008).

How do we isolate and culture MSCs? With some specific differences related to the different sources and protocols established in different laboratories, isolation of MSCs broadly entails some common steps: (1) collection of the starting material: for example by aspiration, as in the case of MSCs from BM (Meirelles and Nardi 2009), or by mechanical separation of the tissue, as in the case of MSCs from human amniotic membrane which requires separation from the rest of the placenta (Bailo et al. 2004; Soncini et al. 2007); (2) treatment of the starting material in order to allow its dissociation and the consequent separation of MSCs from other cells and/or components of the tissue (Soncini et al. 2007; Meirelles and Nardi 2009; de Girolamo et al. 2013); (3) suspension of the collected cells in an appropriate medium.

Current research has produced isolation protocols and culture conditions, in compliance with Good Manufacturing Practice (GMP) to ensure the reproducibility, efficacy and safety of the MSCs, to allow their introduction in clinical use for therapeutic purposes. When placed in plastic culture with the appropriate medium, freshly isolated MSCs rapidly adhere and present with a spindle/fibroblastoid-like shape and start to proliferate. At or near confluence, cells are generally trypsin-treated, then expanded in culture. The time needed to reach confluence, the expandability rate and the cell yield after expansion vary widely and are significantly influenced by the source of isolation, the age and condition of the donor and the intrinsic/inherent properties of the cells, but also by the ex-vivo manipulation of the source material, the harvesting technique and the culture conditions (e.g. density of plating, oxygen tension, temperature, media composition and additives) (Phinney et al. 1999; Bernardo et al. 2009; de Girolamo et al. 2013). The variability/heterogeneity is further complicated by the fact that the MSC isolation and expansion protocols are generally laboratory-specific and pose important challenges for clinical translation as detailed in Sect. 11.4.1.

One of the features that make MSCs so attractive for the research and clinical use is their ability to undergo *multilineage differentiation* in media supplemented with specific growth factors, hormones and/or other additives that stimulate differentiation. The morphological changes and the expression of various lineage-specific genes, and the acquired ability to exert tissue-specific functions, all allow evaluation of the level of in vitro differentiation achieved. Besides their ability to differentiate into cells with characteristics of adipocytes, osteocytes and chondrocytes (Dominici et al. 2006), MSCs have been reported to differentiate towards cells of different lineages, including neurogenic (Phinney and Prockop 2007; Neirinckx et al. 2013a), pancreatic (Domínguez-Bendala and Ricordi 2012; Marappagounder et al. 2013), myogenic (Makino et al. 1999) and hepatic lineages (Sancho-Bru et al. 2009; Dong et al. 2013), although these trans-differentiation properties in vivo are still debated (Phinney and Prockop 2007; Barzilay et al. 2009; Neirinckx et al. 2013a).

Another attractive aspect of MSC biology relates to their *immunogenicity*. MSCs have long been believed to possess poor immunogenicity because of their low or absent expression of HLA class I molecules HLA-A, HLA-B and HLA-C and the absent/negligible expression of HLA class II molecules (HLA-DQ and HLA-DR) and co-stimulatory molecules, although some variabilities reported in terms of the level of expression of these markers in different laboratories and in relation to the presence of inflammatory stimuli (Nauta and Fibbe 2007). The assumption of MSC hypo-immunogenicity encouraged their use in transplantation settings even across a histocompatibility barrier and in immune-competent patients. However, there have been reports challenging the limited immunogenicity of allogeneic MSCs (Eliopoulos et al. 2005; Nauta et al. 2006; Knaän-Shanzer 2013) and there are also conflicting findings about the immunogenicity of differentiated BM-MSCs (Knaän-Shanzer 2013) (see Sect. 11.4.2 for details). Further studies are still needed for a better understanding of the immune status of MSCs. The role of their in vitro and in vivo immunomodulatory properties in the therapeutic potential for the injured brain is discussed in Sect. 11.2.3.

11.2 Possible Mechanisms of Protection by MSCs in TBI

The aim of protective strategies is to limit injury progression and rescue threatened tissue. The sequelae of traumatic brain injury (TBI) can be divided into primary injury (caused by the biomechanical impact) and secondary events starting minutes after TBI and lasting months. These cascades are associated with activation of genomic, cellular and/or biochemical processes that interact in a complex network

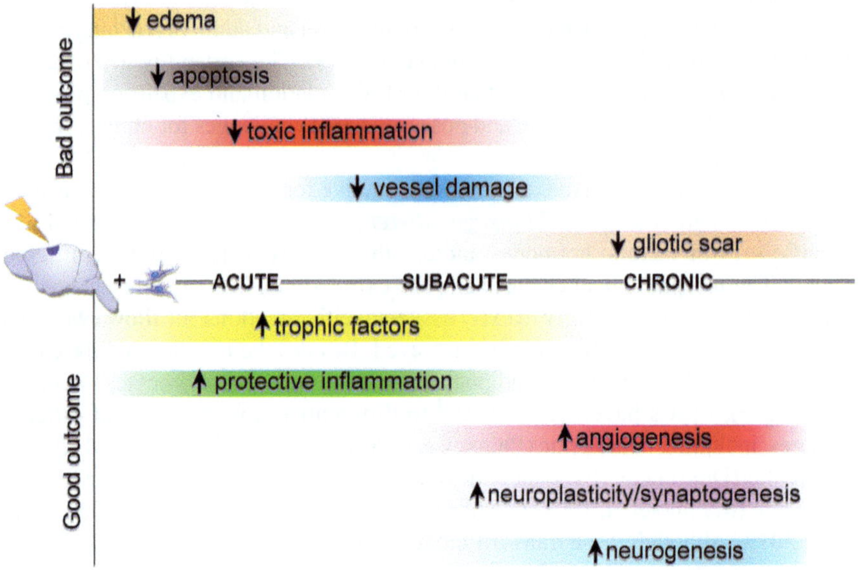

Fig. 11.1 Toxic and protective events affected by MSCs and their time course after TBI

leading to delayed cellular dysfunction and death (McIntosh et al. 1996; DeKosky et al. 1998). Although the extensive nature of these events and the abundance of targets offer the potential for therapeutic intervention, so far no pharmacological treatment is available to reverse the pathologic cellular cascade underlying progression of cell death and to improve the neurobehavioral outcome. MSCs might prove an ideal candidate (Salem and Thiemermann 2010) since they simultaneously affect multiple secondary injury mechanisms (Fig. 11.1).

11.2.1 Trophic Factors

There is experimental evidence of beneficial effects in tissue repair and regeneration after transplantation of MSCs despite a rare/transient presence of transplanted cells in the host tissues. This prompted the proposal that the restorative outcome might be due to mechanisms other than in vivo differentiation of injected cells and replacement of host defective cells (Gnecchi et al. 2005, 2008; Caplan and Dennis 2006; Meirelles and Nardi 2009; de Girolamo et al. 2013; Neirinckx et al. 2013b). Thus a new mechanism was proposed in which the transplanted MSCs release bioactive trophic molecules that might reprogram the surrounding host environment through paracrine actions. In support of this, it has been shown that the conditioned medium generated from the culture of MSCs (i.e. the medium containing the set of molecules secreted by these cells) injected in animal models of disease can recapitulate the beneficial effects of their cellular counterpart in tissue protection and repair (Timmers et al. 2007; van Poll et al. 2008; Di Santo et al. 2009; Cargnoni et al. 2012, 2014).

Several MSC secreted pleiotropic molecules are involved in these effects, including cytokines/chemokines, growth factors, extracellular matrix proteins and tissue remodelling enzymes (Caplan 2009). In vitro studies have described the secretome profile of MSCs isolated from BM, the best characterized source (Kinnaird et al. 2004; Schinköthe et al. 2008; Tate et al. 2010). Among the secreted factors released in vitro under basal conditions, neurotrophins i.e. brain-derived neurotrophic factor (BDNF), glial cell-derived neurotrophic factor (GDNF), nerve growth factor (NGF) and neurotrophin−3 (NT−3) are of great interest for TBI (Arnhold et al. 2006; Crigler et al. 2006). In vivo studies have investigated the effect of MSCs on neurotrophin production after TBI, confirming that these neurotrophins increase after MSC treatment in the injured brain. The question remains, however, whether the infused cells are directly or indirectly responsible for the changes. Commonly used biochemical methods (western blot or ELISA) cannot distinguish between the human (in case of xenotransplant) and the rodent isoforms of neurotrophins, so when measuring these proteins after TBI and MSC treatment, it is not possible to distinguish the contribution of infused MSCs from that of host cells. Thus, increased production could be related to direct secretion by MSCs and/or to an indirect effect of MSCs on endogenous cell populations.

Kim et al. (2010) injected 2 million-human MSCs intravenously 24 h after TBI, modeled by controlled cortical impact (CCI) in rats. At 48 h the homogenized injured hemisphere of TBI MSC rats showed increased amounts of NGF, BDNF and NT−3. NGF plays a key role in neuronal plasticity and prevents neuronal apoptosis (Saito et al. 2004). BDNF is involved in neural development (Ernfors et al. 1995), neurogenesis (Zigova et al. 1998; Benraiss et al. 2001) and in synaptic plasticity processes (Parkhurst et al. 2013). NT−3 supports the survival and differentiation of existing neurons and induces neurite outgrowth (Kumar et al. 1998; Bregman et al. 2002). The survival cellular signaling of these neurotrophins is mediated by binding to the neurotrophic tyrosine kinase receptor (Trk). NGF preferentially binds TrkA, BNDF binds TrkB and NT−3 binds TrkC (Chao 2003). Autophosphorylation of the cytosolic domain induces the phosphorylation of Akt and ERK signal transducers which, in turn, through NF-κB, induces the transcription of genes responsible for cell survival (Yuan and Yankner 2000; Chao 2003). The increased expression of neurotrophins in TBI MSC treated animals found by Kim et al. was in fact related to increased phosphorylation of Akt (but not of ERK) at 2 days, together with reduced cleavage of caspase−3 (the main signal for cell apoptosis) at 8 days. The molecular modifications induced by MSCs resulted in the functional recovery of motor function (Kim et al. 2010).

Our group and others have reported similar results on the increase in neurotrophin production with intravenous (iv) (Mahmood et al. 2004a) or intracerebroventricular (icv) (Chen et al. 2005; Zanier et al. 2011) transplantation of MSCs after TBI, showing increased production of BDNF (Mahmood et al. 2004a; Zanier et al. 2011) or NGF (Mahmood et al. 2004a; Chen et al. 2005) induced by infused/endogenous cells. We used MSCs from human cord blood, observing for the first time an effect on neurotrophin production similar to that described with BM-MSCs (Zanier et al. 2011).

In addition to the direct and indirect production of neurotrophins, a third mechanism for the increase in neurotrophins induced by MSCs was shown by Mahmood et al. (Mahmood et al. 2011). Three millions human MSCs transplanted 7 days after CCI, directly in the lesion cavity or impregnated in scaffolds, increased the production of tissue plasminogen activator (tPA) in the boundary zone 15 days after surgery. tPA regulates the final production of plasmin, a serine protease known for its part in the degradation of fibrin blood clots. In the brain, plasmin plays a pivotal role in the activation of BDNF and NGF, by cleaving pro-BDNF and pro-NGF into active forms (Bruno and Cuello 2006; Barnes and Thomas 2008). Thus, focusing on the increase in neurotrophins, MSCs may: (1) directly produce and/or (2) induce other cells to produce and/or (3) produce, or induce the production of, proteases that can cleave neurotrophin precursors. In principle the paracrine action means that trophic factors alone can be applied, although up to now single-drug therapy has always failed in clinical trials. MSCs act like poly-pharmacy and improve the neurological outcome by creating a growth-promoting niche, this pleiotropic activity more than any single released factor possibly being the key to MSC therapeutic efficacy.

11.2.2 Microvesicles

MSCs may also exert their paracrine effects through the release of extracellular membrane vesicles that mediate the horizontal transfer of signals and molecules from one cell to another even over long distances. MSCs, like other many cell types (Camussi et al. 2010), release circular membrane fragments known as *microvesicles*, either constitutively or in response to activation stimuli. These microvesicles, depending on their size and process of release, are generally classified as exosomes (multivesicular bodies of 30–120 nm and released from the endosomal compartment) or shedding vesicles (100–1000 nm, derived from direct budding of the cell plasma membrane). They are capable of transferring specific proteins, lipids and nucleic acids (mRNA, microRNA and DNA) thus serving as a unique mechanism for the intercellular trafficking of complex biological messages (Verderio 2013). However, the complete biochemical composition of microvesicles remains to be determined and seems to vary depending on the cell source (Camussi et al. 2010). Recent studies have reported on the miRNA (Collino et al. 2010) and protein (Kim et al. 2012) composition of microvesicles from human BM-MSCs, indicating that they include several molecules representative of the multiple properties of MSCs.

In spite of the fact that the physiological role and the biogenesis of MSC-derived microvesicles are still only partially understood, they appear to have a vital role in cell-to-cell communication (Mathivanan et al. 2010). In terms of the microvesicle-mediated therapeutic effects of MSCs, human MSC-derived microvesicles injected in SCID mice with glycerol-induced acute kidney injury (AKI) had the same efficacy as their cellular counterparts on the functional and morphological recovery of AKI (Bruno et al. 2009) and protected SCID mice from lethal cisplatin-induced AKI (Bruno et al. 2012). Exosomes released by human umbilical cord blood MSCs were reported to alleviate carbon tetrachloride (CCl_4)-induced fibrotic liver (Li et al. 2013). Lai et al. have documented the beneficial effects after the injection of MSC-derived microvesicles in animal models with cardiovascular diseases (Lai et al. 2010, 2011, 2013; Arslan et al. 2013). Zhu and collaborators, have reported that human BM-MSC-derived microvesicles were therapeutically effective in C57BL/6 mice with *E.coli* endotoxin-induced acute lung injury (Zhu et al. 2014). Microvesicles have also been used to deliver selected factors and target specific therapeutic signals. For instance, Katsuda and collaborators administered adipose tissue-derived MSC exosomes carrying enzymatically active neprilysin, a trivial beta amyloid-degrading enzyme in Alzheimer rodents (Katsuda et al. 2013). Exosomes were transferred into N2a cells overproducing Aβ and seemed to reduce both secreted and intracellular Aβ levels, suggesting therapeutic possibilities for adipose tissue-derived MSC exosomes in Alzheimer disease.

Alvarez-Erviti and collaborators succeeded in delivering functional siRNA to the mouse brain by systemically injecting targeted exosomes (Alvarez-Erviti et al. 2011). Xin et al. have reported that systemic administration of rat BM-MSC-derived exosomes after acute brain injury in ischemic rats (induced by middle cerebral artery occlusion) significantly improved functional recovery compared with control rats (Xin et al. 2013a) and raised miR−133b in the ipsilateral hemisphere.

It was shown earlier that miR−133b is specifically expressed in midbrain dopaminergic neurons and regulates the production of tyrosine hydroxylase and the dopamine transporter (Dreyer 2010). Employing knock-in and knock-down technologies to up- or down-regulate miR−133b, evidence was obtained that exosomes from MSCs mediate the miR−133b transfer to astrocytes and neurons, which in turn regulates gene expression and is involved in neurite remodeling and functional recovery after stroke (Xin et al. 2013b). The bi-lipid membrane composition enriched with membrane-bound proteins suggests that microvesicles may become a therapeutic agent, homing the injured brain and treating TBI. To our knowledge the effects of microvesicles generated by MSCs after TBI have not been examined and detailed studies are needed to see whether MSC-derived microvesicles mimic the phenotype of their parent cells and provide a protective effect on TBI.

11.2.3 Immunomodulation

After TBI, the inflammatory response plays a major role in the evolution of the brain lesion. The innate immune response is needed for resolution of the damage and wound healing but it is also a key factor in the secondary injury cascades leading to exacerbation of the initial damage. The blood brain barrier (BBB) breakdown allows the passage of inflammatory molecules and cells into the injured parenchyma. The infiltrated leukocytes directly affect neuronal survival, releasing pro-inflammatory cytokines, reactive oxygen species (ROS), nitric oxide (NO) and cytotoxic proteases which may in turn exacerbate neuronal death (Lucas et al. 2006;

Morganti-Kossmann et al. 2007). However, inflammatory mediators and immune cells can also have a neuroprotective effect and promote neurogenesis and lesion repair after central nervous system (CNS) injuries (Correale and Villa 2004). Recruited macrophages and activated microglia have been proposed as beneficial through different mechanisms including glutamate uptake (Nakajima et al. 2008), removal of cell debris (Stoll and Jander 1999) and production of neurotrophins (Batchelor et al. 1999; David and Kroner 2011). Thus strategies to direct the inflammatory response toward a protective phenotype could prove effective for TBI patients.

MSCs exert immunomodulatory properties: several studies have demonstrated that they influence the function of different immune cells in vitro, including proliferation of T cells after stimulation by alloantigens or mitogens, as well as activation of T cells by CD3 and CD28 antibodies. MSCs inhibit the generation and function of monocyte-derived dendritic cells in vitro and may also modulate B-cell functions, affecting the cytotoxic activity of natural killer cells by inhibiting proliferation and cytokine secretion (reviewed in Uccelli et al. 2008).

Despite the general agreement on the MSC's effects on the immune cell functions, little is known about the mechanisms involved. For example, inhibition of T-cell proliferation by MSCs appears to depend on cell-to-cell contact and on the release of soluble factors. MSC-derived molecules that may have immunomodulatory activity on T-cell responses include transforming growth factor-β (TGF-β), indoleamine 2,3-dioxygenase (IDO), prostaglandin E2 (PGE−2), interleukin (IL)-10, NO, miRNA (reviewed in Uccelli et al. 2008), as well as exosomes (see above). However, this aspect of the MSC biology calls for further study. Characterization of the composition and functions of the MSC secretome will undoubtedly provide important information towards a better understanding.

In TBI pathology, the effects of MSCs on immune responses have been investigated by different groups. Zhang et al. (2013) transplanted 4 million syngenic MSCs intravenously 2 h after weight-drop injury in rats, reducing the infiltration of neutrophils (MPO^+) and $CD3^+$ lymphocytes and the infiltration/activation of macrophages/microglia (Iba^+) 3 days after injury. These anti-inflammatory effects were accompanied by a reduction of apoptotic cells in the pericontusional cortex and early improvement of sensorimotor function. Analysis of cytokine levels in the injured cortex showed decreases in the expression of the pro-inflammatory mediators IL−1β, IL−6, IL−17, *tumor necrosis factor*-α (TNF-α) and interferon (INF)-γ in the acute phase (12–72h) and increased expression of the anti-inflammatory cytokines IL−10 and TGF-β1 24–72h after TBI, indicating attenuation of the inflammatory response and a shift toward an anti-inflammatory microenvironment. There was also lower expression of the chemoattractant mediators *monocyte chemotactic protein−1* (MCP−1, also known as CCL2, chemoattractant for monocytes, macrophages and microglia), macrophage inflammatory protein−2 (MIP−2, also known as CXCL2, secreted by monocytes/macrophages is chemotactic for polymorphonuclear leukocytes (Wolpe et al. 1989)) and RANTES (also known as CCL5, chemotactic for T cells, eosinophils and basophils and active in recruiting leukocytes into inflammatory sites), all indicating less recruitment of inflammatory cells from the circulatory system. The authors indicated the tumor necrosis factor-stimulated gene−6

(TSG−6) as one possible mediator of the attenuated innate immune response after MSC treatment. TSG−6 is an anti-inflammatory protein induced by TNFα and IL−1 (Wisniewski and Vilcek 1997) and was up-regulated in MSC-treated rats, together with a decrease in the expression of NF-kB, an important transcription factor that regulates many genes involved in the inflammatory response. Similarly, Watanabe et al. focused on TSG−6 and MSC-mediated protection (Watanabe et al. 2013). CCI injured mice were injected with 10 million MSCs or TSG−6 protein (50 μg/mouse) intravenously 6 h after TBI. Both treatments induced a comparable inhibition of neutrophil infiltration and a decrease in the expression of matrix metalloproteinase (MMP)-9 in the injured cortex at 24 h, leading to less BBB leakage on day three. These data support the idea that MSC are able to attenuate the inflammatory response and one possible mediator is TSG−6.

Another important MSC-mediated immunomodulatory effect is the ability to modulate the activation state of microglia/macrophages (M, resident or infiltrated immune cells in the brain) toward a beneficial and protective phenotype after cerebral insults. There is growing evidence that M play ambivalent roles in inflammation (Lai and Todd 2008; Madinier et al. 2009; Shechter and Schwartz 2013). In response to TBI, M can adopt diverse, complex activation states, enabling them to participate in the cytotoxic response, but also in immune regulation and injury resolution (Kigerl et al. 2009; Longhi et al. 2013; Shechter and Schwartz 2013).

The activation states of M can be classified in two main phenotypes: classical activation, or M1, with pro-inflammatory and detrimental properties, and the alternative one, M2, with beneficial and protective functions. Among M2-activated M, three subsets can be distinguished with different properties: M2a with pro-regenerative functions (mainly growth stimulation and tissue repair), M2b with immunoregulatory phenotype, and M2c involved in debris scavenging and healing functions (David and Kroner 2011). After acute brain injury, both M1 and M2 activation states are present in the injured tissue (Perego et al. 2011; Hu et al. 2012); however, the M2 phenotype vanishes very soon, favoring the balance towards the M1 phenotype which is responsible for the continued production of pro-inflammatory cytokines and exacerbation of injury (Loane and Byrnes 2010). Great attention has been centered on MSC's effects on M activation in different in vitro or in vivo models (Maggini et al. 2010; Ooi et al. 2010; Nakajima et al. 2012; Song et al. 2013), with the aim to skew pro-inflammatory M1 polarization towards M with an M2 immunosuppressive and pro-regenerative profile. After TBI, Walker et al. (2012a) reported that iv injection of human BM-derived multipotent adult progenitor cells (MAPC, 10^7/kg, two injections 2 and 24 h after CCI) raised the percentage of T regulatory cells in the periphery (spleen and blood) and the ratio of M2/M1 microglia in the brain. They reported that direct contact between MAPC and splenocytes is required for modulation of parenchymal microglia and attributed a central role to extraneurologic organs, by which transplanted cells act as "remote bioreactors" that boost systemic anti-inflammatory cytokine production, thereby affecting the resident microglia and the infiltrated macrophages in the peri-injury area (Walker et al. 2012b). Our group has demonstrated that MSCs can also act as a "local bioreactor" in the brain (Zanier et al. 2014). MSCs injected icv 24 h after CCI induced early and lasting acquisition of protective M2 traits by M. In the absence of any treatment the M2 wave seen 3 days after TBI vanished very soon, returning to basal levels 7 days after injury, while after MSC treatment the significant increase of M2 traits at 3 days was still present 7 days after injury. MSC-treated animals also showed a decrease in active phagocytosis of the M populations in the perilesional cortex 7 days after injury.

The role of phagocytosis after an acute brain injury is complex. On one hand, phagocytosis is needed to remove cell debris and dying cells, thus limiting the propagation of danger signals that can exacerbate damage progression ("secondary phagocytosis"). On the other hand, under certain conditions such as inflammation, phagocytosis can target viable neurons, causing their death ("primary phagocytosis"). After injury, such detrimental phagocytic activity may result from exposure of eat-me signals on otherwise viable neurons as a result of subtoxic and reversible insult (Neher et al. 2011, 2012). The ability of MSCs to influence and limit "primary phagocytosis" might be a key pathway to confer protection after TBI.

A direct link between MSC infusion and changes in the activation state of microglia was established in vitro by us and others (Giunti et al. 2012). Primary microglia cultures were exposed to an inflammatory condition, followed 2 h later by the addition of MSCs. Pro-inflammatory stimuli induced microglia M1 polarization, while the co-culture condition was able to skew the M1 acquired phenotype towards M2 polarization.

These data indicate that immunomodulation is a real healing/protective mechanism of action of MSCs after TBI, involving modulation of both the local and systemic inflammatory responses.

11.2.4 Brain Repair

The aim of repair is to rewire or restore the damaged or missing parenchyma, yielding new functional tissue. Brain repair can be considered as the ability of the CNS to remodel itself in response to insults that impair tissue homeostasis. Beside damage, TBI induces, endogenous neurorestorative events (Richardson et al. 2007). The brain retains neurogenic zones with neural stem cells that can differentiate into functional neurons (Kokaia and Lindvall 2003; Zhao et al. 2008). Like neurogenesis, other endogenous restorative processes take place in the traumatized brain and include gliogenesis, angiogenesis, synaptic plasticity and axonal sprouting. All these events: (1) are stimulated by endogenous growth factors, (2) may continue for weeks and months, (3) may contribute to functional and structural recovery. However, these spontaneous brain restorative processes are largely ineffective in counteracting the progression of damage and cumulative increases in injury over time. Providing the injured tissue with a facilitatory milieu that enhances neuroregeneration is an important additional therapeutic strategy. The roles of MSCs in neurogenesis, synaptic plasticity and axonal sprouting and in angiogenesis/vascular repair are discussed under the heading below.

11.2.4.1 Neurogenesis

The beneficial effects on functional and structural damage after MSC infusion in TBI models are not the result of replacement of injured or dead cells with exogenous cells. The first experimental study using MSCs for TBI was published in 2001 by Chopp and collaborators (Lu et al. 2001b) and showed that MSCs infused intravenously 24 h after TBI improved motor function in rats. Analysis of their fate showed that less than 0.005 % of transplanted MSCs migrated to the injured cerebral area and only 5 % expressed the neuronal marker NeuN and 7 % the astrocytic marker glial fibrillary acidic protein (GFAP). The same group then worked to increase the engraftment rate in the traumatically injured area by intra-arterial (ia) (Lu et al. 2001a) or intracerebral (ic) (Mahmood et al. 2002) injection or by culturing MSCs with neurotrophic factors (Lu et al. 2001a). In vivo neuronal differentiation occurred in a negligible number of MSCs, with no relationship between route of administration and degree of efficacy. In the context of TBI no evidence has ever been provided that neurons newly generated from MSCs become functionally active. In some studies functional improvement was induced by MSCs in the absence of any neuronal differentiation (Mahmood et al. 2005; Mori et al. 2005) and there is ample evidence that MSC-treated animals enjoy long lasting protection even when no transplanted cells are found in the brain injured tissue (Chen et al. 2005; Kim et al. 2010; Zanier et al. 2011; Pischiutta et al. 2014). These findings, along with evidence that transplanted MSCs induce early (days-weeks) functional improvement suggest that the beneficial effects are due to stimulation of endogenous neuroreparative processes through a paracrine action, so neuron replacement is not the primary mechanism of MSC-induced protection.

The observation that MSCs may promote endogenous restorative processes through interaction with local neural cells is consistent with several studies showing the induction of local neurogenesis after MSC infusion in acute brain injured rodents. After TBI we found greater expression of doublecortin (DCX, a marker of developing neurons) in the subventricular zone (svz) of MSC-treated than in untreated TBI mice at 5 weeks post-injury (Pischiutta et al. 2014). Similarly, Mahmood et al. found an increased proliferation rate (5-bromo–2-deoxyuridine (BrdU) positive cells) in the svz and subgranular zone (sgz) of traumatized rats treated with MSCs (Mahmood et al. 2004b). This effect was regularly detected after iv, ic or icv MSCs (Mahmood et al. 2004b).

11.2.4.2 Synaptic Plasticity and Axonal Sprouting

Neuroplasticity is a major compensatory mechanism following acute brain injury. Through axonal sprouting, undamaged axons can reconnect neurons whose links were injured or disrupted or can establish new networks with undamaged neurons. Thus adjacent viable regions of the cortex might function vicariously after injury (Nudo 2013). The MSC's ability to foster synaptic processes after TBI was demonstrated by infusing a fluorescent dye (DiI) into the contralateral cortex 5 weeks after injury and measuring its transport from the injection site to the injured hemisphere through the corpus callosum 1 week later (Xiong et al. 2009). MSC-treated animals had a greater axonal fiber length that was directly correlated to performance in behavioral tests, indicating that neuroplastic processes are enhanced by MSCs, and can promote neuronal connectivity by directing axonal projections, neurite outgrowth and elongation in the injured cortex. Reduction of the growth inhibitory molecule Nogo-A is linked to MSC promotion of neuroplastic processes (Mahmood et al. 2013a). Nogo-A is a myelin-derived inhibitor of axonal outgrowth highly expressed in scar tissue after TBI (Yang et al. 2013). Acute glial activation is needed to clear excessive glutamate release and remove dying cells and cellular debris after injury, limiting damage progression (Laird et al. 2008; Loane and Byrnes 2010). However, excessive glial scar in chronic stages inhibits remodeling (Iseki et al. 2012; Lin et al. 2014). Our group has demonstrated that both cord blood (Zanier et al. 2011) and BM (Pischiutta et al. 2014) derived MSCs inhibit glial scar formation around the traumatized area 5 weeks after injury, with smaller lesion volume and better functional recovery. Thus, inhibition of gliotic scar and reduction of inhibitory molecules by MSCs render the damaged tissue more permissive to neuroregenerative and neuroplastic processes, fostering recovery and better outcome. Whether MSCs are directly involved in promoting plasticity of the injured neurons and proliferation of the endogenous cells in the subventricular zone or whether this is due to the interaction between MSCs and glial cells which, in turn, will be induced to secrete neurotrophins, still needs to be clarified.

11.2.4.3 Angiogenesis/Vascular Repair

The human brain accounts for 2% of body weight but receives about 20% of cardiac output and uses about 20% of oxygen. Brain glycogen reserves are very limited so cerebral metabolism depends strictly on blood flow and energy supply. Deficient delivery of oxygen, glucose and nutrients increases susceptibility to neuronal injury. Global and regional cerebral ischemia/hypoperfusion have been observed in animal models of TBI and in brain-injured patients, after analysis of post-mortem tissue (Jones et al. 1981; Graham et al. 1989; Jenkins et al. 1989). Therapeutic strategies to restore blood flow and brain oxygenation are therefore particularly important for the vulnerable perilesional tissue that can be preserved from secondary damage and death if oxygen and glucose delivery is adequate.

Strategies aimed at restoring cerebral blood flow (CBF) after TBI may act on different mechanisms including vascular preservation from secondary damage and angiogenesis/vascular remodeling. MSC effects on CBF have been examined in live animals by magnetic resonance imaging (MRI) (Li et al. 2011). CBF was quantified by arterial spin-labeling done longitudinally on rats up to 6 weeks post-TBI. TBI induced an early, persistent reduction of CBF (<30 mL/100 g/min) in the mechanically damaged and remote regions. MSCs restored and preserved CBF in the brain regions adjacent to and further from the lesion at chronic stages (3–6 weeks), showing their ability to improve hemodynamics and moderate post-TBI hypoperfusion. *Post-mortem* analysis showed MSCs boosted vascular density in the pericontusional area when administered in the acute phase (24 h) (Pischiutta et al. 2014; Mahmood et al. 2007; Qu et al. 2008), or in the sub-acute phase (7 days) (Qu et al. 2009) after TBI and even at 2 months (Bonilla et al. 2009). In principle a rescue effect on pre-existing vessels and/or the promotion of neoangiogenic processes may be responsible for the increase in vessel density, though the effects after delayed administration of MSCs clearly support a regenerative action on brain vasculature. Gene expression microarray analysis on MSCs in vitro detected the expression of genes involved in angiogenic processes that could potentially sustain both neurovascular repair at early stages and neovascularization later (Qu et al. 2011). The vascular network improved when MSCs were transplanted ic (Qu et al. 2008; Bonilla et al. 2009) or icv (Pischiutta et al. 2014) suggesting a paracrine effects of infused MSCs. However, neurovascular protection has also been seen after systemic infusion of MSCs (Menge et al. 2012), presumably due to a different mechanism. In this setting the tissue inhibitor of matrix metalloproteinase−3 (TIMP−3) has been identified as a critical factor. Iv injected MSCs are mainly trapped in the lungs where, interacting with pulmonary endothelial cells, they secrete TIMP−3 which significantly reduces BBB leakage. Thus, MSCs do sustain the brain vasculature and promote angiogenesis, acting as local bioreactors, as well as interacting with extra-neurologic organs, providing a "remote" systemic control.

11.3 Manipulation of MSCs: Engineered Stem Cells, Survival-Promoting Scaffolds

Tissue engineering combines scaffolds (spatially guiding tissue regeneration), transplantation of cells (engrafting and/or providing a favorable regenerative environment) and delivery of different biomolecules (trophic factors, modulators of inflammatory molecules) in the hope that combinatory strategies may be more effective in achieving tissue repair. Scaffolds are supporting elements of various compositions and shapes. For TBI, which causes a mixture of dead, contused and spared tissue, injectable rather than preassembled scaffolds may be preferred. The infusion of injectable scaffolds minimizes the risk of damaging intact surrounding tissue (Delcroix et al. 2010; Saracino et al. 2013). Collagen (Lu et al. 2007; Qu et al. 2009, 2011; Xiong et al. 2009; Mahmood et al. 2011, 2013b, 2014; Guan et al. 2013), fibrin (Yasuda et al. 2010) and alginate (Heile et al. 2009) scaffolds have been seeded with BM-MSCs and transplanted up to 7 days after injury in rodents. Details of the protective/regenerative effects observed in the different studies, mostly from the same leading group, are provided in Table 11.1. On the whole, implanting human MSCs with scaffolds is more effective than implanting the cells alone. The mechanical and molecular 3D environment of the scaffolds improves transplanted MSCs survival and stimulates their expression of trophic molecules, fostering regenerative processes in the injured brain. Whether these effects depend on the scaffolds increasing delivery of donor MSCs at the injury site, or whether

Table 11.1 Manipulation of MSCs with survival-promoting scaffolds for TBI

Reference	TBI: model/recipient	Scaffolds	Gel state (injectable/preformed)	N° and cell origins	Treatment (timing and location)	Outcome	Mechanisms
Lu et al. 2007	CCI/rat	Collagen	Preformed (cylinder)	3×10^6 human BM-MSCs	4d post-injury/lesion cavity	Evaluated up to 5 weeks: spatial learning sensorimotor function lesion volume	Cell migration into the lesion boundary zone; neurotrophic support for the surrounding brain tissue
Xiong et al. 2009 Qu et al. 2011 Mahmood et al. 2011, 2013a, b 2014	CCI/rat	Collagen	Preformed (cylinder)	3×10^6 human BM-MSCs	7d post-injury/lesion cavity	Evaluated up to 5 weeks: spatial learning sensorimotor function neuronal cell loss axonal density	neural fiber length in the injured cortex neurogenesis, angiogenesis and signal transduction tissue plasminogen activator (tPA) and tPA-positive neurons Nogo-A protein
Qu et al. 2009	CCI/mouse	Collagen	Preformed (cylinder)	0.3×10^6 human BM-MSCs	7d post-injury/lesion cavity	Evaluated up to 5 weeks: spatial learning lesion volume	vascular density
Heile et al. 2009	CCI/rat	Alginate	Preformed (capsule)	6.4×10^4human BM-MSCs+ GLP-1 fusion gene transfection	Immediately before-injury ICV	Evaluated up to 2 weeks: neuronal cell loss	GLP-1 level in cerebrospinal fluid

Table 11.1 (continued)

Reference	TBI: model/ recipient	Scaffolds	Gel state (injectable/ preformed)	N° and cell origins	Treatment (timing and location)	Outcome	Mechanisms
Guan et al. 2013	CCI/rat	Collagen	Injectable	3×10^6 human BM-MSCs	7d post-injury/ lesion cavity	Evaluated up to 4 weeks: spatial learning sensorimotor function cell survival	Glucose metabolism in the lesion site
Yasuda et al. 2010	Cold rod to the right cortex/rat	Fibrin matrix	Injectable	10^4mice GFP + BM-MSCs	7d post-injury/ center of the lesion	Evaluated up to 4 weeks: cell survival and migration	GFP + MSC subpopulation expressing-MAP2 and co-localizing with neurons and perivascular cells

scaffolds also modify the molecular properties of MSCs is not clear. Nanostructured scaffolds for neuroregenerative purposes hold promise, but further work is needed to develop fully biodegradable injectable scaffolds, biocompatible for surgical use.

11.4 Challenges Involved in the Therapeutic Use of MSCs

11.4.1 MSC Heterogeneity

The protection observed in experimental models varies widely in different studies. Next to conceptual issues and methodological differences between injury models and laboratories, heterogeneity of MSC populations may contribute to disparate outcomes. This heterogeneity is at various levels. The first is *donor-to-donor heterogeneity*. Every MSC donor intrinsically differs genetically, physiologically, etc., and this may for example affect the patterns of MSC gene expression, differentiation capacity and secretion of bioactive molecules. Donor age also clearly contributes to differences in BM-derived MSCs, though heterogeneity has also been observed in MSCs isolated from age- and sex-matched donors (Phinney 2012; Dimarino et al. 2013). Another level of heterogeneity is between MSC populations isolated from different human tissues. MSCs from different sources cannot be considered entirely equivalent in terms of their immunophenotype, secretory and proteomic profile, differentiation potential and immunomodulatory ability (Phinney 2012). One explanation may relate to the in vivo origin of MSCs from different tissues, which can influence the commitment, phenotype and functions of the cells differently. However, there are only a few studies directly comparing gene and protein expression and potential of cells isolated from different sources and comparison is further complicated because different laboratories use non-standardized isolation and cultivation methods.

A further level of heterogeneity is *intra-population heterogeneity*. MSCs isolated from a specific source still tend to be heterogeneous populations which, when cultured, may contain both undifferentiated stem/progenitor cells and more mature cell types, with different functional abilities (Magatti et al. 2008; Bernardo et al. 2009; Meirelles and Nardi 2009). In the case of BM-MSCs this heterogeneity has been detected using different experimental approaches, including transcriptome or immuno-staining analysis as well as assays aimed at investigating differentiation abilities (Phinney 2012). The question of MSC heterogeneity is even more complex since culture and expansion conditions can introduce experimental artefacts, modifying the expression of natively expressed markers and promoting the expression of new ones, altering the original cellular phenotype and functions. Additional work is urgently needed to identify the specific properties of each MSC sub-population and to understand the determinants of intrinsic MSC heterogeneity. This is vital in order to reduce experimental and clinical variability, predict MSC in vivo potency and develop successful MSC-based treatment transferable to the clinic.

11.4.2 Autologous or Allogenic Transplant and Immunosuppression

The choice of rodent or human MSC sources in the experimental regenerative field is still in its early years. On one hand, rodent MSCs allow syngenic or allotransplants which are the conditions faced in the clinical setting (matched human MSCs transplanted in patients). On the other hand, human MSC candidates for clinical use need to be tested in rodent models to assess their efficacy, long-term effects, and safety and to obtain regulatory approvals. The minimum criteria required to define MSCs are applicable specifically to humans, and cannot be entirely extended to cells isolated from other species. Rodent MSCs differ from human MSCs in marker expression and in some of their general characteristics and potency (Peister et al. 2004; Chamberlain et al. 2007) (see Sect. 11.1). If rodent MSCs are fundamental as proof of concept studies, human MSCs offer a higher predictive value but their efficacy and mechanisms of action need to be fully addressed in the experimental setting in order to move to clinical trials.

Autologous stem cells have been the treatment of choice in TBI trials so far. Harvesting patient-specific MSCs poses timing, logistic and standardization constraints. To interact promptly with pathological pathways of secondary damage and to foster restorative processes, MSCs have to be transplanted in the acute phase. This limits the possibility of autologous transplantation. The transplantation of bank-stored GMP-grade certified MSCs may overcome some of the logistic limitations associated with autologous MSCs in the organ transplant setting and allow institutions without GMP facilities or the capacity to isolate MSCs to participate actively in this field of research. While the autologous MSC product can introduce differences in cell potency related to the patient's age and disease (Pietilä et al. 2012), allogenic cell transplant can be easily standardized and therefore provide more comparable results among different trials (Franquesa et al. 2013). Allotransplant, however, poses the risk of host rejection due to immunological mismatch. Thus, immunosuppression becomes a critical question before cell therapy can move to clinical application.

Data from acute brain injury models, including TBI (Kim et al. 2010; Li et al. 2011; Zanier et al. 2011), stroke (Nomura et al. 2005; Wakabayashi et al. 2010; Xin et al. 2010) and spinal cord injury (Yang et al. 2008; Hu et al. 2010; Cizkova et al. 2011), show the efficacy of allo- and xeno-transplanted MSCs with different immunosuppression strategies (Anderson et al. 2011). MSCs do not appear to retain intrinsic immunogenic properties, do not trigger alloreactivity and can survive and differentiate into allogenic or even xenogenic immunocompetent recipients in vivo (Atoui and Chiu 2012). Thus, MSCs have been proposed as "universal donor cells". However, emerging reports have challenged the limited immunogenicity of allogenic MSCs (Eliopoulos et al. 2005; Nauta et al. 2006; Knaän-Shanzer 2013) and there are also conflicting findings regarding immunogenicity of differentiated BM-derived MSCs (Knaän-Shanzer 2013). MSCs are rejected after xenotransplantation into the ischemic rodent myocardium and immunosuppression is needed to improve their efficacy and survival in the ischemic heart (Grinnemo et al. 2004, 2006). Transplantation of MSCs into the non-injured adult rodent brain can induce an inflammatory response leading to rapid and complete rejection of the transplanted cells, preventing plastic effects (Coyne et al. 2006, 2007). Consequently, the immunological impunity of MSCs in vivo is not fully supported. Immunosuppression in TBI patients clearly has dangerous implications because it increases susceptibility to infection which is directly related to unfavorable outcomes (Stocchetti 2005; Zygun et al. 2006). We have recently analyzed the specific contribution of immunosuppressive treatment to MSC efficacy for TBI, xenotransplanting human MSCs into mouse hosts, a condition that amplifies immunological mismatch-related problems (Pischiutta et al. 2014). MSCs had similar effects in immunosuppressed and immunocompetent mice on all outcome measurements at 1 month post-TBI (i.e. sensorimotor and cognitive deficits, contusion volume, vascular density, gliotic scar, neurogenesis). No signs of early xenogenic rejection (3 days post-injury) were observed and at longer time points (35 days post-injury). MSCs showed similar presence in immunosuppressed and immunocompetent traumatized brains, indicating that after TBI MSCs injected in the acute phase can escape the normal processes of xenogenic rejection. These results constitute a step forward for the development of MSC therapies for TBI, suggesting that MSCs isolated and expanded from donors, tested for their functional abilities and stored as an "off the shelf" medicinal product, may be made immediately available, with no delay to therapy and multiple doses can be given if necessary.

11.4.3 Cell Administration Route

Various routes of administration of MSCs have been employed in experimental models of TBI and protection has been reported after iv, ia, ic or icv infusion. Thus different routes can be taken into consideration after TBI. Below we discuss some of the advantages and limitations of each route. We also suggest that the biology of TBI and its heterogeneity could be important factors in deciding the route of choice.

11.4.3.1 Systemic MSC Administration

Intravenous Delivery. A number of preclinical studies (Mahmood et al. 2005; Harting et al. 2009; Zhang et al. 2013) used iv infusion for cell delivery for TBI. The iv route has two important advantages: (1) it is minimally invasive and, compared to direct transplantation into the central nervous system, overcomes the risks of bleeding and tissue injury; (2) it can be done quickly thus allowing timely treatment. However, some limitations also need to be considered. An initial obstacle to iv delivery is the large proportion of first-pass pulmonary sequestration. Many studies in animal models have shown that iv infusion of MSCs does not yield a large number of cells reaching the organ of interest, because the majority are trapped in the lungs (Schrepfer et al. 2007; Fischer et al. 2009). Pulmonary sequestration is primarly related to the MSC size. Schrepfer et al. showed that the mean size of suspended mouse MSCs (15–19 μm) is bigger than the pulmonary capillaries so most of the iv-injected MSCs are trapped in the capillaries, preventing access to the intended organs (Schrepfer et al. 2007). Besides cell size, the expression of adhesion molecule by MSCs is another important factor in pulmonary sequestration. Pre-clinical studies have shown that MSCs interact with endothelial cells, engaging P-selectin and vascular cell adhesion molecule−1 (VCAM−1) (Rüster et al. 2006). The inactivation of VCAM−1 counter ligand (VLA−4/CD49d) on the MSC surface blocks the MSC-endothelial interaction resulting in a significant increase of MSCs in the arterial system (Fischer et al. 2009).

Syngenic stem cell passage across the pulmonary circulation was investigated in anesthetized Sprague-Dawley rats, using silicone tubing catheters placed in the left internal jugular vein and common carotid artery to measure pulmonary passage of MSCs, MAPCs, neural stem cells (NSCs) and BM derived mononuclear cells (BMMCs) (average diameters respectively 18, 15, 16 and 7 μm) co-labeled with specific nanocrystals and infused iv. The labelled cells in the arterial circulation and in peripheral filter organs (lungs, spleen and kidney) were quantified by flow cytometry and infrared imaging, respectively. MSC pulmonary sequestration was 30 times that of BMMCs (Fischer et al. 2009). Two independent studies found that after TBI, only 0.001 % of iv injected cells were found in the brain or organ systems other than lung parenchyma 2–3 days after injection (Harting et al. 2009; Prockop and Oh 2012). These studies strongly suggest that the efficacy after MSC iv infusion is very likely unrelated to the MSCs reaching the injured tissue but that infused MSCs act as remote "bioreactors" stimulating resident cells in lung (macrophages) and spleen (T-cells) to acquire an anti-inflammatory phenotype (Harting et al. 2009; Prockop and Oh 2012; Walker et al. 2012b) thereby promoting resolution of the brain injury.

Intra-Arterial Delivery. The rationale behind ia administration is to bypass the pulmonary first pass effect, increasing delivery of infused cells to the target tissue. However, microvascular occlusions have been documented and CBF impairment (80–90 % reduction in laser Doppler flow signal) have been shown to occur in 35 % of treated animals (Walczak et al. 2008). More encouraging results were obtained by Lundberg and collaborators (Lundberg et al. 2012) who reported no thromboembolic complications after ia delivery. The presence of human MSCs in the brain was higher compared to iv administration, but the contusion model they used was not associated with any gross neurological symptoms, thus preventing the assessment of stem cell transplantation efficacy.

Additional studies are needed to establish if ia transplantation of MSCs gives more favorable effects than iv injection. Furthermore, while this approach may be particularly interesting in ischemic stroke when an endovascular procedure may already be planned, in TBI patients who have intracranial pressure and perfusion pressure problems, all the complications associated with carotid puncture/manipulation must be carefully evaluated before considering this strategy as promising.

11.4.3.2 Focal MSC Administration

Intracerebroventricular Delivery. Icv cannulation in human TBI is invasive and may have significant complications; however, it is recommended by authoritative guidelines for intracranial pressure monitoring of severe TBI in the intensive care unit (Brain Trauma Foundation et al. 2007; Stocchetti et al. 2008). In these patients, this site would therefore be free from additional surgical complications and offer the advantage of focal administration directly in the region of interest.

In mice icv administration of human umbilical cord blood or BM-MSC induced lasting improvement in sensorimotor and cognitive functions and reduced contusion volume 1 month after TBI (Zanier et al. 2011; Pischiutta et al. 2014). Our data provide evidence that MSCs can also act as a local "bioreactor" in the brain. In our model icv injected MSCs are detected in the ventricles and at the lesion site for up to 5 weeks in TBI mice, but are confined to the ventricles in sham-operated mice (Pischiutta et al. 2014) supporting a local action of MSCs on host tissue. However, further studies are necessary to see whether if direct contact between MSCs and the injured cells in vivo is needed for protection, as previous evidence suggests that MSCs may act through the release of active molecules rather than through cell-to-cell contact (Zanier et al. 2011; Walker et al. 2012b).

Intracerebral Delivery. The rationale for ic stem cell implantation is to maximize the MSCs load at the site of injury. However, increasing evidence that MSC engraftment is not required for therapeutic efficacy challenges this approach. Furthermore, BM-MSC differentiation could trigger immune rejection (Niemeyer et al. 2008; Huang et al. 2010) probably due to a switch in surface-MHC molecule composition during MSC differentiation, as described in the heart (Huang et al. 2010). The invasiveness of the ic approach and the possibility of further tissue damage during cell transplantation make this strategy unlikely in the treatment of TBI at the present time.

The decision on the administration route is therefore fundamental in the definition of a clinical protocol. Issues to be considered are the type of injury, the biodistribution of injected cells and the cell type. Iv administration offers easy access and the potential for broad distribution, but has the disadvantage of a large pulmonary first-pass effect, thus significantly reducing the cells delivered to the arterial circulation. Ia delivery can target the injured tissue better, but it can cause emboli, impairing blood flow and worsening the clinical outcome. Icv delivery may be the choice in a selected group of severe TBI patients.

11.4.4 Treatment Timing and Doses

The optimal timing of therapeutic MSC administration after TBI is a point open for discussion. There is ample evidence of the reciprocal interaction between infused MSCs and endogenous cell population. After TBI, the severity and kinetics of the TBI-related metabolic cellular and molecular cascades are the determinants of the injured microenvironment that, with time, may be more or less permissive to MSC functions. At present, technical aspects related to patient stabilization and identification of the patient-matched allogenic MSCs set the lower limit of the window of treatment at approximately 12–24 h post-TBI. No robust data are available to define the upper limit. However, considering the rapid evolution of secondary damage involving pericontusional cerebral tissue that could be rescued, it seems reasonable to treat the patient as soon as possible.

Preclinical studies in rodents report a wide range of TBI-to-transplantation intervals, with a preference in the acute phase (within 24 h). Both systemic (Lu et al. 2001b; Mahmood et al. 2001a, 2003, 2004a; Harting et al. 2009; Kim et al. 2010; Menge et al. 2012; Watanabe et al. 2013; Zhang et al. 2013; Tajiri et al. 2014) and central (Mahmood et al. 2001b, 2002, 2004b; Chen et al. 2005; Mori et al. 2005; Walker et al. 2010; Zanier et al. 2011; Wang et al. 2013; Pischiutta et al. 2014) administration of MSCs within the first 24 h after injury have resulted in improvements of functional and structural outcome in the chronic stages by multiple mechanisms of protection and repair.

Sub-acute transplantation (day 4–7 after injury) has shown protective effects too after systemic (Li et al. 2011; Matsumoto et al. 2013) or central infusion (Xiong et al. 2009; Mahmood et al. 2011; Qu et al. 2011). Acute and sub-acute transplantation have been recently compared (Han et al. 2013) by transplanting MSCs one or 7 days after injury in the corpus callosum ipsilateral to the injured site. The results show that injection on day 7 produces greater functional and structural improvements 1 month post-TBI than the injection on day one. Confirmatory studies by independent research groups are needed to further explore the therapeutic window.

Repeated administration could be a strategy to obtain a "booster effect", combining acute and sub-acute doses through a multiple delivery system. Acute MSC infusion will allow the interaction with early pathologic pathways and sub-acute administration will replace any MSCs may have been damaged or eliminated, allowing the stimulation of protective and restorative processes. No data are available in TBI so far and this possibility needs to be explored in order to confirm a potential advantage and exclude any secondary side effects.

Finally, treatment in the chronic phases of TBI were investigated by Bonilla and co-workers. They showed that MSC ic infusion into the lesion core 2 months after injury improved sensorimotor deficits and promoted neurorestorative processes (increase of vessel density and endogenous neurogenesis) (Bonilla et al. 2009); iv injection at the same time failed to induce any significant difference between MSC-treated and control animals (Bonilla et al. 2012). These data suggest that at chronic stages MSCs can still act as a local bioreactor promoting endogenous brain restorative processes. A pivotal mechanism of iv MSC infusion is the modulation of the focal and systemic inflammatory environment, which is no longer affected by later treatment.

To conclude, no systematic analysis of transplantation timing and its effect on TBI sequelae has been done yet. More experimental data are needed to clarify the ideal timeframe for MSC transplantation in order to identify the best approach in the clinical setting.

As for the timing of transplantation, there is no a consensus on the ideal MSC dose. Depending on the route of delivery and the animal model used, most experimental studies use the maximum dose that does not affect the animal's health (e.g. obstructing vessels with cellular emboli in case of iv injection) or cell viability (e.g. inducing cell mortality due to constriction in the syringe needle).

Iv injections in rat models go from two (Lu et al. 2001b; Mahmood et al. 2001a, 2004a; Kim et al. 2010), three (Li et al. 2011) or four (Harting et al. 2009; Zhang et al. 2013; Tajiri et al. 2014) million-MSCs, while in mouse models the range is between 300,000 (Qu et al. 2009; Matsumoto et al. 2013) and one (Watanabe et al. 2013) or 2 million (Menge et al. 2012) MSCs.

Smaller amounts are commonly used for cerebral infusion: the icv dose to mice never exceeds 150,000–200,000 MSCs (Chen et al. 2005; Zanier et al. 2011; Pischiutta et al. 2014), while MSCs transplanted directly in the lesion cavity in rats range from 3 to 400,000 (Mori et al. 2005; Lu et al. 2007) to one (Mahmood et al. 2001b, 2002; Walker et al. 2010; Wang et al. 2013) or 5 million (Bonilla et al. 2009). MSCs impregnated into scaffolds transplanted into the lesion core of rats have been seeded at dose of 10,000 cells (Yasuda et al. 2010), 64,000 (Heile et al. 2009) or 3 million (Lu et al. 2007; Xiong et al. 2009; Mahmood et al. 2011, 2013a; Qu et al. 2011; Guan et al. 2013) while the only dose tested in mice was 300,000 MSCs (Qu et al. 2009).

A vast range of MSC doses have been investigated up to now, but few studies have tried to identify a dose-response effect. Mahmood and co-workers directly compared different doses: 1 or 2 million MSCs injected iv in rats 24 h after injury had different effects on sensorimotor function, with significant improvement only after the highest dose (Mahmood et al. 2003). However, they found no dose-dependent effect when infusing 2, 4 or 8 million MSCs in the same experimental setting (Mahmood et al. 2005).

To conclude, there is no agreement on dose-dependent effects of MSC infusion and more studies are needed to clarify the issue. However, translation of results from the experimental setting to the clinical context is rarely possible.

11.5 Clinical Trials of MSCs for TBI Patients

ClinicalTrials.gov (http://clinicaltrials.gov/) is the world's largest registry and result database of publicly and privately supported clinical trials (CTs). It currently lists 160,935 trials in 185 countries. This dataset is maintained by the US National Library of Medicine at the National Institutes of Health. Reviewing the clinical trials in ClinicalTrials.gov, we found 4598 trials in response to the search criteria 'stem cell* OR stromal cell*'. The strategy used to refine the search and to identify only CTs focused on the use of MSCs in acute brain injury is detailed in Fig. 11.2A. Among the 4598 CTs, 81 involved acute neurological conditions and 23 used MSCs (Fig. 11.2). The majority of the CTs were conducted on stroke (12 CTs; 52 %) and in spinal cord-injured patients (10 CTs, 43 %); one (4 %) was on intraparenchymal hemorrhage and none in TBI patients.

To date CTs with adult stem/stromal cells in TBI have all focused on acute interventions with an autologous source, thus excluding the use of MSCs that need

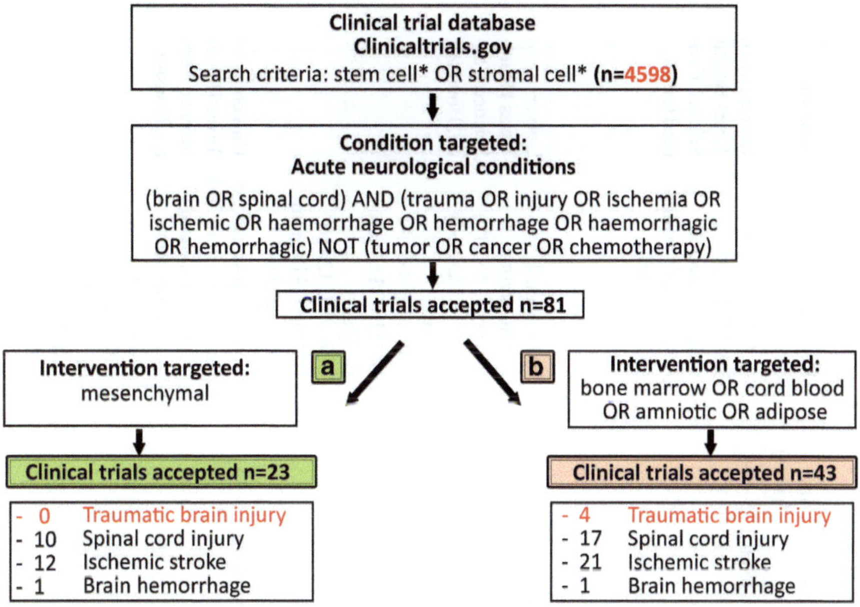

Fig. 11.2 Clinical trials (CTs) registered in ClinicalTrials.gov and focused on stem/stromal cells for acute neurological conditions on 12 February 2014. CTs were first selected by the search criteria 'stem cell* OR stromal cell*'. Then, in the "condition" field acute neurological injuries (e.g. TBI, spinal cord injury, stroke and brain hemorrhage) were targeted. CTs not involving neurological conditions/diseases and acute mechanism of disease were rejected. Last, adding "mesenchymal" in intervention field (A) resulted in 23 trials, but none were focused on TBI. Appling the wider intervention criteria 'bone marrow OR amniotic OR cord blood OR adipose' in the field 'interventions' (B) four TBI CTs were identified

time and manipulation in order to be selected and expanded (Fig. 11.2B). There are four CTs with BM-derived mononuclear cells (BMMCs), a heterogeneous population that includes MSCs together with hematopoietic stem cells, lymphoid cells (lymphocytes, plasma cells), monocytes and macrophages. The main features of these trials are summarized in Table 11.2 and will be discussed briefly since we believe they are of interest even if slightly out of focus. One CT (NCT00254722) has been completed while the other three (NCT01575470, NCT01851083 and NCT02028104) are currently recruiting participants. Three of them are led by the same group (The University of Texas Health Science Center, Houston, USA).

NCT00254722 is a Phase I non-randomized, trial carried out in Houston (Texas USA) from April 2006 to November 2009. It was designed to evaluate the logistics, feasibility and safety of BMMC autologous transplantation in children after TBI. The secondary objective was to investigate whether the late functional outcome was improved by BMMCs compared to age and severity matched concomitant controls, in order to estimate the potential treatment effect size for future trial planning. Ten children (aged 5–14 years) with acute TBI (within 24 h) and a post resuscitation Glasgow Coma Scale of 5–8 were recruited. Patients were treated with 6 million autologous BMMCs/kg body weight delivered intravenously within 48 h after TBI.

Table 11.2 Published and ongoing clinical trials applying stem/stromal cells in TBI

Registration number	Trial name	Purpose	Phase	Start date	Status	Age (years)	Regimen	Sponsor
NCT00254722	Safety of autologous stem cell treatment for traumatic brain injury in children	To determine if bone marrow derived mononuclear cells (BMMNC) autologous transplantation in children after isolated traumatic brain injury is safe and will improve functional outcome	1	November 2005	Completed	5–14	6×10^6 autologous BMMNC/kg body weight delivered intravenously within 48 h after TBI	University of Texas Health Science Center, Houston
NCT01575470	Treatment of severe adult traumatic brain injury using bone marrow mononuclear cells	To determine if bone marrow harvest, BMMNC separation, and reinfusion in adults with acute severe TBI is safe and will improve functional outcome	1/2	March 2012	Recruiting	18–55	Intravenously administration. *Cohort 1:* 6×10^6 BMMNC/kg body weight. *Cohort 2:* 9×10^6 BMMNC/kg body weight. *Cohort 3:* 12×10^6 BMMNC/kilogram body weight.	University of Texas Health Science Center, Houston
NCT01851083	Pediatric autologous bone marrow mononuclear cells for severe traumatic brain injury	To determine the effect of intravenous infusion of autologous BMMNCs on brain structure and neurocognitive/functional outcomes after severe TBI in children	1/2	August 2013	Recruiting	5–15	Bone marrow harvest performed within 36 h of injury, followed by a single intravenous infusion of autologous BMMNC (6×10^6 autologous BMMNCs/kg body weight)	University of Texas Health Science Center, Houston

Table 11.2 (continued)

Registration number	Trial name	Purpose	Phase	Start date	Status	Age (years)	Regimen	Sponsor
NCT02028104	Stem cell therapy in traumatic brain injury	To determine the effect of stem cell therapy on common symptoms in patients with chronic TBI	1	March 2010	Recuiting	6 months-65 years	BMMNC are administered intrathecally in traumatic brain injury patients. Data about dose are not available in ClinicalTrials.gov.	Neurogen Brain and Spine Institute, Mumbay, India

The safety of the procedure was evaluated by monitoring systemic and cerebral hemodynamics during BM harvest; infusion-related toxicity was determined by pediatric logistic organ dysfunction scores, hepatic enzymes, Murray lung injury scores and renal function. One and 6 months post-injury, conventional MRI, neuropsychological and functional outcome measures were obtained. Infusion of BMMCs to acutely treat severe TBI in children appeared to be safe. There were no episodes of harvest-related depression of systemic or cerebral hemodynamics and no detectable infusion-related toxicity. MRI 1 and 6 months post-injury showed no significant decrease in grey matter, white matter or intracranial volume; there was no significant rise in cerebrospinal fluid volume during the study. Assessment of functional and neuropsychological outcome 1 and 6 months post-TBI showed improvements in all scores examined. The dichotomized Glasgow outcome scale at 6 months showed 70 % good to 30 % bad outcome or death, which is similar to other major reports in pediatric severe TBI (Cox et al. 2011). However, although structural preservation and improved functional outcomes were observed, the study was underpowered and not designed to assess therapeutic efficacy.

The safety of the protocol led to the controlled prospective, randomized, blinded phase II ongoing trial (August 2013–June 2018) **NCT01851083**,designed to determine the effect of iv infusion of autologous BMMCs on brain structure and neurocognitive/functional outcomes after severe TBI in children (aged 5–15 years, post-resuscitation Glasgow coma scale, (GCS, 3–8, recruitment within 24 h of injury). BM was harvested within 36 h of injury, followed by a single infusion of BMMCs.

NCT01575470 (March 2012–June 2014) is a Phase I/II trial, designed as a dose-escalation study, consisting of four cohorts of five adult TBI patients cohorts (age 18–55 years, admission GCS 5–8). The investigator's primary hypothesis is that autologous BMMCs transplantation after TBI is safe (harvest-and infusion-related toxicity). The secondary hypothesis is that: (1) functional outcome measures will improve after BMMC infusion, (2) BMMC infusion will reduce BBB permeability, (3) BMMCs are neuroprotective and preserve grey and white matter structures assessed by diffusion tensor MRI.

NCT02028104 (March 2010–January 2015) is a Phase I trial led by the Neurogen Brain and Spine Institute (Mumbay, India). The purpose is to evaluate the effect of stem cell therapy on common symptoms in patients with TBI. BMMCs are administered intrathecally in 6 months to 65-year-old patients. Unlike the other CTs mentioned so far patients with chronic TBI are enrolled in this trial.

Considering the overall analysis of CTs focused on TBI, some observations can be made. First, most of them are safety trials and are not powered to detect functional measures of efficacy. However, valid estimates can be made from these findings to allow controlled phase II trials. Second, none of them uses pure MSCs but rely on an autologous source, using the more heterogeneous population of BMMCs. Finally, the patients are mainly children, who have greater neurologic plasticity with a unique injury pattern compared to adults.

Acknowledgments We thank Eliana Sammali and Federica Marchesi for the artwork.

References

Alvarez-Erviti L, Seow Y, Yin H, Betts C, Lakhal S, Wood MJA (2011) Delivery of siRNA to the mouse brain by systemic injection of targeted exosomes. Nat Biotechnol 29:341–345

Anderson AJ, Haus DL, Hooshmand MJ, Perez H, Sontag CJ, Cummings BJ (2011) Achieving stable human stem cell engraftment and survival in the CNS: is the future of regenerative medicine immunodeficient? Regen Med 6:367–406

Arnhold S, Klein H, Klinz F-J, Absenger Y, Schmidt A, Schinköthe T, Brixius K, Kozlowski J, Desai B, Bloch W, Addicks K (2006) Human bone marrow stroma cells display certain neural characteristics and integrate in the subventricular compartment after injection into the liquor system. Eur J Cell Biol 85:551–565

Arslan F, Lai RC, Smeets MB, Akeroyd L, Choo A, Aguor ENE, Timmers L, van Rijen HV, Doevendans PA, Pasterkamp G, Lim SK, de Kleijn DP (2013) Mesenchymal stem cell-derived exosomes increase ATP levels, decrease oxidative stress and activate PI3K/Akt pathway to enhance myocardial viability and prevent adverse remodeling after myocardial ischemia/reperfusion injury. Stem Cell Res 10:301–312

Atoui R, Chiu RCJ (2012) Concise review: immunomodulatory properties of mesenchymal stem cells in cellular transplantation: update, controversies, and unknowns. Stem Cells Transl Med 1:200–205

Bailo M, Soncini M, Vertua E, Signoroni PB, Sanzone S, Lombardi G, Arienti D, Calamani F, Zatti D, Paul P, Albertini A, Zorzi F, Cavagnini A, Candotti F, Wengler GS, Parolini O (2004) Engraftment potential of human amnion and chorion cells derived from term placenta. Transplantation 78:1439–1448

Barnes P, Thomas KL (2008) Proteolysis of proBDNF is a key regulator in the formation of memory. PloS ONE 3:e3248

Barry F, Murphy M, Coleman C, O'Brien T, Rae M, Duffy K (2008) Mesenchymal stem cells: characterization, therapeutic evaluation and manufacturing. Eur Cell Mater 16(Suppl. 3):2

Barzilay R, Melamed E, Offen D (2009) Introducing transcription factors to multipotent mesenchymal stem cells: making transdifferentiation possible. Stem Cells (Dayt Ohio) 27:2509–2515

Batchelor PE, Liberatore GT, Wong JY, Porritt MJ, Frerichs F, Donnan GA, Howells DW (1999) Activated macrophages and microglia induce dopaminergic sprouting in the injured striatum and express brain-derived neurotrophic factor and glial cell line-derived neurotrophic factor. J Neurosci 19:1708–1716

Benraiss A, Chmielnicki E, Lerner K, Roh D, Goldman SA (2001) Adenoviral brain-derived neurotrophic factor induces both neostriatal and olfactory neuronal recruitment from endogenous progenitor cells in the adult forebrain. J Neurosci 21:6718–6731

Bernardo ME, Locatelli F, Fibbe WE (2009) Mesenchymal stromal cells. Ann N Y Acad Sci 1176:101–117

Bianco P, Robey PG, Simmons PJ (2008) Mesenchymal stem cells: revisiting history, concepts, and assays. Cell Stem Cell 2:313–319

Bonilla C, Zurita M, Otero L, Aguayo C, Vaquero J (2009) Delayed intralesional transplantation of bone marrow stromal cells increases endogenous neurogenesis and promotes functional recovery after severe traumatic brain injury. Brain Inj BI 23:760–769

Bonilla C, Zurita M, Otero L, Aguayo C, Rico MA, Rodríguez A, Vaquero J (2012) Failure of delayed intravenous administration of bone marrow stromal cells after traumatic brain injury. J Neurotrauma 29:394–400

Bourin P, Bunnell BA, Casteilla L, Dominici M, Katz AJ, March KL, Redl H, Rubin JP, Yoshimura K, Gimble JM (2013) Stromal cells from the adipose tissue-derived stromal vascular fraction and culture expanded adipose tissue-derived stromal/stem cells: a joint statement of the International Federation for Adipose Therapeutics and Science (IFATS) and the International Society for Cellular Therapy (ISCT). Cytotherapy 15:641–648

Brain Trauma Foundation et al (2007) Guidelines for the management of severe traumatic brain injury. VII. Intracranial pressure monitoring technology. J Neurotrauma 24(Suppl 1):S45–54

Bregman BS, Coumans J-V, Dai HN, Kuhn PL, Lynskey J, McAtee M, Sandhu F (2002) Transplants and neurotrophic factors increase regeneration and recovery of function after spinal cord injury. Prog Brain Res 137:257–273

Bruno MA, Cuello AC (2006) Activity-dependent release of precursor nerve growth factor, conversion to mature nerve growth factor, and its degradation by a protease cascade. Proc Natl Acad Sci U S A 103:6735–6740

Bruno S, Grange C, Deregibus MC, Calogero RA, Saviozzi S, Collino F, Morando L, Busca A, Falda M, Bussolati B, Tetta C, Camussi G (2009) Mesenchymal stem cell-derived microvesicles protect against acute tubular injury. J Am Soc Nephrol JASN 20:1053–1067

Bruno S, Grange C, Collino F, Deregibus MC, Cantaluppi V, Biancone L, Tetta C, Camussi G (2012) Microvesicles derived from mesenchymal stem cells enhance survival in a lethal model of acute kidney injury. PloS ONE 7:e33115

Campagnoli C, Roberts IA, Kumar S, Bennett PR, Bellantuono I, Fisk NM (2001) Identification of mesenchymal stem/progenitor cells in human first-trimester fetal blood, liver, and bone marrow. Blood 98:2396–2402

Camussi G, Deregibus MC, Bruno S, Cantaluppi V, Biancone L (2010) Exosomes/microvesicles as a mechanism of cell-to-cell communication. Kidney Int 78:838–848

Caplan AI (2009) Why are MSCs therapeutic? New data: new insight. J Pathol 217:318–324

Caplan AI, Dennis JE (2006) Mesenchymal stem cells as trophic mediators. J Cell Biochem 98:1076–1084

Cargnoni A, Ressel L, Rossi D, Poli A, Arienti D, Lombardi G, Parolini O (2012) Conditioned medium from amniotic mesenchymal tissue cells reduces progression of bleomycin-induced lung fibrosis. Cytotherapy 14:153–161

Cargnoni A, Piccinelli EC, Ressel L, Rossi D, Magatti M, Toschi I, Cesari V, Albertini M, Mazzola S, Parolini O (2014) Conditioned medium from amniotic membrane-derived cells prevents lung fibrosis and preserves blood gas exchanges in bleomycin-injured mice-specificity of the effects and insights into possible mechanisms. Cytotherapy 16:17–32

Chamberlain G, Fox J, Ashton B, Middleton J (2007) Concise review: mesenchymal stem cells: their phenotype, differentiation capacity, immunological features, and potential for homing. Stem Cells (Dayt Ohio) 25(11):2739–2749

Chao MV (2003) Neurotrophins and their receptors: a convergence point for many signalling pathways. Nat Rev Neurosci 4:299–309

Chen Q, Long Y, Yuan X, Zou L, Sun J, Chen S, Perez-Polo JR, Yang K (2005) Protective effects of bone marrow stromal cell transplantation in injured rodent brain: synthesis of neurotrophic factors. J Neurosci Res 80:611–619

Cizkova D, Novotna I, Slovinska L, Vanicky I, Jergova S, Rosocha J, Radonak J (2011) Repetitive intrathecal catheter delivery of bone marrow mesenchymal stromal cells improves functional recovery in a rat model of contusive spinal cord injury. J Neurotrauma 28:1951–1961

Collino F, Deregibus MC, Bruno S, Sterpone L, Aghemo G, Viltono L, Tetta C, Camussi G (2010) Microvesicles derived from adult human bone marrow and tissue specific mesenchymal stem cells shuttle selected pattern of miRNAs. PloS ONE 5:e11803

Correale J, Villa A (2004) The neuroprotective role of inflammation in nervous system injuries. J Neurol 251:1304–1316

Cox CS Jr, Baumgartner JE, Harting MT, Worth LL, Walker PA, Shah SK, Ewing-Cobbs L, Hasan KM, Day M-C, Lee D, Jimenez F, Gee A (2011) Autologous bone marrow mononuclear cell therapy for severe traumatic brain injury in children. Neurosurgery 68:588–600

Coyne TM, Marcus AJ, Woodbury D, Black IB (2006) Marrow stromal cells transplanted to the adult brain are rejected by an inflammatory response and transfer donor labels to host neurons and glia. Stem Cells (Dayt Ohio) 24:2483–2492

Coyne TM, Marcus AJ, Reynolds K, Black IB, Woodbury D (2007) Disparate host response and donor survival after the transplantation of mesenchymal or neuroectodermal cells to the intact rodent brain. Transplantation 84:1507–1516

Crigler L, Robey RC, Asawachaicharn A, Gaupp D, Phinney DG (2006) Human mesenchymal stem cell subpopulations express a variety of neuro-regulatory molecules and promote neuronal cell survival and neuritogenesis. Exp Neurol 198:54–64

Da Silva Meirelles L, Chagastelles PC, Nardi NB (2006) Mesenchymal stem cells reside in virtually all post-natal organs and tissues. J Cell Sci 119:2204–2213

David S, Kroner A (2011) Repertoire of microglial and macrophage responses after spinal cord injury. Nat Rev Neurosci 12:388–399

De Bari C, Dell'Accio F, Tylzanowski P, Luyten FP (2001) Multipotent mesenchymal stem cells from adult human synovial membrane. Arthritis Rheum 44:1928–1942

De Girolamo L, Lucarelli E, Alessandri G, Avanzini MA, Bernardo ME, Biagi E, Brini AT, D'Amico G, Fagioli F, Ferrero I, Locatelli F, Maccario R, Marazzi M, Parolini O, Pessina A, Torre ML, Italian Mesenchymal Stem Cell Group (2013) Mesenchymal stem/stromal cells: a new cells as drugs' paradigm. Efficacy and critical aspects in cell therapy. Curr Pharm Des 19:2459–2473

DeKosky ST, Kochanek PM, Clark RS, Ciallella JR, Dixon CE (1998) Secondary injury after head trauma: subacute and long-term mechanisms. Semin Clin Neuropsychiatry 3:176–185

Delcroix GJ-R, Schiller PC, Benoit J-P, Montero-Menei CN (2010) Adult cell therapy for brain neuronal damages and the role of tissue engineering. Biomaterials 31:2105–2120

Di Santo S, Yang Z, Wyler von Ballmoos M, Voelzmann J, Diehm N, Baumgartner I, Kalka C (2009) Novel cell-free strategy for therapeutic angiogenesis: in vitro generated conditioned medium can replace progenitor cell transplantation. PloS ONE 4:e5643

Dimarino AM, Caplan AI, Bonfield TL (2013) Mesenchymal Stem Cells in Tissue Repair. Front Immunol 4:201

Domínguez-Bendala J, Ricordi C (2012) Present and future cell therapies for pancreatic beta cell replenishment. World J Gastroenterol WJG 18:6876–6884

Dominici M, Le Blanc K, Mueller I, Slaper-Cortenbach I, Marini F, Krause D, Deans R, Keating A, Prockop D, Horwitz E (2006) Minimal criteria for defining multipotent mesenchymal stromal cells. The International Society for Cellular Therapy position statement. Cytotherapy 8:315–317

Dong X, Pan R, Zhang H, Yang C, Shao J, Xiang L (2013) Modification of histone acetylation facilitates hepatic differentiation of human bone marrow mesenchymal stem cells. PloS ONE 8:e63405

Dreyer J-L (2010) New insights into the roles of microRNAs in drug addiction and neuroplasticity. Genome Med 2:92

Eliopoulos N, Stagg J, Lejeune L, Pommey S, Galipeau J (2005) Allogeneic marrow stromal cells are immune rejected by MHC class I- and class II-mismatched recipient mice. Blood 106:4057–4065

Erices A, Conget P, Minguell JJ (2000) Mesenchymal progenitor cells in human umbilical cord blood. Br J Haematol 109:235–242

Ernfors P, Kucera J, Lee KF, Loring J, Jaenisch R (1995) Studies on the physiological role of brain-derived neurotrophic factor and neurotrophin-3 in knockout mice. Int J Dev Biol 39:799–807

Fischer UM, Harting MT, Jimenez F, Monzon-Posadas WO, Xue H, Savitz SI, Laine GA, Cox CS Jr (2009) Pulmonary passage is a major obstacle for intravenous stem cell delivery: the pulmonary first-pass effect. Stem Cells Dev 18:683–692

Franquesa M et al (2013) Mesenchymal Stem Cells in Solid Organ Transplantation (MiSOT) fourth meeting: lessons learned from first clinical trials. Transplantation 96:234–238

Friedenstein A (1990) Osteogenic stem cells in bone marrow. In: Heersche JNM, Kanis JA (eds) Bone and mineral research. Elsevier, Amsterdam

Fukumoto T, Sperling JW, Sanyal A, Fitzsimmons JS, Reinholz GG, Conover CA, O'Driscoll SW (2003) Combined effects of insulin-like growth factor-1 and transforming growth factor-beta1 on periosteal mesenchymal cells during chondrogenesis in vitro. Osteoarthritis Cartilage 11(1):55–64

Giunti D, Parodi B, Usai C, Vergani L, Casazza S, Bruzzone S, Mancardi G, Uccelli A (2012) Mesenchymal stem cells shape microglia effector functions through the release of CX3CL1. Stem Cells (Dayt Ohio) 30:2044–2053

Gnecchi M, He H, Liang OD, Melo LG, Morello F, Mu H, Noiseux N, Zhang L, Pratt RE, Ingwall JS, Dzau VJ (2005) Paracrine action accounts for marked protection of ischemic heart by Akt-modified mesenchymal stem cells. Nat Med 11:367–368

Gnecchi M, Zhang Z, Ni A, Dzau VJ (2008) Paracrine mechanisms in adult stem cell signaling and therapy. Circ Res 103:1204–1219

Graham DI, Ford I, Adams JH, Doyle D, Teasdale GM, Lawrence AE, McLellan DR (1989) Ischaemic brain damage is still common in fatal non-missile head injury. J Neurol Neurosurg Psychiatry 52:346–350

Grinnemo KH, Månsson A, Dellgren G, Klingberg D, Wardell E, Drvota V, Tammik C, Holgersson J, Ringdén O, Sylvén C, Le Blanc K (2004) Xenoreactivity and engraftment of human mesenchymal stem cells transplanted into infarcted rat myocardium. J Thorac Cardiovasc Surg 127:1293–1300

Grinnemo KH, Månsson-Broberg A, Leblanc K, Corbascio M, Wärdell E, Siddiqui AJ, Hao X, Sylvén C, Dellgren G (2006) Human mesenchymal stem cells do not differentiate into cardiomyocytes in a cardiac ischemic xenomodel. Ann Med 38:144–153

Guan J, Zhu Z, Zhao RC, Xiao Z, Wu C, Han Q, Chen L, Tong W, Zhang J, Han Q, Gao J, Feng M, Bao X, Dai J, Wang R (2013) Transplantation of human mesenchymal stem cells loaded on collagen scaffolds for the treatment of traumatic brain injury in rats. Biomaterials 34:5937–5946

Han EY, Chun MH, Kim ST, Lim D (2013) Injection time-dependent effect of adult human bone marrow stromal cell transplantation in a rat model of severe traumatic brain injury. Curr Stem Cell Res Ther 8:172–181

Haniffa MA, Collin MP, Buckley CD, Dazzi F (2009) Mesenchymal stem cells: the fibroblasts' new clothes? Haematologica 94:258–263

Harting MT, Jimenez F, Xue H, Fischer UM, Baumgartner J, Dash PK, Cox CS (2009) Intravenous mesenchymal stem cell therapy for traumatic brain injury. J Neurosurg 110:1189–1197

Heile AMB, Wallrapp C, Klinge PM, Samii A, Kassem M, Silverberg G, Brinker T (2009) Cerebral transplantation of encapsulated mesenchymal stem cells improves cellular pathology after experimental traumatic brain injury. Neurosci Lett 463:176–181

Horwitz EM, Le Blanc K, Dominici M, Mueller I, Slaper-Cortenbach I, Marini FC, Deans RJ, Krause DS, Keating A, International Society for Cellular Therapy (2005) Clarification of the nomenclature for MSC: the international society for cellular therapy position statement. Cytotherapy 7:393–395

Hu S-L, Luo H-S, Li J-T, Xia Y-Z, Li L, Zhang L-J, Meng H, Cui G-Y, Chen Z, Wu N, Lin J-K, Zhu G, Feng H (2010) Functional recovery in acute traumatic spinal cord injury after transplantation of human umbilical cord mesenchymal stem cells. Crit Care Med 38:2181–2189

Hu X, Li P, Guo Y, Wang H, Leak RK, Chen S, Gao Y, Chen J (2012) Microglia/macrophage polarization dynamics reveal novel mechanism of injury expansion after focal cerebral ischemia. Stroke J Cereb Circ 43:3063–3070

Huang X-P, Sun Z, Miyagi Y, McDonald Kinkaid H, Zhang L, Weisel RD, Li R-K (2010) Differentiation of allogeneic mesenchymal stem cells induces immunogenicity and limits their long-term benefits for myocardial repair. Circulation 122:2419–2429

In't Anker PS, Scherjon SA, Kleijburg-van der Keur C, Noort WA, Claas FHJ, Willemze R, Fibbe WE, Kanhai HHH (2003) Amniotic fluid as a novel source of mesenchymal stem cells for therapeutic transplantation. Blood 102:1548–1549

Iseki K, Hagino S, Nikaido T, Zhang Y, Mori T, Yokoya S, Hozumi Y, Goto K, Wanaka A, Tase C (2012) Gliosis-specific transcription factor OASIS coincides with proteoglycan core protein genes in the glial scar and inhibits neurite outgrowth. Biomed Res Tokyo Jpn 33:345–353

Jankowski RJ, Deasy BM, Huard J (2002) Muscle-derived stem cells. Gene Ther 9:642–647

Jenkins LW, Moszynski K, Lyeth BG, Lewelt W, DeWitt DS, Allen A, Dixon CE, Povlishock JT, Majewski TJ, Clifton GL (1989) Increased vulnerability of the mildly traumatized rat brain to cerebral ischemia: the use of controlled secondary ischemia as a research tool to identify common or different mechanisms contributing to mechanical and ischemic brain injury. Brain Res 477:211–224

Jones TH, Morawetz RB, Crowell RM, Marcoux FW, FitzGibbon SJ, DeGirolami U, Ojemann RG (1981) Thresholds of focal cerebral ischemia in awake monkeys. J Neurosurg 54:773–782

Katsuda T, Kosaka N, Takeshita F, Ochiya T (2013) The therapeutic potential of mesenchymal stem cell-derived extracellular vesicles. Proteomics 13:1637–1653

Kern S, Eichler H, Stoeve J, Klüter H, Bieback K (2006) Comparative analysis of mesenchymal stem cells from bone marrow, umbilical cord blood, or adipose tissue. Stem Cells (Dayt Ohio) 24:1294–1301

Kigerl KA, Gensel JC, Ankeny DP, Alexander JK, Donnelly DJ, Popovich PG (2009) Identification of two distinct macrophage subsets with divergent effects causing either neurotoxicity or regeneration in the injured mouse spinal cord. J Neurosci Off J Soc Neurosci 29:13435–13444

Kim H-J, Lee J-H, Kim S-H (2010) Therapeutic effects of human mesenchymal stem cells on traumatic brain injury in rats: secretion of neurotrophic factors and inhibition of apoptosis. J Neurotrauma 27:131–138

Kim H-S, Choi D-Y, Yun SJ, Choi S-M, Kang JW, Jung JW, Hwang D, Kim KP, Kim D-W (2012) Proteomic analysis of microvesicles derived from human mesenchymal stem cells. J Proteome Res 11:839–849

Kinnaird T, Stabile E, Burnett MS, Lee CW, Barr S, Fuchs S, Epstein SE (2004) Marrow-derived stromal cells express genes encoding a broad spectrum of arteriogenic cytokines and promote in vitro and in vivo arteriogenesis through paracrine mechanisms. Circ Res 94:678–685

Knaän-Shanzer S (2013) The immune status of mesenchymal stem cells and its relevance for therapeutic application. Stem Cells (Dayt Ohio)

Kokaia Z, Lindvall O (2003) Neurogenesis after ischaemic brain insults. Curr Opin Neurobiol 13:127–132

Kumar S, Kahn MA, Dinh L, de Vellis J (1998) NT-3-mediated TrkC receptor activation promotes proliferation and cell survival of rodent progenitor oligodendrocyte cells in vitro and in vivo. J Neurosci Res 54:754–765

Lai AY, Todd KG (2008) Differential regulation of trophic and proinflammatory microglial effectors is dependent on severity of neuronal injury. Glia 56:259–270

Lai RC, Arslan F, Lee MM, Sze NSK, Choo A, Chen TS, Salto-Tellez M, Timmers L, Lee CN, El Oakley RM, Pasterkamp G, de Kleijn DPV, Lim SK (2010) Exosome secreted by MSC reduces myocardial ischemia/reperfusion injury. Stem Cell Res 4:214–222

Lai RC, Chen TS, Lim SK (2011) Mesenchymal stem cell exosome: a novel stem cell-based therapy for cardiovascular disease. Regen Med 6:481–492

Lai RC, Yeo RWY, Tan KH, Lim SK (2013) Mesenchymal stem cell exosome ameliorates reperfusion injury through proteomic complementation. Regen Med 8:197–209

Laird MD, Vender JR, Dhandapani KM (2008) Opposing roles for reactive astrocytes following traumatic brain injury. Neurosignals 16:154–164

Li L, Jiang Q, Qu CS, Ding GL, Li QJ, Wang SY, Lee JH, Lu M, Mahmood A, Chopp M (2011) Transplantation of marrow stromal cells restores cerebral blood flow and reduces cerebral atrophy in rats with traumatic brain injury: in vivo MRI study. J Neurotrauma 28:535–545

Li T, Yan Y, Wang B, Qian H, Zhang X, Shen L, Wang M, Zhou Y, Zhu W, Li W, Xu W (2013) Exosomes derived from human umbilical cord mesenchymal stem cells alleviate liver fibrosis. Stem Cells Dev 22:845–854

Lin B, Xu Y, Zhang B, He Y, Yan Y, He M-C (2014) MEK inhibition reduces glial scar formation and promotes the recovery of sensorimotor function in rats following spinal cord injury. Exp Ther Med 7:66–72

Loane DJ, Byrnes KR (2010) Role of microglia in neurotrauma. Neurother J Am Soc Exp Neurother 7:366–377

Longhi L, Perego C, Ortolano F, Aresi S, Fumagalli S, Zanier ER, Stocchetti N, De Simoni M-G (2013) Tumor necrosis factor in traumatic brain injury: effects of genetic deletion of p55 or p75 receptor. J Cereb Blood Flow Metab 33(8):1182–1189

Lu D, Li Y, Wang L, Chen J, Mahmood A, Chopp M (2001a) Intraarterial administration of marrow stromal cells in a rat model of traumatic brain injury. J Neurotrauma 18:813–819

Lu D, Mahmood A, Wang L, Li Y, Lu M, Chopp M (2001b) Adult bone marrow stromal cells administered intravenously to rats after traumatic brain injury migrate into brain and improve neurological outcome. Neuroreport 12:559–563

Lu D, Mahmood A, Qu C, Hong X, Kaplan D, Chopp M (2007) Collagen scaffolds populated with human marrow stromal cells reduce lesion volume and improve functional outcome after traumatic brain injury. Neurosurgery 61:596–602; discussion 602–603

Lucas S-M, Rothwell NJ, Gibson RM (2006) The role of inflammation in CNS injury and disease. Br J Pharmacol 147(Suppl 1):S232–240

Lundberg J, Södersten E, Sundström E, Le Blanc K, Andersson T, Hermanson O, Holmin S (2012) Targeted intra-arterial transplantation of stem cells to the injured CNS is more effective than intravenous administration: engraftment is dependent on cell type and adhesion molecule expression. Cell Transplant 21:333–343

Madinier A, Bertrand N, Mossiat C, Prigent-Tessier A, Beley A, Marie C, Garnier P (2009) Microglial involvement in neuroplastic changes following focal brain ischemia in rats. PloS ONE 4:e8101

Magatti M, De Munari S, Vertua E, Gibelli L, Wengler GS, Parolini O (2008) Human amnion mesenchyme harbors cells with allogeneic T-cell suppression and stimulation capabilities. Stem Cells (Dayt Ohio) 26:182–192

Maggini J, Mirkin G, Bognanni I, Holmberg J, Piazzón IM, Nepomnaschy I, Costa H, Cañones C, Raiden S, Vermeulen M, Geffner JR (2010) Mouse bone marrow-derived mesenchymal stromal cells turn activated macrophages into a regulatory-like profile. PloS ONE 5:e9252

Mahmood A, Lu D, Wang L, Li Y, Lu M, Chopp M (2001a) Treatment of traumatic brain injury in female rats with intravenous administration of bone marrow stromal cells. Neurosurgery 49:1196–1203; discussion 1203–1204

Mahmood A, Lu D, Yi L, Chen JL, Chopp M (2001b) Intracranial bone marrow transplantation after traumatic brain injury improving functional outcome in adult rats. J Neurosurg 94:589–595

Mahmood A, Lu D, Wang L, Chopp M (2002) Intracerebral transplantation of marrow stromal cells cultured with neurotrophic factors promotes functional recovery in adult rats subjected to traumatic brain injury. J Neurotrauma 19:1609–1617

Mahmood A, Lu D, Lu M, Chopp M (2003) Treatment of traumatic brain injury in adult rats with intravenous administration of human bone marrow stromal cells. Neurosurgery 53:697–702; discussion 702–703

Mahmood A, Lu D, Chopp M (2004a) Intravenous administration of marrow stromal cells (MSCs) increases the expression of growth factors in rat brain after traumatic brain injury. J Neurotrauma 21:33–39

Mahmood A, Lu D, Chopp M (2004b) Marrow stromal cell transplantation after traumatic brain injury promotes cellular proliferation within the brain. Neurosurgery 55:1185–1193

Mahmood A, Lu D, Qu C, Goussev A, Chopp M (2005) Human marrow stromal cell treatment provides long-lasting benefit after traumatic brain injury in rats. Neurosurgery 57:1026–1031; discussion 1026–1031

Mahmood A, Lu D, Qu C, Goussev A, Chopp M (2007) Treatment of traumatic brain injury with a combination therapy of marrow stromal cells and atorvastatin in rats. Neurosurgery 60:546–553; discussion 553–554

Mahmood A, Qu C, Ning R, Wu H, Goussev A, Xiong Y, Irtenkauf S, Li Y, Chopp M (2011) Treatment of TBI with collagen scaffolds and human marrow stromal cells increases the expression of tissue plasminogen activator. J Neurotrauma 28:1199–1207

Mahmood A, Wu H, Qu C, Mahmood S, Xiong Y, Kaplan D, Chopp M (2013a) Down-regulation of Nogo-A by collagen scaffolds impregnated with bone marrow stromal cell treatment after traumatic brain injury promotes axonal regeneration in rats. Brain Res

Mahmood A, Wu H, Qu C, Xiong Y, Chopp M (2013b) Effects of treating traumatic brain injury with collagen scaffolds and human bone marrow stromal cells on sprouting of corticospinal tract axons into the denervated side of the spinal cord. J Neurosurg 118:381–389

Mahmood A, Wu H, Qu C, Mahmood S, Xiong Y, Kaplan D, Chopp M (2014) Down-regulation of Nogo-A by collagen scaffolds impregnated with bone marrow stromal cell treatment after traumatic brain injury promotes axonal regeneration in rats. Brain Res 1542:41–48

Makino S, Fukuda K, Miyoshi S, Konishi F, Kodama H, Pan J, Sano M, Takahashi T, Hori S, Abe H, Hata J, Umezawa A, Ogawa S (1999) Cardiomyocytes can be generated from marrow stromal cells in vitro. J Clin Invest 103:697–705

Marappagounder D, Somasundaram I, Dorairaj S, Sankaran RJ (2013) Differentiation of mesenchymal stem cells derived from human bone marrow and subcutaneous adipose tissue into pancreatic islet-like clusters in vitro. Cell Mol Biol Lett 18:75–88

Mathivanan S, Ji H, Simpson RJ (2010) Exosomes: extracellular organelles important in intercellular communication. J Proteomics 73:1907–1920

Matsumoto M, Imura T, Fukazawa T, Sun Y, Takeda M, Kajiume T, Kawahara Y, Yuge L (2013) Electrical stimulation enhances neurogenin2 expression through β-catenin signaling pathway of mouse bone marrow stromal cells and intensifies the effect of cell transplantation on brain injury. Neurosci Lett 533:71–76

McIntosh TK, Smith DH, Meaney DF, Kotapka MJ, Gennarelli TA, Graham DI (1996) Neuropathological sequelae of traumatic brain injury: relationship to neurochemical and biomechanical mechanisms. Lab Investig J Tech Methods Pathol 74:315–342

Meirelles L da S, Nardi NB (2009) Methodology, biology and clinical applications of mesenchymal stem cells. Front Biosci Landmark Ed 14:4281–4298

Meng X, Ichim TE, Zhong J, Rogers A, Yin Z, Jackson J, Wang H, Ge W, Bogin V, Chan KW, Thébaud B, Riordan NH (2007) Endometrial regenerative cells: a novel stem cell population. J Transl Med 5:57

Menge T, Zhao Y, Zhao J, Wataha K, Gerber M, Zhang J, Letourneau P, Redell J, Shen L, Wang J, Peng Z, Xue H, Kozar R, Cox CS Jr, Khakoo AY, Holcomb JB, Dash PK, Pati S (2012) Mesenchymal stem cells regulate blood-brain barrier integrity through TIMP3 release after traumatic brain injury. Sci Transl Med 4:161ra150

Morganti-Kossmann MC, Satgunaseelan L, Bye N, Kossmann T (2007) Modulation of immune response by head injury. Injury 38:1392–1400

Mori K, Iwata J, Miyazaki M, Nakao Y, Maeda M (2005) Functional recovery of neuronal activity in rat whisker-barrel cortex sensory pathway from freezing injury after transplantation of adult bone marrow stromal cells. J Cereb Blood Flow Metab 25:887–898

Murray IR, West CC, Hardy WR, James AW, Park TS, Nguyen A, Tawonsawatruk T, Lazzari L, Soo C, Péault B (2013) Natural history of mesenchymal stem cells, from vessel walls to culture vessels. Cell Mol Life Sci CMLS

Musina RA, Belyavski AV, Tarusova OV, Solovyova EV, Sukhikh GT (2008) Endometrial mesenchymal stem cells isolated from the menstrual blood. Bull Exp Biol Med 145:539–543

Nakajima K, Yamamoto S, Kohsaka S, Kurihara T (2008) Neuronal stimulation leading to upregulation of glutamate transporter-1 (GLT-1) in rat microglia in vitro. Neurosci Lett 436:331–334

Nakajima H, Uchida K, Guerrero AR, Watanabe S, Sugita D, Takeura N, Yoshida A, Long G, Wright KT, Johnson WEB, Baba H (2012) Transplantation of mesenchymal stem cells promotes an alternative pathway of macrophage activation and functional recovery after spinal cord injury. J Neurotrauma 29:1614–1625

Nauta AJ, Fibbe WE (2007) Immunomodulatory properties of mesenchymal stromal cells. Blood 110:3499–3506

Nauta AJ, Westerhuis G, Kruisselbrink AB, Lurvink EGA, Willemze R, Fibbe WE (2006) Donor-derived mesenchymal stem cells are immunogenic in an allogeneic host and stimulate donor graft rejection in a nonmyeloablative setting. Blood 108:2114–2120

Neher JJ, Neniskyte U, Zhao J-W, Bal-Price A, Tolkovsky AM, Brown GC (2011) Inhibition of microglial phagocytosis is sufficient to prevent inflammatory neuronal death. J Immunol 186:4973–4983

Neher JJ, Neniskyte U, Brown GC (2012) Primary phagocytosis of neurons by inflamed microglia: potential roles in neurodegeneration. Front Pharmacol 3:27

Neirinckx V, Coste C, Rogister B, Wislet-Gendebien S (2013a) Concise review: adult mesenchymal stem cells, adult neural crest stem cells, and therapy of neurological pathologies: a state of play. Stem Cells Transl Med 2:284–296

Neirinckx V, Marquet A, Coste C, Rogister B, Wislet-Gendebien S (2013b) Adult bone marrow neural crest stem cells and mesenchymal stem cells are not able to replace lost neurons in acute MPTP-lesioned mice. PloS ONE 8:e64723

Niemeyer P, Vohrer J, Schmal H, Kasten P, Fellenberg J, Suedkamp NP, Mehlhorn AT (2008) Survival of human mesenchymal stromal cells from bone marrow and adipose tissue after xenogenic transplantation in immunocompetent mice. Cytotherapy 10:784–795

Nomura T, Honmou O, Harada K, Houkin K, Hamada H, Kocsis JD (2005) I.V. infusion of brain-derived neurotrophic factor gene-modified human mesenchymal stem cells protects against injury in a cerebral ischemia model in adult rat. Neuroscience 136:161–169

Nudo RJ (2013) Recovery after brain injury: mechanisms and principles. Front Hum Neurosci 7:887

Ooi YY, Ramasamy R, Rahmat Z, Subramaiam H, Tan SW, Abdullah M, Israf DA, Vidyadaran S (2010) Bone marrow-derived mesenchymal stem cells modulate BV2 microglia responses to lipopolysaccharide. Int Immunopharmacol 10:1532–1540

Parkhurst CN, Yang G, Ninan I, Savas JN, Yates JR 3rd, Lafaille JJ, Hempstead BL, Littman DR, Gan W-B (2013) Microglia promote learning-dependent synapse formation through brain-derived neurotrophic factor. Cell 155:1596–1609

Parolini O et al (2008) Concise review: isolation and characterization of cells from human term placenta: outcome of the first international workshop on placenta derived stem cells. Stem Cells (Dayt Ohio) 26:300–311

Peister A, Mellad JA, Larson BL, Hall BM, Gibson LF, Prockop DJ (2004) Adult stem cells from bone marrow (MSCs) isolated from different strains of inbred mice vary in surface epitopes, rates of proliferation, and differentiation potential. Blood 103:1662–1668

Perego C, Fumagalli S, De Simoni M-G (2011) Temporal pattern of expression and colocalization of microglia/macrophage phenotype markers following brain ischemic injury in mice. J Neuroinflammation 8:174

Phinney DG (2012) Functional heterogeneity of mesenchymal stem cells: implications for cell therapy. J Cell Biochem 113:2806–2812

Phinney DG, Prockop DJ (2007) Concise review: mesenchymal stem/multipotent stromal cells: the state of transdifferentiation and modes of tissue repair–current views. Stem Cells (Dayt Ohio) 25:2896–2902

Phinney DG, Kopen G, Righter W, Webster S, Tremain N, Prockop DJ (1999) Donor variation in the growth properties and osteogenic potential of human marrow stromal cells. J Cell Biochem 75:424–436

Pietilä M, Palomäki S, Lehtonen S, Ritamo I, Valmu L, Nystedt J, Laitinen S, Leskelä H-V, Sormunen R, Pesälä J, Nordström K, Vepsäläinen A, Lehenkari P (2012) Mitochondrial function and energy metabolism in umbilical cord blood- and bone marrow-derived mesenchymal stem cells. Stem Cells Dev 21:575–588

Pischiutta F, D'Amico G, Dander E, Biondi A, Biagi E, Citerio G, De Simoni M-G, Zanier ER (2014) Immunosuppression does not affect human bone marrow mesenchymal stromal cell efficacy after transplantation in traumatized mice brain. Neuropharmacology 79:119–126

Prockop DJ, Oh JY (2012) Mesenchymal stem/stromal cells (MSCs): role as guardians of inflammation. Mol Ther J Am Soc Gene Ther 20:14–20

Qu C, Mahmood A, Lu D, Goussev A, Xiong Y, Chopp M (2008) Treatment of traumatic brain injury in mice with marrow stromal cells. Brain Res 1208:234–239

Qu C, Xiong Y, Mahmood A, Kaplan DL, Goussev A, Ning R, Chopp M (2009) Treatment of traumatic brain injury in mice with bone marrow stromal cell-impregnated collagen scaffolds. J Neurosurg 111:658–665

Qu C, Mahmood A, Liu XS, Xiong Y, Wang L, Wu H, Li B, Zhang ZG, Kaplan DL, Chopp M (2011) The treatment of TBI with human marrow stromal cells impregnated into collagen scaffold: functional outcome and gene expression profile. Brain Res 1371:129–139

Richardson RM, Sun D, Bullock MR (2007) Neurogenesis after traumatic brain injury. Neurosurg Clin N Am 18:169–181, xi

Rodriguez A-M, Elabd C, Amri E-Z, Ailhaud G, Dani C (2005) The human adipose tissue is a source of multipotent stem cells. Biochimie 87:125–128

Rogers I, Casper RF (2004) Umbilical cord blood stem cells. Best Pract Res Clin Obstet Gynaecol 18:893–908

Rossignoli F, Caselli A, Grisendi G, Piccinno S, Burns JS, Murgia A, Veronesi E, Loschi P, Masini C, Conte P, Paolucci P, Horwiz EM, Dominici M (2013) Isolation, characterization, and transduction of endometrial decidual tissue multipotent mesenchymal stromal/stem cells from menstrual blood. Biomed Res Int 2013:901821

Rüster B, Göttig S, Ludwig RJ, Bistrian R, Müller S, Seifried E, Gille J, Henschler R (2006) Mesenchymal stem cells display coordinated rolling and adhesion behavior on endothelial cells. Blood 108:3938–3944

Saito A, Narasimhan P, Hayashi T, Okuno S, Ferrand-Drake M, Chan PH (2004) Neuroprotective role of a proline-rich Akt substrate in apoptotic neuronal cell death after stroke: relationships with nerve growth factor. J Neurosci 24:1584–1593

Salem HK, Thiemermann C (2010) Mesenchymal stromal cells: current understanding and clinical status. Stem Cells (Dayt Ohio) 28:585–596

Salingcarnboriboon R, Yoshitake H, Tsuji K, Obinata M, Amagasa T, Nifuji A, Noda M (2003) Establishment of tendon-derived cell lines exhibiting pluripotent mesenchymal stem cell-like property. Exp Cell Res 287:289–300

Sancho-Bru P, Najimi M, Caruso M, Pauwelyn K, Cantz T, Forbes S, Roskams T, Ott M, Gehling U, Sokal E, Verfaillie CM, Muraca M (2009) Stem and progenitor cells for liver repopulation: can we standardise the process from bench to bedside? Gut 58:594–603

Saracino GAA, Cigognini D, Silva D, Caprini A, Gelain F (2013) Nanomaterials design and tests for neural tissue engineering. Chem Soc Rev 42:225–262

Schinköthe T, Bloch W, Schmidt A (2008) In vitro secreting profile of human mesenchymal stem cells. Stem Cells Dev 17:199–206

Schrepfer S, Deuse T, Reichenspurner H, Fischbein MP, Robbins RC, Pelletier MP (2007) Stem cell transplantation: the lung barrier. Transplant Proc 39:573–576

Shechter R, Schwartz M (2013) Harnessing monocyte-derived macrophages to control central nervous system pathologies: no longer "if" but "how." J Pathol 229:332–346

Shi S, Gronthos S (2003) Perivascular niche of postnatal mesenchymal stem cells in human bone marrow and dental pulp. J Bone Miner Res 18:696–704

Soncini M, Vertua E, Gibelli L, Zorzi F, Denegri M, Albertini A, Wengler GS, Parolini O (2007) Isolation and characterization of mesenchymal cells from human fetal membranes. J Tissue Eng Regen Med 1:296–305

Song D, Ohtaki H, Tsumuraya T, Miyamoto K, Shibato J, Rakwal R, Xu Z, Hiraizumi Y, Inoue T, Shioda S (2013) The anti-inflammatory property of human bone marrow-derived mesenchymal stem/stromal cells is preserved in late-passage cultures. J Neuroimmunol 263:55–63

Stocchetti N (2005) Brain and sepsis: functional impairment, structural damage, and markers. Anesth Analg 101:1463–1464

Stocchetti N, Longhi L, Zanier ER (2008) Intracranial pressure monitoring for traumatic brain injury: available evidence and clinical implications. Minerva Anestesiol 74:197–203

Stoll G, Jander S (1999) The role of microglia and macrophages in the pathophysiology of the CNS. Prog Neurobiol 58:233–247

Tajiri N, Acosta SA, Shahaduzzaman M, Ishikawa H, Shinozuka K, Pabon M, Hernandez-Ontiveros D, Kim DW, Metcalf C, Staples M, Dailey T, Vasconcellos J, Franyuti G, Gould L, Patel N, Cooper D, Kaneko Y, Borlongan CV, Bickford PC (2014) Intravenous transplants of human adipose-derived stem cell protect the brain from traumatic brain injury-induced neurodegeneration and motor and cognitive impairments: cell graft biodistribution and soluble factors in young and aged rats. J Neurosci 34:313–326

Tate CC, Fonck C, McGrogan M, Case CC (2010) Human mesenchymal stromal cells and their derivative, SB623 cells, rescue neural cells via trophic support following in vitro ischemia. Cell Transplant 19:973–984

Timmers L, Lim SK, Arslan F, Armstrong JS, Hoefer IE, Doevendans PA, Piek JJ, El Oakley RM, Choo A, Lee CN, Pasterkamp G, de Kleijn DPV (2007) Reduction of myocardial infarct size by human mesenchymal stem cell conditioned medium. Stem Cell Res 1:129–137

Uccelli A, Moretta L, Pistoia V (2008) Mesenchymal stem cells in health and disease. Nat Rev Immunol 8:726–736

Van Poll D, Parekkadan B, Cho CH, Berthiaume F, Nahmias Y, Tilles AW, Yarmush ML (2008) Mesenchymal stem cell-derived molecules directly modulate hepatocellular death and regeneration in vitro and in vivo. Hepatol Baltim Md 47:1634–1643

Verderio C (2013) Extracellular membrane microvesicles and nanotubes in the brain: understanding their nature, their function in cell-to-cell communication, their role in transcellular spreading of pathological agents and their therapeutic potential. Front Physiol 4:163

Villaron EM, Almeida J, López-Holgado N, Alcoceba M, Sánchez-Abarca LI, Sanchez-Guijo FM, Alberca M, Pérez-Simon JA, San Miguel JF, Del Cañizo MC (2004) Mesenchymal stem cells are present in peripheral blood and can engraft after allogeneic hematopoietic stem cell transplantation. Haematologica 89:1421–1427

Wagner W, Wein F, Seckinger A, Frankhauser M, Wirkner U, Krause U, Blake J, Schwager C, Eckstein V, Ansorge W, Ho AD (2005) Comparative characteristics of mesenchymal stem cells from human bone marrow, adipose tissue, and umbilical cord blood. Exp Hematol 33:1402–1416

Wakabayashi K, Nagai A, Sheikh AM, Shiota Y, Narantuya D, Watanabe T, Masuda J, Kobayashi S, Kim SU, Yamaguchi S (2010) Transplantation of human mesenchymal stem cells promotes functional improvement and increased expression of neurotrophic factors in a rat focal cerebral ischemia model. J Neurosci Res 88:1017–1025

Walczak P, Zhang J, Gilad AA, Kedziorek DA, Ruiz-Cabello J, Young RG, Pittenger MF, van Zijl PCM, Huang J, Bulte JWM (2008) Dual-modality monitoring of targeted intraarterial delivery of mesenchymal stem cells after transient ischemia. Stroke J Cereb Circ 39:1569–1574

Walker PA, Harting MT, Jimenez F, Shah SK, Pati S, Dash PK, Cox CS Jr (2010) Direct intrathecal implantation of mesenchymal stromal cells leads to enhanced neuroprotection via an NFkappaB-mediated increase in interleukin–6 production. Stem Cells Dev 19:867–876

Walker PA, Bedi S, Shah SK, Jimenez F, Xue H, Hamilton JA, Smith P, Thomas CP, Mays RW, Pati S, Cox CS (2012a) Intravenous multipotent adult progenitor cell therapy after traumatic brain injury: modulation of the resident microglia population. J Neuroinflammation 9:228

Walker PA, Shah SK, Jimenez F, Aroom KR, Harting MT, Cox CS Jr (2012b) Bone marrow-derived stromal cell therapy for traumatic brain injury is neuroprotective via stimulation of non-neurologic organ systems. Surgery 152:790–793

Wang Z, Yao W, Deng Q, Zhang X, Zhang J (2013) Protective effects of BDNF overexpression bone marrow stromal cell transplantation in rat models of traumatic brain injury. J Mol Neurosci 49:409–416

Watanabe J, Shetty AK, Hattiangady B, Kim D-K, Foraker JE, Nishida H, Prockop DJ (2013) Administration of TSG-6 improves memory after traumatic brain injury in mice. Neurobiol Dis 59:86–99

Wisniewski HG, Vilcek J (1997) TSG-6: an IL-1/TNF-inducible protein with anti-inflammatory activity. Cytokine Growth Factor Rev 8:143–156

Wolpe SD, Sherry B, Juers D, Davatelis G, Yurt RW, Cerami A (1989) Identification and characterization of macrophage inflammatory protein 2. Proc Natl Acad Sci U S A 86:612–616

Xin H, Li Y, Shen LH, Liu X, Wang X, Zhang J, Pourabdollah-Nejad D S, Zhang C, Zhang L, Jiang H, Zhang ZG, Chopp M (2010) Increasing tPA activity in astrocytes induced by multipotent mesenchymal stromal cells facilitate neurite outgrowth after stroke in the mouse. PloS ONE 5:e9027

Xin H, Li Y, Cui Y, Yang JJ, Zhang ZG, Chopp M (2013a) Systemic administration of exosomes released from mesenchymal stromal cells promote functional recovery and neurovascular plasticity after stroke in rats. J Cereb Blood Flow Metab 33:1711–1715

Xin H, Li Y, Liu Z, Wang X, Shang X, Cui Y, Gang Zhang Z, Chopp M (2013b) Mir-133b Promotes Neural Plasticity and Functional Recovery after Treatment of Stroke with Multipotent Mesenchymal Stromal Cells in Rats Via Transfer of Exosome-Enriched Extracellular Particles. Stem Cells (Dayt Ohio)

Xiong Y, Qu C, Mahmood A, Liu Z, Ning R, Li Y, Kaplan DL, Schallert T, Chopp M (2009) Delayed transplantation of human marrow stromal cell-seeded scaffolds increases transcallosal neural fiber length, angiogenesis, and hippocampal neuronal survival and improves functional outcome after traumatic brain injury in rats. Brain Res 1263:183–191

Xu Y, Malladi P, Wagner DR, Longaker MT (2005) Adipose-derived mesenchymal cells as a potential cell source for skeletal regeneration. Curr Opin Mol Ther 7:300–305

Yang C-C, Shih Y-H, Ko M-H, Hsu S-Y, Cheng H, Fu Y-S (2008) Transplantation of human umbilical mesenchymal stem cells from Wharton's jelly after complete transection of the rat spinal cord. PloS ONE 3:e3336

Yang J, Han Y, Ye W, Liu F, Zhuang K, Wu G (2013) Alpha tocopherol treatment reduces the expression of Nogo-A and NgR in rat brain after traumatic brain injury. J Surg Res 182:e69–77

Yasuda H, Kuroda S, Shichinohe H, Kamei S, Kawamura R, Iwasaki Y (2010) Effect of biodegradable fibrin scaffold on survival, migration, and differentiation of transplanted bone marrow stromal cells after cortical injury in rats. J Neurosurg 112:336–344

Yuan J, Yankner BA (2000) Apoptosis in the nervous system. Nature 407:802–809

Zanier ER, Montinaro M, Vigano M, Villa P, Fumagalli S, Pischiutta F, Longhi L, Leoni ML, Rebulla P, Stocchetti N, Lazzari L, De Simoni M-G (2011) Human umbilical cord blood mesenchymal stem cells protect mice brain after trauma. Crit Care Med 39:2501–2510

Zanier RE, Pischiutta F, Riganti L, Marchesi F, Turola E, Fumagalli S, Perego C, Parotto E, Vinci P, Veglianese P, D'Amico G, Verderio C, De Simoni M-G (2014) Bone marrow mesenchymal stromal cells drive protective M2 microglia polarization after brain trauma. Neurother J Am Soc Exp Neurother 11:679–695

Zhang R, Liu Y, Yan K, Chen L, Chen X-R, Li P, Chen F-F, Jiang X-D (2013) Anti-inflammatory and immunomodulatory mechanisms of mesenchymal stem cell transplantation in experimental traumatic brain injury. J Neuroinflammation 10:106

Zhao C, Deng W, Gage FH (2008) Mechanisms and functional implications of adult neurogenesis. Cell 132:645–660

Zhu Y-G, Feng X-M, Abbott J, Fang X-H, Hao Q, Monsel A, Qu J-M, Matthay MA, Lee JW (2014) Human Mesenchymal Stem Cell Microvesicles for Treatment of Escherichia coli Endotoxin-Induced Acute Lung Injury in Mice. Stem Cells (Dayt Ohio) 32:116–125

Zigova T, Pencea V, Wiegand SJ, Luskin MB (1998) Intraventricular administration of BDNF increases the number of newly generated neurons in the adult olfactory bulb. Mol Cell Neurosci 11:234–245

Zuk PA, Zhu M, Mizuno H, Huang J, Futrell JW, Katz AJ, Benhaim P, Lorenz HP, Hedrick MH (2001) Multilineage cells from human adipose tissue: implications for cell-based therapies. Tissue Eng 7:211–228

Zygun DA, Zuege DJ, Boiteau PJE, Laupland KB, Henderson EA, Kortbeek JB, Doig CJ (2006) Ventilator-associated pneumonia in severe traumatic brain injury. Neurocrit Care 5:108–114

Chapter 12
Transplantation of Embryonic Stem Cells in Traumatic Brain Injury

Peter Riess, Marek Molcanyi, Edmund A. M. Neugebauer and Ewa K. Stuermer

Contents

Abstract Traumatic Brain Injury (TBI) is one of the leading causes of death and disability worldwide. Experimental research during the past decade has increased our understanding of the pathophysiology of TBI and allowed us to develop neuroprotective pharmacological therapies. However, to date, no therapeutic approach has been proven effective in reversing the pathologic cellular sequelae underlying the progression of cell loss and in improving neurobehavioral outcome. Embryonic stem cells (ESC) are pluripotent cells with the ability to differentiate into any brain-tissue-specific cell type. Therefore ESC may possess great therapeutic potential in brain injury. Therefore ESC transplantation has begun to be evaluated in several models of experimental TBI, with promising results. The following is a compendium of these new and exciting studies, including a critical discussion of the rationale and hazards associated with ESC transplantation techniques in experimental TBI research.

P. Riess (✉)
Department of Trauma and Orthopedic Surgery, HELIOS Klinikum Bad Berleburg, Teaching Hospital University Marburg, An der Gontardslust 7, 57319 Bad Berleburg, Germany
e-mail: peter.riess@helios-kliniken.de

M. Molcanyi
Institute for Neurophysiology, University of Cologne, Robert Koch Str. 39, 50931 Cologne, Germany

E. A. M. Neugebauer · E. K. Stuermer
Institute for Research in Operative Medicine, University of Witten/Herdecke, Ostmerheimer Str. 200, 51109 Köln, Germany

L.-R. Zhao, J. H. Zhang (eds.), *Cellular Therapy for Stroke and CNS Injuries*, Springer Series in Translational Stroke Research, DOI 10.1007/978-3-319-11481-1_12

Abbreviations

TBI	Traumatic Brain Injury
ESC	Embryonic stem cells
BBB	Blood-brain barrier
CCI	Controlled Cortical Impact
CNS	Central nervous system
FP	Lateral fluid percussion brain injury
IFN	Gamma interferon-gamma
TNFalpha	Tumor necrosis factor-alpha

12.1 Introduction

Traumatic Brain Injury (TBI) is one of the leading causes of death and disability worldwide (Bruns et al. 2003). Thus, it is a highly relevant medical and socio-economic problem of modern society. During the last two decades, acute pre- and in hospital care, time management, diagnostic procedures, and rehabilitation strategies have substantially improved the level of care and outcome following TBI (Maegele et al. 2007). But still, to date, no therapeutic approach has been proven effective in reversing the pathologic cellular sequelae underlying the progression of cell loss and in improving neurobehavioral outcome. As the brain has limited capacity for self-repair, restorative approaches with focus on replacement and repair of dysfunctional and dead cells by transplantation of embryonic stem cells (ESC) following TBI has been studied. The therapeutic potential of these cells has been examined in experimental brain injury using a variety of approaches. An overview of current pre-clinical knowledge with respect to ESC replacement into the experimentally injured brain is presented.

12.2 Traumatic Brain Injury—A Disease Characterized by Cell Loss

The pathologic sequelae after TBI are separated into the following: The primary injury with immediate and nonreversible mechanical damage to the central nervous system (CNS) that occurs at the moment of impact, and secondary or delayed injury: An unclarified pathologic cascade that is initiated at the moment of the traumatic insult and progresses up to months or even years. This secondary injury to the CNS is a complex network of interacting structural, functional, cellular, and molecular changes, including breakdown of the blood-brain barrier (BBB), edema formation with brain swelling, impairment of energy metabolism, changes in cerebral perfusion, ionic dyshomeostasis, activation of destructive neurochemicals and enzymes, generation of free radicals, and genomic changes. Alone or in combination these events may lead to delayed cell death (Laurer et al. 1999).

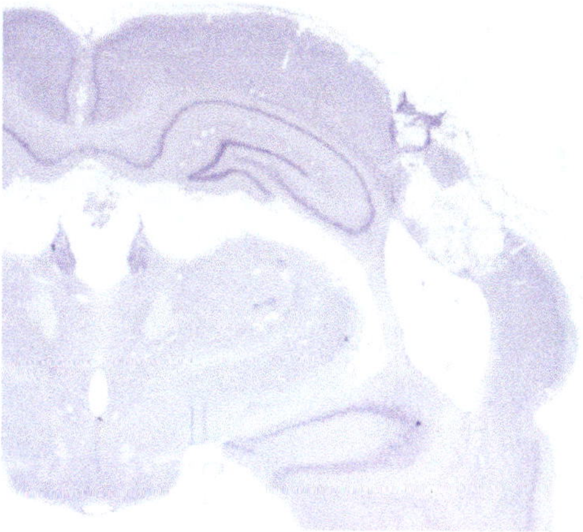

Fig. 12.1 Cresyl-Violet-Staining: Histological evaluation at 6 weeks after lateral Fluid-Percussion brain injury. Ventriculomegaly, shrinkage of the hippocampal cell layer, and necrosis is shown

Today consensus is reached that the stimulation of regenerative potentials within the injured adult central nervous system requires one or more of the following processes (Schouten et al. 2004):

- Cellular replacement,
- delivery of neurotrophic factors,
- removal of growth inhibition,
- promotion of axonal guidance,
- adequate intracellular signaling,
- bridging and artificial substrates, and
- modulation of the host immunoresponse.

The development of clinically relevant experimental models of TBI has greatly increased our understanding of the pathophysiology of TBI. Behavioral evaluation following both clinical and experimental TBI in rodent models demonstrates that long term impairments of cognitive and neuromotor function occur. It is proven that enduring cognitive, neurobehavioral, and histopathological changes persist for up to 1 year following severe experimental brain injury in rats (Pierce et al. 1998). The extent of cognitive dysfunction following experimental TBI appears to be related to both injury severity and neuronal loss in the CA3 and dentate hilar regions of the hippocampus (Hicks et al. 1993). While mild experimental TBI in rat only induces cognitive deficits associated with regional neuronal loss in the hippocampus, severe lateral fluid percussion (FP) injury in the rat (McIntosh et al. 1989) resulted in neurological motor deficits with an initial but incomplete recovery, persisting up to 1 year. Yet cognitive deficits endured up to 1 year post-injury. Histological evaluation demonstrated ongoing axonal degeneration in the striatum, corpus callosum, and injured cortex continuing up to 1 year and for up to 6 months post-injury in the thalamus (Pierce et al. 1998). In addition, ventriculomegaly, thalamic degeneration, shrinkage of the hippocampal pyramidal cell layer, progressive bilateral neuronal death in the dentate gyrus, reactive astrocytosis, and progressive atrophy of the cortex, thalamus, hippocampus, and septum have been reported up to 1 year following lateral Fluid Percussion brain injury in rats (Fig. 12.1) (Bramlett et al. 2002).

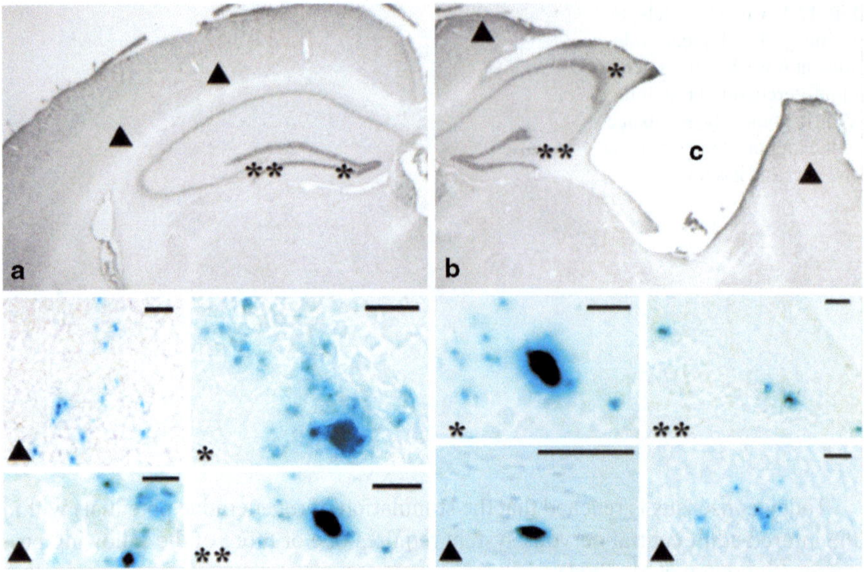

Fig. 12.2 Detection of neural stem cells (NSCs) (C17-2) applying X-gal staining after 13 weeks following transplantation. NSCs were transplanted either contralateral—into the healthy hemisphere- (**a**) or ipsilateral—into the injured hemisphere- (**b**). After transplantation into the contralateral hemisphere (**a**), NSCs were localized in the cortex (▲), dentate gyrus (*) or CA3 (**) region of the hippocampus. After ipsilateral hemispheric transplantation (**b**), X-gal- cells were located in the cortex (▲) surrounding the injury cavity (C), the granule cell layer of the dentate gyrus (**), and the CA3 region (*) of the hippocampus. (Adapted from Riess et al. 2002)

Using another TBI inducing model, the Controlled Cortical Impact (CCI) in mice and rats, histological findings could be confirmed by the detection of substantial tissue loss in the region of impact and selective hippocampal neuronal cell loss have also been observed (Fig. 12.2), and have been related to behavioral deficits seen in this model (Dixon et al. 1999; Riess et al. 2002).

As the brain has limited capacity for self-repair, mature neurons have no ability to regenerate, and endogenous progenitor cells, although present in multiple locations in the adult mammalian brain, appear to have limited ability to generate new functional neurons in response to injury. For this reason, there is great interest in the possibility of repairing the nervous system by transplanting cells that can replace those lost due to the damage (Bjorklund et al. 2002), or to manipulate endogenous progenitor cells to increase their neurogenic potential or their ability to facilitate regeneration. On the base of current neuropathological knowledge of TBI, a variety of potentially beneficial cellular replacement strategies should be considered. For further information see Schouten et al. (Schouten et al. 2004). One of these options is the use of embryonic stem cells. They show a great potential in the experimental phase on cellular levels as well as in animal trials, but ethical obstacles could be not dismissed in human.

12.3 Embryonic Stem Cells for Transplantation

Embryonic stem cells (ESCs) are nontransformed, pluripotent cells that are derived directly from the inner cell mass of the blastocyst (Evans et al. 1981). These cells are able to participate fully in embryonic development when they were reintroduced into the blastocyst. In vitro, ESCs give rise to cell types of all three primary germ layers in a way that recapitulates events of embryogenesis (Doetschman et al. 1985).

ESCs cultured under specific conditions will differentiate into neuons, astrocytes, or oligodendrocytes, depending on specific factors (Schouten et al. 2004). Due to their ability to differentiate into any tissue-specific cell type, ESCs may therefore possess great therapeutic potential in brain injury. Because TBI is associated with a massive loss of multiple cell types due to primary mechanical tissue disruption, bleeding, and secondary insults such as edema and rise of intracranial pressure leading to cell necrosis. In such case, an entire tissue segment including neurons, glia and vascular structures has to be treated by cells, able to differentiate into all lost cell types (Schouten et al 2004; Riess et al 2007; Molcanyi et al. 2013). Furthermore, ESCs provide a unique cellular system for experimental dissection of lineage specification and determination. Moreover, understanding and controlling ESC differentiation is an important step toward harnessing their potential to differentiate in any cell type of need for biomedical purposes. ESCs have been isolated from mice, monkeys, and humans (Schouten et al. 2004).

12.4 Host Environment Impairs ESC Survival But Induces ESC Differentiation and Trophic Factor Release

It has been suggested that stem cells hold great potential for the repair of the damaged nervous system. Migration and differentiation of stem cell derived precursors or progenitors seem to be accompanied by an improvement of neurological motor function as well as sensorimotor functional recovery (Riess et al. 2002, 2007). Despite these promising reports, it has to be noted that the number of surviving and differentiating cells after implantation is mostly reduced and therefore cannot completely explain the functional recovery reported. The cell survival and differentiation seems to be not just defined but also restricted by the host environment. This assumption is underlined by reports describing the post-traumatic brain as a hostile environment due to the onset of an acute inflammatory response associated with an activation of immune competent cells, the release of immune mediators, the breakdown of the blood-brain barrier as well as the infiltration of peripheral blood cells and neurotoxic components (Kelly et al. 2004; Lenzlinger et al. 2001). Furthermore, the cellular immune responses, particularly the recruitment of macrophages has been shown to be responsible for loss of cells due to extensive phagocytosis of the transplanted ESCs following TBI (Molcanyi et al. 2007). This potential detrimental effect of the post traumatic brain environment was also shown *in vitro.* Inflammatory cytokines released following brain trauma such as interferon-gamma

(IFN gamma) or tumor necrosis factor-alpha (TNFalpha) seem to inhibit the generation of neurospheres and to exhibit cytotoxic effects on neural stem cells (Wong et al. 2004).

Hence, pathophysiological changes associated with TBI may affect the survival, migration and differentiation of transplanted ESCs (Bentz et al. 2010) examined the effect of trauma associated environmental alterations on stem cell survival and differentiation. They added tissue extract after FP-brain injury (TBE) or healthy rat brains (HBE) to undifferentiated murine embryonic stem cells (CGR8) cultured in feeder-free conditions. Time-dependent survival, proliferation and differentiation of murine ESCs were examined over a period of 7 days. Hereby omission of serum from the culture medium induced neural differentiation of ESCs, as indicated by a significant time dependent down-regulation of oct −4 with a concomitant up-regulation of nestin after 7 days. Pronounced cell loss largely due to apoptotic cell death was observed additionally. In TBE treated cells, on the other hand, a significant amplification of apoptotic cell death, enhancement of nestin and MAP2 expression and marked morphological changes such as axonal-like outgrowth was observed within 3 days of conditioning. Treatment of ESCs with HBE resulted in less pronounced neuronal differentiation processes. Axonal-like outgrowth was not noticed. The pronounced expression of nestin was enhanced in the CGR8 cells following 3 days of incubation with TBE (Fig. 12.3). Furthermore TBE treated cells also expressed Map2. Cells treated with HBE or untreated cells also expressed nestin and Map2. However, analysis of fluorescence intensity revealed significant more intensive nestin and Map2 signals in TBE treated cells than in untreated cells or cells incubated with HBE (Fig. 12.3). The expression of nestin and Map2 confirmed the observed tendency of cells, in particular of TBE treated cells, towards neuronal development.

In this experiment, the authors suggest that during the early acute phase of traumatic brain injury the cerebral environment is disposed to detrimental as well as potent protective signals that seem to rapidly induce neurogenic processes.

In another experiment (Bentz et al. 2007) compared the capacity of two different ES cell lines to secrete neurotrophins in response to cerebral tissue extract derived from healthy or FP brain injury in rats. The intrinsic capacity of the embryonic cell lines BAC7 (feeder cell-dependent cultivation) to release brain-derived neurotrophic factor (BDNF) or neurotrophin-3 (NT-3) exceeded the release of these factors by CGR8 cells (feeder cell-free growth) by factors of 10 and 4, respectively. Nerve growth factor (NGF) was secreted only by BAC7 cells. Conditioning of cell lines with cerebral tissue extract derived from healthy or brain injured rat brains resulted in a significant time-dependent increase in BDNF release in both cell lines. The increase in BDNF release by BAC7 cells was more pronounced when cells were incubated with brain extract derived from injured brain. Neutrophin-3 and NGF release was inhibited when cell lines were exposed to cerebral tissue extract. The magnitude of the response to cerebral tissue extract was dependent on the intrinsic capacity of the cell lines to release neurotrophins. In this experiment the authors were able to show, that significant variations in the intrinsic capability of different stem cell lines to produce neurotrophic factors exist. In this experiment the authors

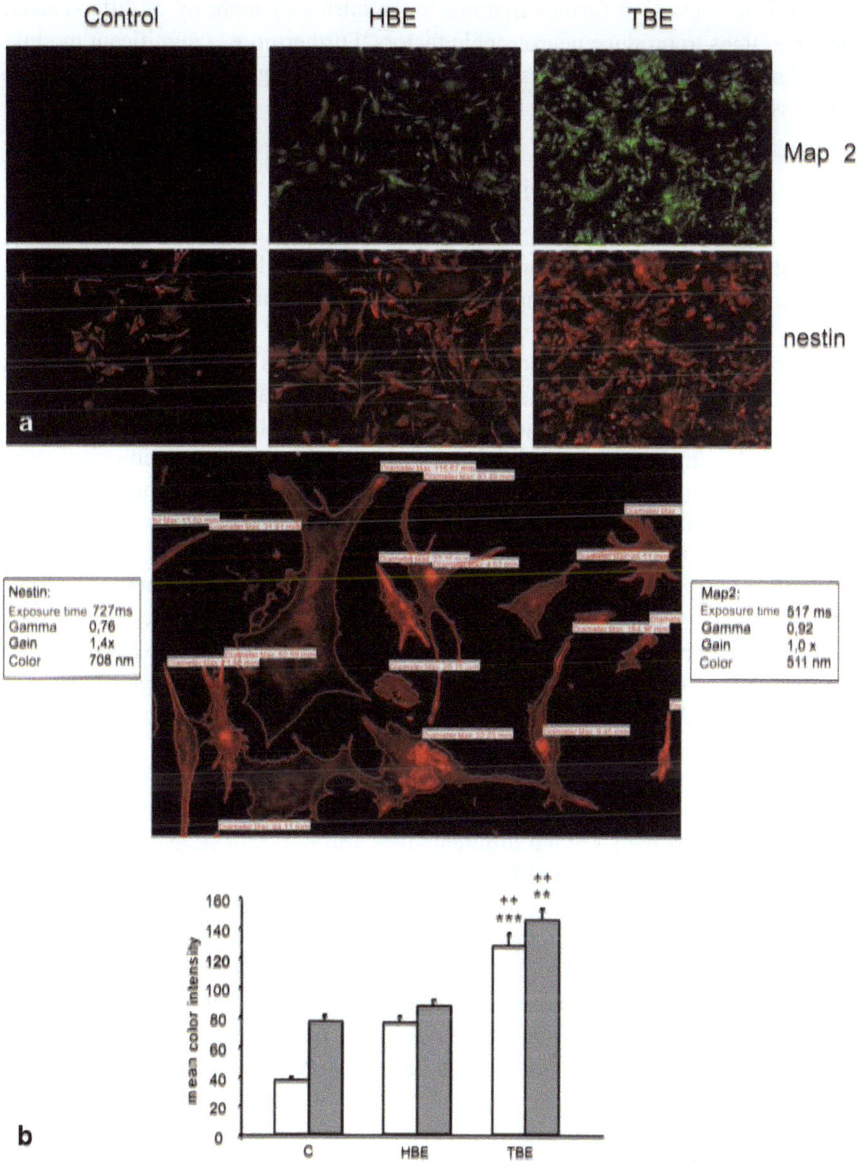

Fig. 12.3 Brain-extract modulation of stem cell differentiation at day 3 following conditioning. CGR8 stem cells were conditioned with brain extract derived from traumatized rats (TBE) or brain extract from healthy animals (HBE). Alternatively, cells were grown in serum-free medium (differentiation control) or under normal/standard conditions (+FCS, +lif). **a** Expression of *nestin* (*red*), and Map2 (*green*) was assessed by immunocytochemistry. **b** Quantification of optical density of nestin (*white bars*) or MAP2 (*grey bars*) signals were evaluated. Data are presented as mean ± s.e.m of raw data. Significant difference from control ($^{**}=p<0.01$; $^{***}=p<0.001$) or HBE conditioned cells ($++=p<0.01$). (Adapted from Bentz K et al. 2010)

were able to show significant variations in the intrinsic capability of differ-190 ent stem cell lines to produce neurotrophic factors. Furthermore, a significant modulation of neurotrophic factor release was observed following conditioning of cell lines with tissue extract derived from rat brains.

12.5 ESC Transplantation Following TBI

The cell loss due to TBI based on the primary (mechanical) direct trauma and on the developing oedema leading to apoptosis and necrosis of multiple cell types. Therefore the choice of cell population utilized in replacement therapies after TBI might be critical. Pluripotent ESCs derived from the inner mass of the blastocyst are able to differentiate into any tissue-specific cell type and may therefore possess great therapeutic potential in brain injury, since a variety of cell types are damaged or destroyed following cerebral trauma. In accordance with this assumption pluripotent murine embryonic stem cells (D3 ES cell line) have been shown to survive and differentiate following transplantation into rat brains in an experimental stroke model (Erdö et al. 2003).

Furthermore, enhanced green fluorescent protein (eGFP)-transfected D3 ESCs have been shown to migrate along the corpus callosum to the ventricular walls and to populate the border zone of the damaged brain tissue. They were also found on the hemisphere opposite to the implantation site, indicating the highly migratory behavior of implanted ESCs (Hoehn et al. 2002). In another murine model of TBI (induced by injecting phosphate-buffered saline into the left frontal and right caudal cortex) (Srivastava et al. 2006) analyzed the feasibility of ESC transplantation with the focus on the migration in response to lesions induced in brain tissues, and the mechanism of their in vivo differentiation into neighboring neural cells. They demonstrate that undifferentiated ESCs migrate within 1 week after injection to the damaged regions of brain tissue, engraft, and proliferate. Behavioral assessment was not performed. They conclude that damaged brain tissue provides a niche that attracts ESCs to migrate and proliferate. It has to be taken into consideration that the highly proliferative characteristics (self-renewal) of ESCs combined with the ability to differentiate into all embryonic germinal layers (pluripotency) present a potential threat of tumor development (teratoma, teratocarcinoma) when they are transplanted into the adult CNS. Tumorigenesis has been observed after implantation of undifferentiated human ESCs into healthy rat brains, giving rise to teratomas and malignant teratocarcinomas (Thomson et al. 1998). Accordingly, Erdö et al. (2003) compared the tumorigenic outcome after implantation of D3 ESCs in a homologous (mouse to mouse) vs. xenogeneic (mouse to rat) stroke model. In injured and healthy mouse brains, both transplanted undifferentiated and pre-differentiated murine ESCs produced highly malignant teratomas, while mouse ESCs xenotransplanted into injured rat brain migrated towards the lesion and differentiated into neurons at the border zone of the ischemic infarct. This suggests that tumorigenesis may be related to the host animal rather than to the differentiation status of the implanted cells.

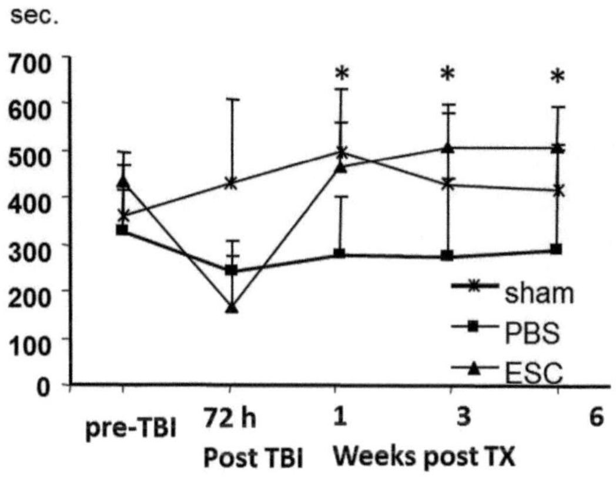

Fig. 12.4 Evaluation of time-dependent modification in locomotive motor function using the rotarod test. *Sham-operated animals. ▲ animals treated with embryonic stem (*ES*) cells following fluid percussion injury; ■ control animals treated with phosphate-buffered saline (*PBS*). Times of two performances on the Rotarod were calculated and are expressed as sum ± SE. $^{*}p \leq 0.5$ when injured ES-cell transplanted animals are compared to their respective PBS-injected animals. (Adapted from Riess et al. 2007)

Based upon these findings of Erdö undifferentiated murine ESCs of the D3 line stably transfected with the pCX -(ß-act-) were implanted into the ipsilateral cortex of rats (Fig. 12.5a) 72 h after lateral fluid-percussion brain injury of moderate severity to examine their effects on neurofunctional recovery. The chosen transplantation paradigm based upon reports demonstrating that the initial inflammatory response decreases 3 days after trauma to the brain and the development of astroglial scar begins to build thereafter (Okano 2002). The relatively early time point was chosen in order to avoid the peak of any inflammatory reaction and to allow for the migration and differentiation of stem cells that might be obstructed by the later formation of astroglial scar.

Neuromotor function was assessed at 1, 3, and 6 weeks after transplantation using a Rotatrod and a Composite Neuroscore test. During this 6 weeks of observation transplanted animals showed a significant improvement in the Rotarod Test (Fig. 12.4) and in the Composite Neuroscore Test as compared to controls.

At Five days after transplantation (TX), ESCs were detected as cell clusters in 100 % of transplanted animals (Fig. 12.5b). At 7 weeks following transplantation, only a few transplanted cells were detected in one animal (Fig. 12.5c).

There was also tumor formation detected in this animal (Fig. 12.6), although a tumor-suppressive effect of xenotransplantation has been suggested (Erdö et al. 2003). Neither differentiation nor migration of cells was observed at any time point. Due to the unexpected tumor formation, Molcanyi et al. (2009) critically investigated tumorigenesis and possible mechanisms of tumor-suppressive effects following xenotransplantation of D-3 murine ESCs into uninjured adult rat brains lacking any preliminary inflammatory potential. In this experiment in 5 out of 8 healthy animals tumor formation was observed within 2 weeks of ESC implantation (Figs. 12.7

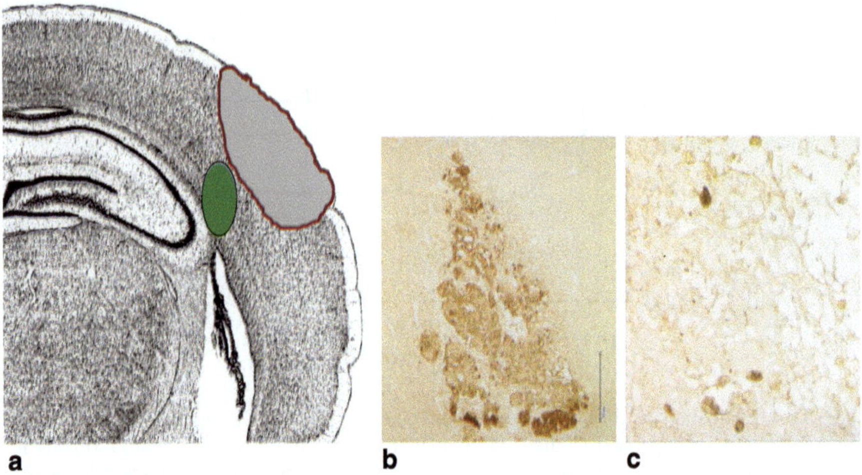

Fig. 12.5 **a**, **b**, **c** Schematic representation of injury and transplantation site. Immunohistochemical analysis of embryonic stem (*ES*) cell location at 1 week in cluster formation, and 7 weeks post-transplantation. (Adapted from Riess et al. 2007)

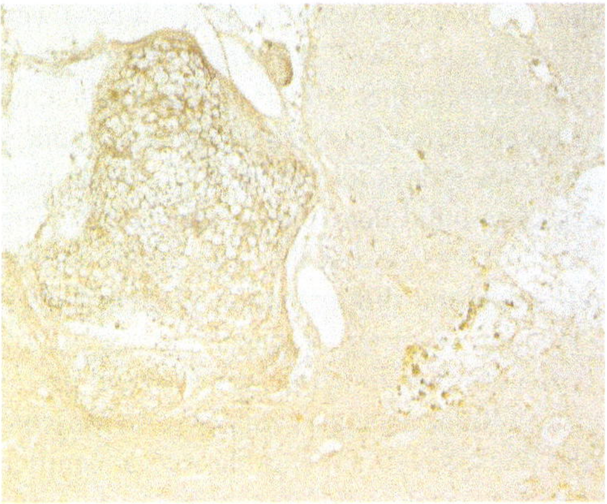

Fig. 12.6 Seven weeks after ESC-Transplantation histological evaluation reveals tumor formation (chondroma). (Adapted from Riess et al. 2007)

and 12.8), indicating the tumorsuppressive effect mainly occur after transplantation of ESC s into injured rat brains. Tumor-suppressive effects, reflected by Erdö could possibly be ascribed to immunomodulatory activity of macrophages scavenging the tumorigenic fraction of the implanted cells (Fig. 12.9) (Molcanyi et al. 2009). Macrophage activation could also be additionally triggered by potential impurity stem cell graft caused by feeder cells, the stem cells have been initially raised on (Molcanyi et al. 2014). In the same time, macrophage populations may become a major source of methodological pitfalls interfering with correct identification of implanted stem cells (Molcanyi et al. 2013).

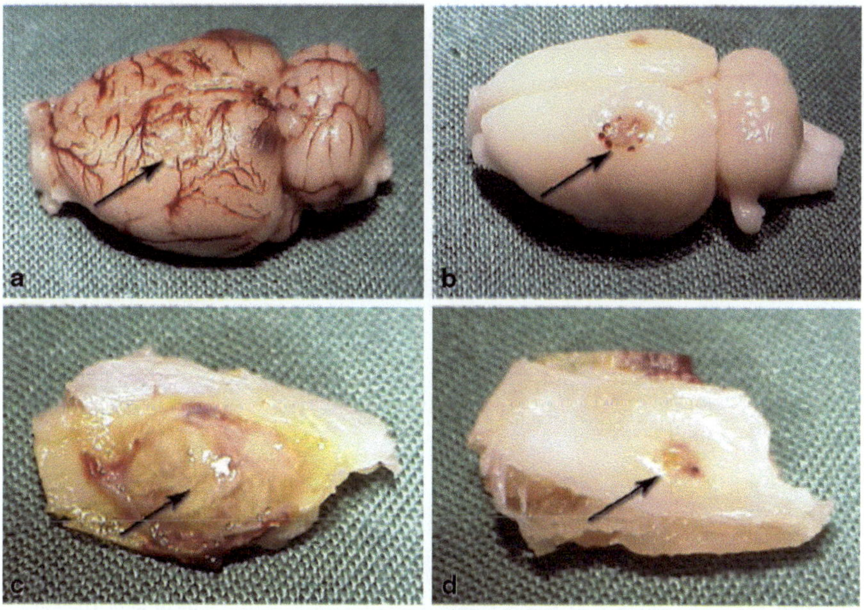

Fig. 12.7 Autopsy view—pathological changes in brains that received ESC grafts. **a** Brain without transcardial perfusion, because the animal died 8 days after ESC transplantation. Rough area with firm surface texture, lacking superficial vasculature was later histologically identified as tumor. **b** Brain perfused with PFA solution (2 weeks after ESC transplanation) shows typical pathological signs of tumor, including the vascular reorganization at the border between tumor and healthy tissue. **c** Tumor formation found on top of the skull, communicating with the brain tumor lying underneath via craniotomy opening. **d** Craniotomy opening (view from inside of scull). (Adapted from Molcanyi et al. 2009)

This is in accordance with studies suggesting that low incidence of tumorigenesis following ESC xenotransplantation may be due, in part, to the removal of implanted cells by activated inflammatory cells in the injured brain (Molcanyi et al. 2007).

Molcanyi examined the time dependent fate of the ESCs following ipsi- and contralateral transplantation into rat brains following FP brain injury. Double-staining for GFP and macrophage antigens revealed stem cell clusters embedded and phagocytosed by infiltrated and activated macrophages, indicating the loss of implanted stem cells was due to an early posttraumatic inflammatory response (Fig. 12.9). When ESCs were implanted into completely intact healthy brains macrophage infiltration was less pronounced. The authors therefore suggested that the massive macrophage infiltration at graft sites might be ascribed to the combined stimulus exhibited by the FP brain injury and the substantial conglomeration of ESCs. Macrophage activation could also be additionally triggered by potential impurity stem cell graft caused by feeder cells, the stem cells have been initially raised on (Molcanyi et al. 2014). In the same time, macrophage populations may become a major source of methodological pitfalls interfering with correct identification of implanted stem cells (Molcanyi et al. 2013).

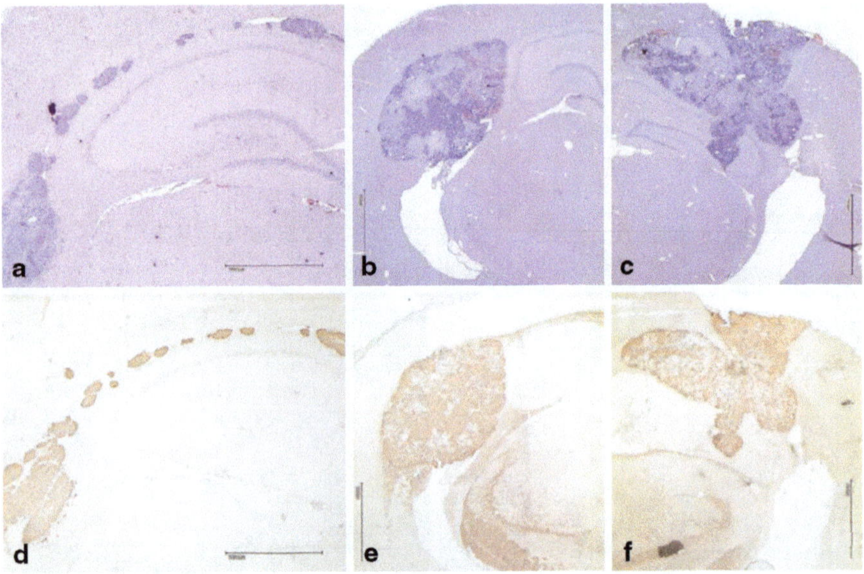

Fig. 12.8 Gross morphology of tumor formations. **a** Coronal section stained with HE reveals the main cell cluster implanted in the lateral parahippocampal cortex and several smaller clusters spreading along corpus callosum (8 day survival). **b** Tumor formation, stained by HE, exerting pressure on the hippocampus and lateral ventricle (2 week survival). **c** Tumor formation, stained by HE, showing an infiltrative growth into hippocampus and cortex (2 week survival). **d**, **e**, **f** Corresponding adjacent sections labelled with anti-GFP antibody confirmed the tumor structures to be arising from implanted GFP-transfected ESCs. (Adapted from Molcanyi et al. 2009)

Fig. 12.9 Phagocytosis of the implanted ES cells (appear *red*—anti-GFP-Cy3) by macrophages (*green*—anti-ED -1-FITC) -confocal imaging. High magnification of the *red* stem cell localized inside of a *green* phagocyting macrophage (*thin arrow*). (Adapted from Molcanyi et al. 2007)

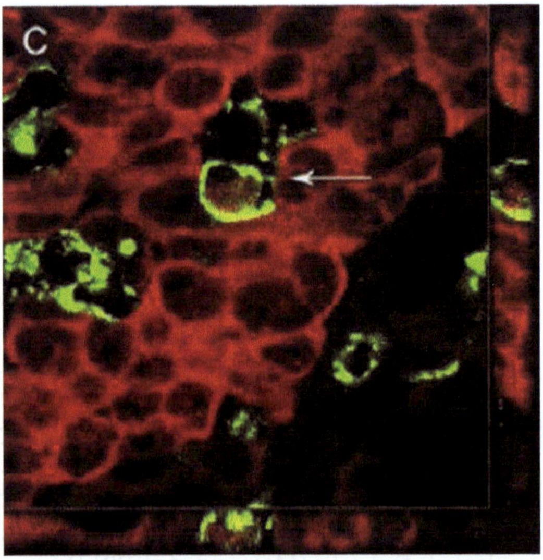

Previous studies have also shown that the grafting procedure itself, i.e. needle insertion followed by pressure exerted by cell graft infusion, may provoke a notable cellular response in terms of macrophage invasion and microglia activation. The extensive loss of ESCs and the apparent lack of differentiation and migration of these cells following implantation into injured rat brains indicate that the significant improvement of motor function that was observed within 5 days after implantation is not attributed to the functional integration of ESCs into the neuronal cerebral network. *In vitro* studies demonstrated that the mechanism underlying stem-cell-mediated functional improvement might be partially due to the release of trophic factors by implanted cells. Incubation of D3 ESCs with extract derived from injured rat hemispheres resulted in a rapid time-dependent and significant release of BDNF into the medium (see previous chapter: Host environment induces cell differentiation and neurotrophic factor release, but also impairs ESCs survival), indicating that similar to the observations made by Chen et al. (2002), the improvement of neuronal function following transplantation of embryonic stem cells might be triggered by the release of protective neurotrophic factors.

Conclusions

The development of therapeutic strategies for neuro-protection and regeneration following TBI has been an active field of research over the past decades. By focusing on the posttraumatic cellular pathophysiological sequelae, our understanding of mechanisms leading to brain damage after TBI has greatly increased, and more neuroprotective agents have been developed and tested in both experimental and clinical settings. The use of ESCs in cell replacement therapy has received a great deal of scientific and public interest in the recent years. This is due to the remarkable pace at which paradigm-changing discoveries have been made regarding the neurogenic potential of embryonic, fetal, and adult stem cells.

Unfortunately, in studies on day 5 after implantation, surviving undifferentiated embryonic stem cells were detected in large clusters but 7 weeks later only a few GFP-positive ESCs were detectable at the implantation site. (Molcanyi et al. 2007). This lack of survival and/or integration of ESCs may be related to the severity of injury as described before (Shindo et al. 2006).

Additionally the cell survival and differentiation seems to be not just defined but also restricted by the host environment. This assumption is underlined by reports describing that the post-traumatic brain is a hostile environment due to the onset of an acute inflammatory response associated with an activation of immunocompetent cells, the release of immune mediators, the breakdown of the blood-brain barrier as well as the infiltration of peripheral blood cells and neurotoxic components (Kelly et al. 2004; Lenzlinger et al. 2001).

Furthermore, the cellular immune responses, particularly the recruitment of macrophages has been shown to be responsible for loss of cells by extensive phagocytosis of the implanted ESCs following TBI (Molcanyi et al. 2007). The observed functional improvements must therefore be regulated by mechanisms that are closely associated with the transplantation of ESCs but are independent of stem cell integration and differentiation. Plausible mechanisms may include the production and/or secretion of trophic factors by ESCs. Tissue culture experiments revealed, that post-traumatic host environment induces rapid ESC differentiation and neurotrophic factor release, but also impairs the survival of ESCs (Bentz et al. 2010).

Tumorigenesis is also an important issue in cell transplantation. It has to be taken into consideration that the highly proliferative characteristics of ESCs combined with the pluripotency present a potential threat of tumor development when they are transplanted into the CNS. Xenogeneic transplantation of undifferentiated murine ESCs into an intact healthy rat brain leads to the onset of tumorigenesis. The absence of tumor formation, observed in other studies, could partially be explained in a successful removal of tumorigenic cell pool by phagocytosis, the prevention of tumorigenesis via xenotransplantation as described before seems unreliable.

The repair of the traumatically injured brain is challenging due to the complex network of interacting structural, functional, cellular, and molecular changes following TBI leading to cell death. The restoration of brain function after TBI requires more than cellular replacement. An understanding of graft integration and the mechanisms underlying transplant-induced behavioral recovery is essential in the development of more effective therapeutic approaches to CNS injury. In addition, combined therapies including cellular replacement, growth factor infusion, gene therapy, environmental enrichment, and manipulation of plasticity are likely to improve the already encouraging data concerning the development of clinically effective transplantation based treatments for the traumatically injured CNS.

The effects of neurotrophic factors expressed by transplanted ESCs following transplantation to the injured brain seem to be crucial. Also important for the therapeutic effects of ESCs are the secondary effects as e.g. inflammatory responses caused by ESC transplantation. Furthermore phagocytosis of the ESCs is important to prevent the host brain of tumor formation. These findings provide evidence for the therapeutic potential of ESC transplantation after TBI in rats, but also raise serious safety concerns for human trails. However, the therapeutic use of undifferentiated ESCs is seriously limited due to the risk of tumor development, and in addition the clinical transplantation of ESC-derivatives is complicated by ethical and immunological concerns.

References

Bentz K, Molcanyi M, Riess P, Elbers A, Pohl E, Sachinidis A, Hescheler J, Neugebauer E, Schäfer U (2007) Embryonic stem cells produce neurotrophins in response to cerebral tissue extract: cell line-dependent differences. J Neurosci Res 85(5):1057–1064

Bentz K, Molcanyi M, Schneider A, Riess P, Maegele M, Bosche B, Hampl JA, Hescheler J, Patz S, Schäfer U (2010) Extract derived from rat brains in the acute phase following traumatic brain injury impairs survival of undifferentiated stem cells and induces rapid differentiation of surviving cells. Cell Physiol Biochem 26(6):821–830

Bjorklund A, Lindvall O (2000) Cell replacement therapies for central nervous system disorders. Nat Neurosci 3:537–544

Bramlett HM, Dietrich WD (2002) Quantitative structural changes in white and gray matter 1 year following traumatic brain injury in rats. Acta Neuropathol 103(6)607–614

Bruns J, Hauser W (2003) The epidemiology of traumatic brain injury: a review. Epilepsia 44(Suppl 10):2–10

Chen X, Katakowski M, Li Y, Lu D et al (2002) Human bone marrow stromal cell cultures conditioned by traumatic brain tissue extracts: growth factor production. J Neurosci Res 69:687–691

Dixon CE, Kochanek PM, Yan HQ (1999) One-year study of spatial memory performance, brain morphology and cholinergic markers after moderate controlled cortical impact in rats. J Neurotrauma 16:109–122

Doetschman TC, Eistetter H, Katz M (1985) The in vitro development of blastocyst-derived embryonic stem cell lines: formation of visceral yolksac, blood islands and myocardium. J Embryol Exp Morphol 87:27–45

Erdö, T, Buhrle C, Blunk J, Hoehn M et al (2003) Host-dependent tumorigenesis of embryonic stem cell transplantation in experimental stroke. J Cereb Blood Flow Metab. 23:780–785

Evans MJ, Kaufman MH (1981) Establishment in culture of pluripotential cells from mouse embryos. Nature 292:154–156

Hicks RR, Smith DH, Lowenstein DH, Saint Marie RL, McIntosh TK (1993) Mild experimental brain injury in the rat induces cognitive deficits associated with regional neuronal loss in the hippocampus. J Neurotrauma 10:405–414

Hoehn M, Kustermann E, Blunk J, Wiedermann D et al (2002) Monitoring of implanted stem cell migration in vivo: a highly resolved in vivo magnetic resonance imaging investigation of experimental stroke in rat. Proc Natl Acad Sci U S A 99:16267–16272

Kelly S, Bliss TM, Shah AK, Sim GH, Ma M, Foo WC, Masel J, Yenari MA, Weissmann IL, Uchida N, Palmer T, Steinberg GK (2004) Transplanted human fetalneural stem cells survive, migrate, and differentiate in ischemic rat cerebralcortex. Proc Natl Acad Sci U S A 101:11839–11844

Laurer HL, McIntosh TK (1999) Experimental models of brain trauma. Curr Opin Neurol 12:715–721

Lenzlinger PM, Morganti-Kossmann MC, Laurer HL, McIntosh TK (2001) The duality of the inflammatory response to traumatic brain injury. Mol Neurobiol 24:169–181

Löhr M, Tzouras G, Molcanyi M, Ernestus RI, Hampl JA, Fischer JH, Sahin K, Arendt T, Härtig W (2008) Degeneration of cholinergic rat basal forebrain neurons after experimental subarachnoid hemorrhage. Neurosurgery 63:336–344; discussion 344–345

Maegele M, Engel D et al (2007) Incidence and outcome of traumatic brain injury in an urban area in Western Europe over 10 years. Eur Surg Res 39:372–379

McIntosh TK, Vink R, Noble L, Yamakami I, Faden AI (1989) Traumatic brain injury in the rat: characterization of a lateral fluid percussion model. Neuroscience 28:233–244

Molcanyi M, Riess P, Bentz K, Maegele M, Hescheler J, Schäfke B, Trapp T, Neugebauer E, Klug N, Schäfer U (2007) Trauma-associated inflammatory response impairs embryonic stem cell survival and integration after implantation into injured rat brain. J Neutotrauma 24:625–37

Molcanyi M, Riess P, Haj-Yasein NN, Bentz K, Loehr M, Kuchta J, Zivcak J, Stenzel W, Miletic H, Hescheler J, Neugebauer E, Hampl JA, Ernestus RI, Schafer U (2009) Developmental potential of the murine embryonic stem cells transplanted into the healthy rat brain–novel insights into tumorigenesis. Cell Physiol Biochem 24(1–2):87–94

Molcanyi M, Bosche B, Kraitsy K, Patz S, Zivcak J, Riess P, Majdoub FE, Hescheler J, Goldbrunner R, Schäfer U (2013) Pitfalls and fallacies interfering with correct identification of embryonic stem cells implanted into the brain after experimental traumatic brain injury. J Neurosci Methods 30(21):60–70

Molcanyi M, Zare Mehrjardi N, Schaefer U, Haj-Yasein NN, Brockmann M, Penner M, Riess P, Reinshagen C, Rieger B, Hannes T, Hescheler J, Bosche B (2014) Impurity of stem cell graft by murine embryonic fibroblasts—implications for cell-based therapy of the central nervous system. Front Cell Neurosci 5(8):257

Okano H (2002) Stem cell biology of the central nervous system. J Neurosci Res 69:698–707
Pierce JES, Smith DH, Trojanowski JQ, McIntosh TK (1998) Enduring cognitive, neurobehavioral, and histopathological changes persist for up to one year following severe experimental brain injury in rats. Neuroscience 87:359–369
Riess P, Zhang C, Saatman KE, Laurer HL, Longhi LG, Raghupathi R, Lenzlinger PM, Lifshitz J, Boockvar J, Neugebauer E, Snyder EY, McIntosh TK (2002) Transplanted neural stem cells survive, differentiate, and improve neurological motor function after experimental traumatic brain injury. Neurosurgery 51:1043–1054
Riess P, Molcanyi M, Bentz K, Maegele M, Simanski C, Carlitscheck C, Schneider A, Hescheler J, Bouillon B, Schäfer U, Neugebauer E. (2007) Embryonic stem cell transplantation after experimental traumatic brain injury dramatically improves neurological outcome, but may cause tumors. J Neurotrauma 24(1):216–225
Schouten J, Fulp C, Royo NC, Saatman KE et al (2004) A review and rationale for the use of cellular transplantation as a therapeutic strategy for traumatic brain injury. J Neurotrauma 21:1501–1538
Shindo T, Matsumoto Y, Wang Q, Kawai N, Tamiya T, Nagao S (2006) Differences in the neuronal stem cells survival, neuronal differentiation and neurological improvement after transplantation of neural stem cells between mild and severe experimental traumatic brain injury. J Med Invest 53:42–51
Srivastava S, Shenouda S, Mishra R, Carrier E (2006) Transplanted embryonic stem cells successfully survive, proliferate, and migrate to damaged regions of the mouse brain. Stem Cells 24:1689–1694
Thomson JA, Itskovitz-Eldor J, Shapiro SS, Waknitz MA et al (1982) Embryonic stem cell lines derived from human blastocysts. Science 282:1145–1147
Wong G, Goldshmit Y, Turnley AM (2004) Interferon-gamma but not TNF alpha promotesneuronal differentiation and neuriteoutgrowth of murine adult neural stemcells. Exp Neurol 187:171–177

Chapter 13
Transplantation of Olfactory Ensheathing Cells in Spinal Cord Injury

Johana Tello Velasquez, Jenny A. K. Ekberg and James A. St John

Contents

J. A. St John (✉) · J. Tello Velasquez · J. A. K. Ekberg
Eskitis Institute for Drug Discovery, Griffith University, Brisbane, Australia
e-mail: j.stjohn@griffith.edu.au

J. A. K. Ekberg
School of Biomedical Sciences, Queensland University of Technology, Brisbane, Australia

L.-R. Zhao, J. H. Zhang (eds.), *Cellular Therapy for Stroke and CNS Injuries*,
Springer Series in Translational Stroke Research, DOI 10.1007/978-3-319-11481-1_13

Abstract Spinal cord injury is characterized by massive cellular and axonal loss, a neurotoxic environment, inhibitory molecules and physical barriers that hamper nerve regeneration and reconnection leading to chronic paralysis. Transplantation of different types of cells is one of the strategies being examined in order to restore the lost cell populations and to re-establish a permissive environment for nerve regeneration. The mammalian olfactory system is one of the few zones in the body where neurogenesis occurs during the lifetime of the organism, with olfactory neurons being replaced daily with their axons elongating from the peripheral nervous system into the central nervous system to re-establish functional connections. The regenerative ability of this system is largely attributed to the presence of a unique group of cells called olfactory ensheathing cells (OECs). OECs have emerged as an encouraging cell candidate for transplantation therapies to repair the injured spinal cord with multiple animal models showing significant functional improvements and several human trials establishing that the procedure is safe and feasible. Even though the results are promising with some animal models showing remarkable restoration of function, the variability amongst studies in terms of outcome assessments, cell purity, cell culture and transplantation protocols make it difficult to reach firm conclusions about the effectiveness of OEC transplant therapy to treat the injured spinal cord. These variations need to be addressed in order to achieve a more realistic understanding of how the benefits of OEC transplantation enhance the therapeutic outcomes.

Abbreviations

BDNF	Brain-derived neurotrophic factor
cAMP	Cyclic adenosine monophosphate
CNS	Central nervous system
CP	Cribriform plate
CSPG	Chondroitin sulfate proteoglycan
DAPI	4',6-diamidino-2-phenylindole
dBcAMP	Dibutyryl cyclic adenosine monophosphate
FGF	Fibroblast growth factor
GDNF	Glial cell-derived neurotrophic factor
GFAP	Glial fibrillary acidic protein
GL	Glomerular layer
HNK-1	Human natural killer-1
IGF	Insulin-like growth factor
LP	Lamina propria
MAG	Myelin associated glycoprotein
NFL	Nerve fibre layer
NGF	Nerve growth factor
Nogo	Neurite outgrowth inhibitory protein
NPY	Neuropeptide Y
NT4	Neurotrophin 4
NT5	Neurotrophin 5

OE	Olfactory epithelium
OEC	Olfactory ensheathing cell
OMgp	Oligodendrocyte-myelin glycoprotein
p75NTR	p75 low-affinity neurotrophin receptor
PNS	Peripheral nervous system
SCI	Spinal cord injury
TROY	TNFRSF expressed on the mouse embryo
VEGF	Vascular endothelial growth factor

13.1 Spinal Cord Injury

Spinal cord injury results in large-scale neuronal loss with very limited capacity for regeneration, leading to chronic paralysis. The main factor hampering recovery is the inability of regenerating spinal cord axons to reach their target (reviewed by Leal-Filho 2011). The pathophysiology of spinal cord injury is divided in two stages: the primary and secondary lesion. The primary lesion is caused by the direct mechanical trauma, i.e. laceration, contusion or compression, resulting in structural disturbances, death of neurons and damage to neural connections. This is followed by ischemia and microvascular damage (Beattie et al. 1997), as well as excessive extracellular glutamate as a consequence of neuronal cell death (Hermann et al. 2001). The high concentration of glutamate and other excitatory amino acids lead to further progressive cell death via excitotoxicity and free radical production (Byrnes et al. 2009). Altogether, the cell death, oxidative stress and inflammatory responses result in massive neuronal and glial cell death (Hulsebosch 2002; Jones et al. 2003).

The secondary lesion is characterized by continued inflammatory immune responses, including cytokine and interleukin secretion by macrophages and neutrophils (Bolton 2005). Furthermore, axon demyelination, as a consequence of oligodendrocyte cell death, results in loss of axonal conduction and subsequently a loss in synaptic communication. Activated astrocytes migrate to the injury site to degrade axonal debris and remove toxic chemicals, but the vast network of activated astrocytes creates a glial scar; a compact structure that becomes a barrier preventing regenerating axons from reaching their target (Bunge et al. 1960; Matthews et al. 1979). Thus, the complex damage resulting from the initial nerve injury leads to an environment that hampers or even completely inhibits neuronal regeneration (Fig. 13.1) (Hulsebosch 2002; Leal-Filho 2011).

Current therapies for spinal cord injury do not lead to significant neural regeneration and functional recovery. Most of these therapies have aimed to minimize the post-traumatic cell damage but fail to achieve the re-establishment of neuronal connections. Drug therapy is generally applied immediately following trauma to treat inflammation and initial degeneration (reviewed by Stahel et al. 2012; Batzofin et al. 2013; Hurlbert et al. 2013). This treatment is often followed by long term therapies aimed at promoting axonal growth and neutralising the toxic environment at the injury site. A major factor hampering axonal regeneration

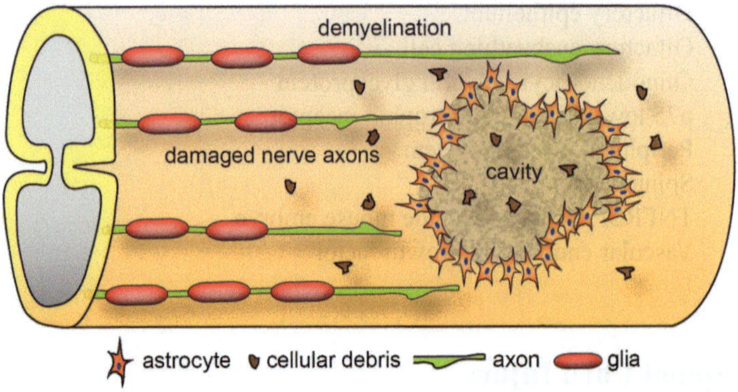

Fig. 13.1 Detrimental effects of secondary spinal cord injury events at the cellular and molecular level. Neurons are shown in *green*; glia in *red*. Injury to the spinal cord leads to (1) apoptosis and cell death (neurons and glial cells) with accumulation of cell debris and the generation of a toxic environment due to high extracellular glutamate concentrations and the presence of free radicals; (2) formation of a glial scar which creates a physical barrier consisting of activated astrocytes. The resultant effect is that damaged axons are unable to regenerate and communicate due to presence of inhibitory molecules, loss of myelin and the physical barrier

following spinal cord injury is the down-regulation of endogenous neurotrophins and one method that has shown promise is the injection of neurotrophins at the site of injury to replace the lost endogenous neurotrophins (Hulsebosch 2002). Peripheral glial cells can produce many growth factors and hence transplanting glia to the injury site is an even more promising approach as these cells can integrate with endogenous cells and scar tissue, producing a more long-term growth-promoting environment (Yan et al. 2001; Feron et al. 2005; Cao et al. 2007; Centenaro et al. 2011). Similarly, pluripotent stem cells can be transplanted to the injury site, potentially resulting in neuronal regeneration and production of glial cells. This method is still experimental, but has resulted in promising functional outcomes in animals (reviewed by Antonic et al. 2013). Further, manipulation of gene expression to block production of growth-inhibitory and toxic molecules has also resulted in some promising functional outcomes (reviewed by Leal-Filho 2011). Overall, however, while these therapeutic interventions have led to some positive outcomes, to date, none have produced a significant functional recovery in humans (Lim and Tow 2007; Leal-Filho 2011).

13.2 Endogenous Glial Cells and Their Role in Spinal Cord Injury

Glial cells are the most abundant cells in the nervous system. They are closely associated with neurons and were previously described simply as supportive nervous tissue. A deeper understanding of glial cell biology, however, has demonstrated that glial cells exhibit a multitude of complex roles and are essential for the development and function of the entire nervous system (Jessen 2006). Glial cells are a heterogeneous population of cells that differ in developmental origin, molecular composition, structure and specific behaviour, and exist together with neurons and other cells in an integrated and co-dependent system (Chung and Barres 2012). Throughout the nervous system, glial cells have crucial roles in axonal extension and guidance, protection against mechanical, chemical and oxidative injury, as well as preservation of the electrical and chemical balance of all neurons (Ndubaku and de Bellard 2008).

Glial cells can be broadly classified as being either central nervous system glia or peripheral nervous system glia. In the mature central nervous system (CNS), there are two major types of glial cells of neural origin; astrocytes and oligodendrocytes. Other types of CNS glial cells exist that originate from non-neuronal precursors; microglia constitute part of the innate immune system and originate from macrophage lineages (Chugani et al. 1991). In the peripheral nervous system (PNS), Schwann cells constitute the main glial cell type, with the exception of the olfactory nervous system, which is populated by specialized glia termed olfactory ensheathing cells (OECs).

Astrocytes play a critical role in the function and homeostasis of the CNS. They are required for the formation and maintenance of the blood-brain barrier, provide support for axonal extension and play an active role in neuronal signalling by exchange of ions and production of neurotransmitters, as well as cell adhesion and synapse signalling molecules (Kriegstein and Gotz 2003). Astrocyte–neuron interactions are known to secure the survival and normal function of neurons (Jessen 2004). Numerous studies have demonstrated that astrocytes play important neuroprotective roles, in neurodegenerative disorders (reviewed by Singh et al. 2011; Cabezas et al. 2012) and they have the ability to promote neuronal survival by protecting against reactive oxygen species and other stressors (Lopez et al. 2007).

After spinal cord injury, astrocytes respond rapidly by migrating to the injury site, where they proliferate and form a compact structure, a glial scar, to preserve the blood-brain barrier, protecting the CNS and maintaining the adequate ionic environment necessary for nerve function. However, the glial scar eventually becomes a physical barrier that stops damaged axons from regenerating and reconnecting (Fig. 13.1) (Leal-Filho 2011). Furthermore, astrocytes respond to neuronal injury by increasing their proliferation and by secreting glycoproteins such as chondroitin sulfate proteoglycans (CSPG), which act to inhibit axon elongation (Table 13.1) (Qiu et al. 2002; Su et al. 2009).

Oligodendrocytes are morphologically similar to astrocytes, albeit with fewer and smaller branched processes. They play different roles in the modulation of neuronal function as well as the regulation of proliferation, survival and differentiation of neurons (Jauregui-Huerta et al. 2010) The most important role of oligodendrocytes, however, is to myelinate axons. The myelin sheet provides electrical insulation around the nerve fibres, speeding the transmission of electrical signals (Jessen 2004). The myelin layer also protects the axons by creating a "safe chamber", resembling a growth-promoting channel through which

Table 13.1 Glial cell response to spinal cord injury

Type of Glia	Response in spinal cord injury event	Reference
Astrocytes	Removal of toxic chemicals (glutamate). Proliferation and secretion of neuroprotective but growth-inhibitory factors. Formation of glial scar	(Qiu et al. 2002; Su et al. 2009)
Oligodendrocytes	Massive death due to high glutamate concentrations. Production of glycoproteins with Nogo-receptor affinity that will suppress myelin production	(Jones et al. 2003; Jessen 2004; Arevalo et al. 2010)
Microglia	Initially phagocytosing debris and producing neuroprotective factors. Over time become neurotoxic and growth-inhibitory due to constant activation	(Chatzipanteli et al. 2002; Pearse et al. 2003; Block and Hong 2005; Kigerl et al. 2009)
Schwann cells	Cells de-differentiate to an immature state, lose their myelin sheath conformation and migrate from the periphery into the injury site in the CNS, where they participate in endogenous repair processes by expression of neurotrophic factors	(Farbman and Squinto 1985; Jessen 2004; Oudega and Xu 2006)

the axon extends. After spinal cord injury, populations of oligodendrocytes are rapidly affected by high levels of glutamate and massive cell death follows. Oligodendrocytes that do survive produce neurite outgrowth inhibitor (Nogo), myelin-associated glycoprotein (MAG), and oligodendrocyte-myelin glycoprotein (OMgp) (Table 13.1), proteins that bind to the Nogo receptor, repressing myelin production and affecting axonal outgrowth and neuronal synapses (Jones et al. 2003). Microglia, which are essentially macrophages present within the CNS, respond to injury by migrating to the injury site, where they phagocytose debris, secrete a range of both pro- and anti-inflammatory cytokines and growth factors which initially have a neuroprotective effect. Over time, however, microglia near and in the injury site respond to the constant prolonged activation by secreting molecules that are growth-inhibitory or toxic, thus repressing axonal regeneration (Chatzipanteli et al. 2002; Pearse et al. 2003; Block and Hong 2005). The majority of the activated microglia transition to the M1 type, which can directly induce neuronal death (Kigerl et al. 2009; Gao et al. 2013). Thus, together with the glial scar, the local environment at a CNS injury site inhibits long-term neuronal extension and regeneration.

The PNS differs dramatically from the CNS in terms of capability to regenerate itself after injury. In contrast to central nerves, peripheral neurons in general regenerate after injury, unless large nerves have been completely severed. Schwann cells play an active role in repair of peripheral damaged nerves as a consequence of their ability to differentiate, migrate, proliferate, secrete growth factors, and produce myelin. Schwann cells are classified as either myelinating or non-myelinating. Myelinating Schwann cells enwrap individual peripheral axons, forming the myelin sheath, whereas the non-myelinating type have metabolic and mechanical support functions (Jessen 2004). After spinal cord injury, Schwann cells migrate from the periphery into the injury site within in the CNS, and participate in endogenous repair processes (Table 13.1). They re-enter the cell cycle, lose their myelinating phenotype and de-differentiate into an immature state, and begin to express trophic factors and cell adhesion molecules that provide a more favourable environment for axon regeneration and extension (Oudega and Xu 2006).

One approach to improve the outcomes after spinal cord injury is to transplant glial cells into the injury site to reduce inflammation, and which will help form a glial bridge across the injury site and thereby promote axon extension. The glial of the PNS system, Schwann cells and OECs, have been trialled in animal models and in humans with various outcomes. The OECs have unique characteristics that may confer an advantage over other glial cell types for transplant therapies.

13.3 The Mammalian Olfactory Nervous System

The mammalian olfactory nervous system is one of the few regions in the CNS in which neurogenesis continuously occurs during the lifetime of the organism (Mackay-Sim and Kittel 1991a, b). The primary sensory neurons of the olfactory system line the dorsal/caudal nasal epithelium and are directly exposed to the environment (Fig. 13.2). The neurons are subjected to attack and destruction by bacterial (St John et al. 2014) and viral pathogens as well as toxins within the air and thus need to be replaced throughout life. Whilst the average life-span of olfactory neurons has not been clearly determined in humans, mouse olfactory neurons generally live for one to three months. Neurons that degenerate are rapidly replaced by new neurons arising from progenitor cells that line in the basal layer of the olfactory mucosa (Mackay-Sim and Kittel 1991b), a process that occurs throughout life (Ramon-Cueto and Santos-Benito 2001).

The primary olfactory system comprises the olfactory mucosa and the bundles of olfactory nerves that project into the olfactory bulb. Stem cells that line basal layer of the olfactory epithelium give rise to the primary olfactory sensory neurons which migrate apically to populate the olfactory epithelium (Fig. 13.2). Olfactory sensory neurons have a bipolar morphology with a single dendrite extending onto the surface of the epithelium and a single axon projecting to and terminating in the olfactory bulb. Each olfactory neuron expresses a single odorant receptor type with the neurons mosaically distributed throughout the epithelium, but the axons of the same odorant receptor type converge to the same targets within the olfactory bulb (Vassar et al. 1994; Mombaerts et al. 1996). To reach their targets in the olfactory bulb, the axons of the olfactory sensory neurons project through the lamina propria that underlies the olfactory epithelium and pass through the bony cribriform plate to enter the nerve fibre layer which is the outer layer of the olfactory bulb and within the CNS. Thus, new axons must constantly traverse the PNS-CNS border and find their correct targets inside the olfactory bulb (Valverde et al. 1992; Tennent and Chuah 1996; Chehrehasa et al. 2010). The constant ability of olfactory

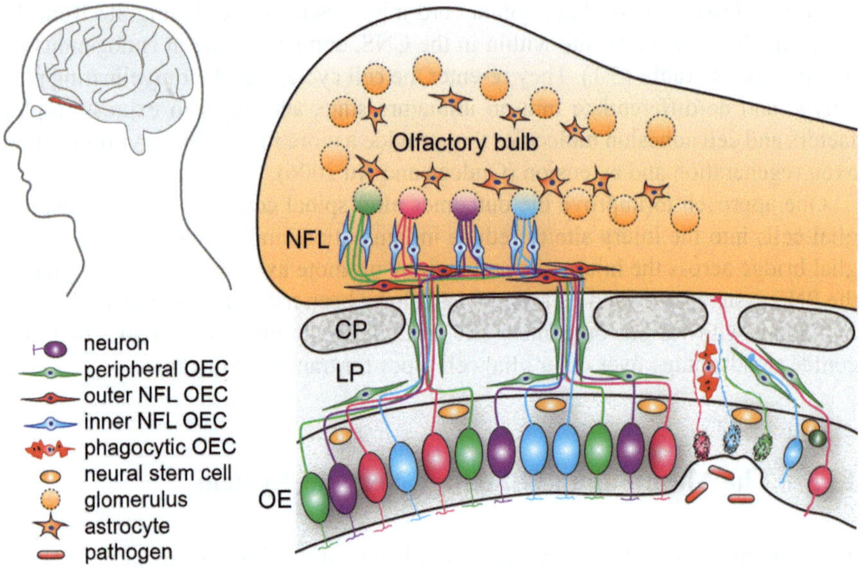

Fig. 13.2 Anatomical organisation of the olfactory system. Primary olfactory sensory neurons lie within the olfactory epithelium (*OE*). Their axons project through the cribriform plate (*CP*) and enter the olfactory bulb where they terminate in their target glomeruli. *OEC*s within the lamina propria (*LP*) encase the bundles of numerous different axons as they project to the olfactory bulb. In the outer nerve fibre layer (*NFL*) of the olfactory bulb, the OECs (*red*) aid the defasciculation and sorting of the different axons. In the inner layer of the nerve fibre layer the *OEC*s (*blue*) assist with the refasciculation and targeting of similar axons to their targets. Astrocytes form a barrier around the glomeruli (*dashed circles*). The olfactory sensory neurons within the *OE* are subjected to toxic molecules within the inhaled air and pathogens such as bacteria and viruses which can result in the death of the neurons (spotted neurons). *OEC*s phagocytose the debris from the degenerated axons. Stem cells lining the basal layer of the *OE* replenish the neuron population which project axons through channels maintained by the *OEC*s

neurons to regenerate and the unique ability of olfactory axons to extend across the PNS-CNS boundary are attributed to the presence of the glia of the olfactory system, called OECs.

13.3.1 Olfactory Ensheathing Cells—the Glia of the Olfactory System

OECs arise from neural crest (Barraud et al. 2010) and they are constantly in close contact with the axons of olfactory neurons all the way from the nasal epithelium to the outer layer of the olfactory bulb. OECs ensheathe the axons of olfactory neurons by the extension of cytoplasmic processes (Chuah and Zheng 1992; Tennent and Chuah 1996) followed by the fasciculation of the axons into larger bundles which ultimately join to form the olfactory nerve (Whitesides and LaMantia 1996).

In contrast to Schwann cells, which in the process of myelination enwrap one single axon (Fig. 13.3), OECs ensheathe bundles of multiple axons by projecting extensive thin cytoplasmic processes around and between the numerous axons within the fascicles (Fig. 13.2; 13.3).

OECs also have a role in promoting axon growth and are known to secrete numerous axon growth promoting factors, such as nerve growth factor, brain derived neurotrophic factor and neuregulins (Boruch et al. 2001). During development, OECs proliferate and migrate ahead of axons or surround the growth cones of axons (Tennent and Chuah 1996; Chehrehasa et al. 2010). Loss of OECs from the olfactory nerve during development results in poor axon growth and targeting (Barraud et al. 2013) which demonstrates that OECs are crucial to the growth and maintenance of axons.

OECs are also thought to be crucial for regeneration during normal turnover of olfactory sensory neurons or after large-scale infection by bacteria and viruses, or major injury. Bacterial infection can lead to the death of olfactory sensory neurons and subsequently their axons (Fig. 13.2; St John et al. 2014), or injury can directly lead to the destruction of the axons (Graziadei et al. 1978; Chehrehasa et al. 2010). The debris from the degenerated axons must be removed but unlike other areas of the body where cells of the immune system usually clear away debris, in the olfactory system this function primarily relies on the OECs (Su et al. 2013). OECs have been shown to continuously phagocytose debris arising from the degenerating axons that occurs during normal turnover of neurons or after widespread injury (Wewetzer et al. 2005; Su et al. 2013). OECs are also able to phagocytose bacteria and thereby protect the olfactory pathway from infection (Wewetzer et al. 2005; Leung et al. 2008; Panni et al. 2013).

OECs form a three-dimensional structure resembling a tunnel through which the axons extend (Fig. 13.3; Li et al. 2005). These structures remain intact even after olfactory axons have degenerated completely following large-scale injury to the olfactory epithelium (Li et al. 2005). However, after large scale injury, OECs can proliferate not only locally around the injury but also from precursors that are present in the olfactory mucosa after which they then migrate along the olfactory nerve (Chehrehasa et al. 2012). By maintaining open channels through which regenerating axons can extend and by responding to injury by proliferating and migrating to the region of need, the OECs with their axon growth-promoting properties provide the structure and support needed for the continuous successful regeneration of the olfactory system.

OECs produce numerous growth factors such as fibroblast growth factor (FGF), insulin-like growth factor (IGF), vascular endothelial growth factor (VEGF), as well as neurotrophic factors such as nerve growth factor (NGF), brain-derived neurotrophic factor (BDNF), glial cell-derived neurotrophic factor (GDNF), neurotrophin 4 (NT4) and NT5; as well as extracellular matrix and cell adhesion molecules including laminin, collagen, galectin-1, heparin sulfate proteoglycans, glial-derived nexin and N-cadherin (Doucette 1990; Doucette and Devon 1993; Chuah and Teague 1999; Kafitz and Greer 1999; Tisay and Key 1999, Boruch et al. 2001; Woodhall et al. 2001; Woodhall et al. 2003; Chuah

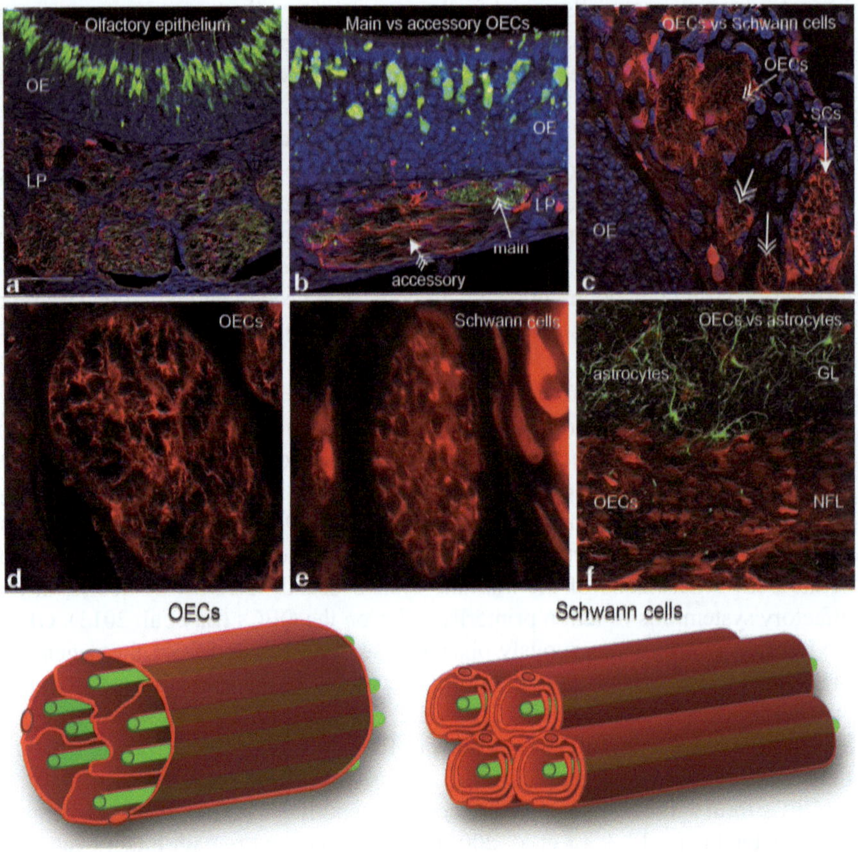

Fig. 13.3 The different populations of glia in the olfactory system. Panels a-f show coronal sections through the olfactory system of a transgenic reporter mouse (OMP-ZsGreen X S100ß-DsRed; Windus et al. 2007; Ekberg et al. 2011) that expresses ZsGreen fluorescent protein in olfactory neurons and DsRed fluorescent protein in glia. **a** The primary olfactory neurons (*green*) within the olfactory epithelium (*OE*) project axons into the lamina propria (*LP*) where they coalesce into fascicles wrapped up by *OEC*s (*red*). Nuclei (*blue*) are stained with DAPI. **b** Along parts of the septum in the nasal cavity, the axon fascicles of the main olfactory system are adjacent to axon fascicles of the accessory (pheromone) olfactory system. **c** Branches of the trigeminal nerve also innervate the nasal cavity; the trigeminal nerve axons are encased by Schwann cells with the trigeminal nerves running adjacent to the main olfactory nerve fascicles that are encased by *OEC*s (*double-headed arrows*). **d** The cell bodies of the olfactory glia are mainly restricted to the periphery of the axon fascicles with their processes permeating the central regions of the axon fascicle. **e** Schwann cells of the trigeminal nerve form tube-like encasing of individual axons. **f** In the olfactory bulb, *OEC*s (*red*) in the nerve fibre layer (*NFL*) form a barrier with the astrocytes (*green*; GFAP immunostaining) in the glomerular layer (*GL*). **g** Schematic of the ensheathment of olfactory axons by *OEC*s. The cell bodies of *OEC*s are largely restricted to the exterior and the processes of the *OEC*s penetrate the internal areas of the nerve bundle where they surround numerous olfactory axons. **h** Schematic of Schwann cell ensheathment of other peripheral nerves in which individual axons are myelinated and encased by Schwann cells. Scale bar is 65 μm in a; 50 μm in b; 30 μm in c; 20 μm in d; 15 μm in e; 40 μm in f

et al. 2004; Chung et al. 2004; Vincent et al. 2005b; Mackay-Sim and St John 2011). These OEC-derived factors are likely to play an important role in nerve repair and regeneration processes as well as neutralization of toxic cell environments due to the excess of free radicals and neurotransmitters such as glutamate (Doucette 1995; Ramon-Cueto 2000; Woodhall et al. 2001; Woodhall et al. 2003; Ramon-Cueto 2011).

13.3.2 Differences Between Olfactory Ensheathing Cells and Schwann Cells

Originally, OECs were referred to as Schwann cells of the olfactory system (Doucette 1984), but their distinctive characteristics separated them from other glial cell types to such an extent that they were classified as an individual glial type. OECs possess features of both CNS and PNS glia in terms of morphology and molecular profile, consistent with their location to both the central and peripheral part of the olfactory nervous system and their ability to cross the PNS-CNS interface. Developmentally, OECs and Schwann cells are of neural crest origin (Barraud et al. 2010), in contrast to astrocytes, which arise from radial glia of neuroepithelial origin (Kriegstein and Gotz 2003). OECs are known to express a number of different proteins found in either Schwann cells or astrocytes. For example non-myelinating Schwann cells and OECs (except those in the inner nerve fibre layer of the olfactory bulb) present immunoreactivity for the p75 low-affinity neurotrophin receptor ($p75^{NTR}$) (Ramon-Cueto 2000).

Whilst similarities between Schwann cells and OECs are evident, one particularly important difference exists in the ability to interact with astrocytes. In contrast to Schwann cells, OECs interact freely with astrocytes, without causing detrimental effect on the astrocyte population (Lakatos et al. 2000). This specific feature is of great interest for nerve regeneration therapies where both populations (OECs and astrocytes) interact at an injury site (Chuah et al. 2011). When OECs are confronted with astrocytes in spinal cord injury sites, astrocyte processes, which form the glial scar, alter their morphology to create a bridging pathway with OECs that allow severed axons to extend across the lesion establishing functional connections (Ramer et al. 2004; Li et al. 2012).

In contrast to Schwann cells, OECs migrate ahead of the regenerating axons, extending their processes to provide a cellular pathway that facilitate axonal extension and adhesion (Tennent and Chuah 1996; Chehrehasa et al. 2010). OECs increase their migration ability by the formation of bigger and thicker processes (Valverde et al. 1992), maintaining a continuous ensheathment of the axons during the regeneration process and leading to enhanced axon growth (Chehrehasa et al. 2010). The capacity of OECs to promote olfactory system renewal and regeneration, as well as their capacity to bridge, enter, and interact with cells of injured host tissue, constitute key factors contributing to the increasing interest in the use of transplanted OECs as therapeutic candidates in spinal cord injury treatments.

13.4 Use of Glial Cells in the Treatment of Spinal Cord Injuries

Re-establishment of nerve connections after spinal cord injury depends of the ability of axons to extend along a pathway to reach their targets. This living pathway consists of glial cells, which provide a dynamic channel through which axons can extend towards their targets (Ramer et al. 2004; Li et al. 2012). After spinal cord injury, the severed nerves are able to survive and sprout locally. However, they are unable to elongate and re-establish the connections, primarily because the glial pathway is altered, blocked and sometimes completely lost. Consequently, a primary objective in the treatment of spinal cord injury is re-establishment of the glial pathway. Transplantation of glial cells into the injury site is therefore a promising therapeutic approach for repair spinal cord injury (Oudega and Xu 2006).

Glial cell transplantation addresses many of the challenges that must be overcome for successful functional improvement, including (1) re-establishment of a growth-promoting environment, (2) replacement of lost cell populations (neurons and glia), and (3) facilitation and promotion of axonal regeneration and extension. Pioneering studies have established that transplantation of glial cells can improve axonal repair, enhance re-growth of damaged nerve cells and improve functional recovery (Yan et al. 2001; Santos-Benito and Ramon-Cueto 2003). Additionally, glial cells have the potential to produce neurotrophic molecules that activate axon regeneration and extension (Jones et al. 2003; Feron et al. 2005).

Different types of glial cells have been investigated as treatment for spinal cord injury including Schwann cells from peripheral nerves and OECs. Schwann cells have been trialled for transplantation due to the important role they play in axon regeneration and myelination. Schwann cells transplanted to the damaged spinal cord can stimulate regeneration of damaged neurons, presumably due to the production of neurotrophic factors (Park et al. 2010), and can also enhance axon remyelination and extension (Lavdas et al. 2010; Flora et al. 2013). However, the axonal regeneration has thus far been limited to restricted areas because Schwann cells have failed to migrate considerable distances into the injured tissue (Lankford et al. 2008). The limited migration is most likely due to unfavorable interaction between the transplanted Schwann cells and host astrocytes (Li et al. 2012). Additionally, Schwann cells have been reported to inhibit myelination by the secretion of connective tissue growth factor, whereas OECs do not (Lamond and Barnett 2013). Thus, Schwann cells may not be the optimal cell type for transplantation therapies due to their poor migration properties, inability to freely interact with endogenous glia and expression of inhibitory molecules.

13.4.1 Transplanted OECs in Spinal Injury Models

Implantation of OECs to promote repair after spinal cord injury have been performed in a variety of spinal cord injury animal models (Table 13.2). Most of these

Table 13.2 Recent examples of in vivo transplantation of olfactory ensheathing cells

First Author (year)	Species	Transplanted cells/tissue	Purity of OECs	Main Outcomes
Richter (2005)	Mouse	Lamina propria Olfactory Bulb	92 % p75 positive cells	Stimulation of outgrowth of axon sprouting. Enhanced angiogenesis LP derived OECs superior ability to migrate
Collazos-Castro (2005)	Rats	Olfactory Bulb derived OECs	90 % p75 positive cells	Partial improvement of motor function. No improvement in axonal regeneration
Feron (2005) Mackay-Sim (2008)	Human	Lamina propria derived OECs	76–88 % p75 positive cells 95 % S100 and GFAP positive cells	Transplantation is feasible and is safe up to 3 years of post-implantation. No deterioration in neurological or functional level
Lu (2006)	Rats	Lamina propria derived OECs	97 % p75 positive cells	Partial Improvement on axonal regeneration. No difference compared with fibroblast
Huang (2006)	Human embryos	Olfactory bulb derived cells	n/a	Neurological functional improvement after transplantation
Lima (2006)	Humans	Whole layer olfactory mucosa	n/a	Transplantation is feasible, relatively safe, and potentially beneficial. Recovery of bladder sensation and improvement in motor function scores
Toft (2007)	Rats	Olfactory Bulb	98 % p75 Positive cells	Improvement on spinal cord function in sensory pathways
Yamamoto (2009)	Rats	Olfactory mucosa	Mixed culture (5 % p75 positive)	Restored directed fore-paw retrieval but not axon regeneration observed
Munoz-Quiles (2009)	Rats	Olfactory Bulb	n/a	Progressive improvement in motor function and axonal regeneration
Salehi (2009)	Rats	Embryonic stem cells + Olfactory Bulb OECs	95 % p75 positive cells	Neural regeneration, neuron survival and partial functional recovery
Chhabra (2009)	Humans	Whole layer olfactory mucosa	n/a	Procedure is relatively safe and feasible. No efficacy could be demonstrated

Table 13.2 (continued)

First Author (year)	Species	Transplanted cells/tissue	Purity of OECs	Main Outcomes
Aoki (2010)	Rats	Whole layer olfactory mucosa	n/a	Partial improvement in motor function and axonal regeneration
Ma (2010)	Rats	neurotrophin-3 genetically modified Olfactory bulb derived OECs	95 % p75 and S100ß positive cells	Effective improvement of axonal regeneration and motor function OECs able to produce NT-3 in vivo
Amemori (2010)	Rats	Lamina propria derived OECs + cAMP infusion	75 % p75 positive cells	Improvement motor function Axon regeneration, sprouting and branching Reduce astrocytic hypertrophy
Ziegler (2011)	Rats	Olfactory bulb	n/a p75 Immuno purification	Improvement in hind limb function, injection of OEG facilitated regeneration of axons across a complete mid-thoracic spinal cord transection.
Centenaro (2011)	Rats	Lamina propria derived OECs	n/a	Discrete motor improvement, improved tissue sparing and sprouting
Zhang (2011)	Rats	Lamina propria derived OECs	n/a	Disappearance of the lesion cavity and integration of repaired tissue Activation of host Schwann cells, improved myelination
Novikova (2011)	Rats	Olfactory bulb derived OECs	93–95 %	Improvement in motor function and neuronal regeneration Aged cell are less effective
Tharion (2011)	Rats	Olfactory lamina propria derived OECS + Olfactory Nerve Fibroblasts (ONL)	Around 50 % p75 positive cells + 50 % fibronectin positive cells (ONL)	Improvement in diaphragm activities Motor function and axonal regeneration improved Disappearance of the lesion cavity

Table 13.2 (continued)

First Author (year)	Species	Transplanted cells/tissue	Purity of OECs	Main Outcomes
Stamegna (2011)	Rats	Lamina propria derived OECs	90 % p75 positive cells	Improvement in axonal regeneration and motor function. Improvement in diaphragm and phrenic nerve activities
Granger (2012)	Dogs	Olfactory mucosa	50 % p75 positive cells	Significantly better fore–hind coordination. Effects are likely to be on local intraspinal circuitry
Huang (2012)	Human embryos	Olfactory bulb derived cells	n/a	Treatment feasible after more than 3 years. Improvement of neurological function. No adverse effects noted
Tabakow (2013)	Humans	Olfactory mucosa	20–50 % S100 positive cells	Improvement of neurological function observed. No adverse effects noted
Mayeur (2013)	Rats	Lamina propria and olfactory bulb	>97 % p75	Improved axon regrowth and reduced scar formation
Toft (2013)	Rats	Olfactory bulb	>97 %	Reduced astrocytic hypertrophy
Torres-Espin (2014)	Rats	Olfactory bulb derived cells	75 % p75/S100 positive cells	Protection/preservation of tissue around injury site

studies have shown that in rodents with spinal cord injury, OEC transplantation promotes the regeneration of axons (Bartolomei and Greer 2000; Gudino-Cabrera et al. 2000; Ramer et al. 2004) and improves functional restoration of breathing and climbing ability (Li et al. 2003; Su and He 2010; Stamegna et al. 2011). In a study of spinal-injured dogs, the transplantation of OECs and fibroblasts restored significant movement in some dogs probably through the restoration of local circuitry, which clearly demonstrates that the procedure has high potential (Granger et al. 2012). Some studies, however, concluded that transplantation of OECs into injured spinal tract did not lead to any detectable difference in axonal extension and functional outcomes (Collazos-Castro et al. 2005; Lu et al. 2006; Chhabra et al. 2009). Discrepancies between results may be attributed to different variables such as the exact nature of spinal cord injury, OEC isolation and culture protocols, transplantation procedure (site of the OECS transplantation and the time after injury that the transplantation is performed) and methodology used to assess outcome.

Numerous studies have reported significant functional improvements following transplantation of OECs. In rodents, OEC transplantation improved or preserved local circuitry as detected electrophysiologically (Toft et al. 2007), and has improved fore-paw movement (Yamamoto et al. 2009), motor function (Munoz-Quiles et al. 2009; Centenaro et al. 2011; Stamegna et al. 2011), hindlimb function (Salehi et al. 2009; Amemori et al. 2010; Aoki et al. 2010; Ziegler et al. 2011), plantar function (Amemori et al. 2010), locomotor function (Ma et al. 2010) and diaphragm activity (Tharion et al. 2011). Thus depending on the type of injury and the assessment of particular functional tests, OEC transplantation can improve functional outcomes.

Corresponding anatomical outcomes after OEC transplantation have also been detected with the majority of studies focusing on axon growth and reduction in cavity formation and astrocytic scarring (Amemori et al. 2010; Centenaro et al. 2011; Tharion et al. 2011; Zhang et al. 2011; Toft et al. 2013; Torres-Espin et al. 2014). The reduction in cavity formation and astrocytic scarring are considered important features as they preserve the axonal circuitry and reduce the formation of physical barriers that prevent axonal growth. In these situations, OECs could be considered to indirectly promoting axon growth as they are enhancing the local environment to make it subsequently amenable to axon growth across the injury site.

However, anatomical improvements are not always detected despite functional recovery being observed. Significant restoration of fore-paw retrieval but not axon regeneration (Yamamoto et al. 2009) may indicate local circuitry changes outside of the immediate injury site as has been suggested for restoration of function in dogs (Granger et al. 2012) or may indicate improved preservation of function of the spared axons.

The timing of transplantation of OECs may also be important for anatomical and functional outcomes. Acute transplantation of OECs may be preferable as the potential ability of OECs to reduce cavity formation and reduce the creation of an inhibitory environment would likely promote axon regeneration. Indeed, acute transplantation of OECs has shown higher functional and anatomical outcomes compared to delayed transplant (Lopez-Vales et al. 2006). However, other studies have not detected significant differences in outcomes for acute versus delayed transplantation of OECs (Munoz-Quiles et al. 2009; Centenaro et al. 2011; Torres-Espin et al. 2014). Despite this, it would seem reasonable that differences in efficacy and outcomes of acute versus delayed treatment may be revealed if improvements in the experimental conditions were made and the purity of the cell preparations and co-transplanted cells was optimized (as discussed below).

What are the mechanisms by which OECs promote regeneration of the spinal cord? OECs transplanted into spinal cord lesions promote regeneration of axons across the lesion, remyelination of axons (although the mechanism remains unknown) and resumption of numerous cellular functions (Ramon-Cueto 2000; Santos-Benito and Ramon-Cueto 2003; Ramer et al. 2004; Li et al. 2012). In particular, OECs provide a physical "bridge" which assists directional regeneration of severed axons across injury sites (Ramon-Cueto and Nieto-Sampedro 1994; Boruch et al. 2001; Resnick et al. 2003) which is consistent with the role of OECs within the injured olfactory system where it has been shown that OECs migrate ahead of axons and that superior axon regeneration occurs when a substantial OEC environment has been created (Chehrehasa et al. 2010). Different factors such as GDNF and NGF have been identified to regulate migration of OECs (Cao et al. 2004; Windus et al. 2007). For example, when OECS are genetically modified to express higher levels of GDNF they display a superior ability to promote axonal growth (Cao et al. 2004). On a cellular level, GDNF acts by stimulating the behaviour of the peripheral lamellipodial waves, increasing cell-cell contact and contact-dependent migration (Windus et al. 2007). Indeed, it has been demonstrated that the motility of OECs in vitro is directly responsible for the apparent motility of axons due to the preference for axons to adhere to OECs (Windus et al. 2011). Thus the migration of OECs across the spinal cord injury site creates a physical glial bridge rich in axon growth promoting factors which together is likely to enhance axon regeneration.

OECs are also thought to be superior to other glial cell types in their interactions with astrocytes. OECs do not tend to form barriers with astrocytes but instead intermix with the astrocytes and thereby reduce the formation of the astrocytic scar (Ramon-Cueto and Avila 1998; Lakatos et al. 2000; Fairless and Barnett 2005; Santos-Silva et al. 2007). Again, this is consistent with their in vivo role in the olfactory system where OECs of the inner nerve fibre layer interact extensively with the astrocytes that populate the more internal layers of the olfactory bulb (Fig. 13.3f; Lakatos et al. 2000).

The phagocytic activity of OECs (Wewetzer et al. 2005; Su et al. 2013) may also improve the local environment of the injury site by removing cellular debris and thereby reducing the inhibitory factors that prevent axon regeneration. It has been clearly shown that OECs phagocytose axon debris within the olfactory system and in vitro (Wewetzer et al. 2005; Su et al. 2013), as well as within the injured spinal cord (Lankford et al. 2008).

In summary, the likely benefits of transplantation of OECs are that they migrate across the injury site to form a glial bridge rich in axon growth promoting factors, they reduce cavity formation, they interact with astrocytes to reduce scar formation, and they likely phagocytose axon debris from the damaged cells to rapidly create an environment favourable for axon regeneration (Fig. 13.4).

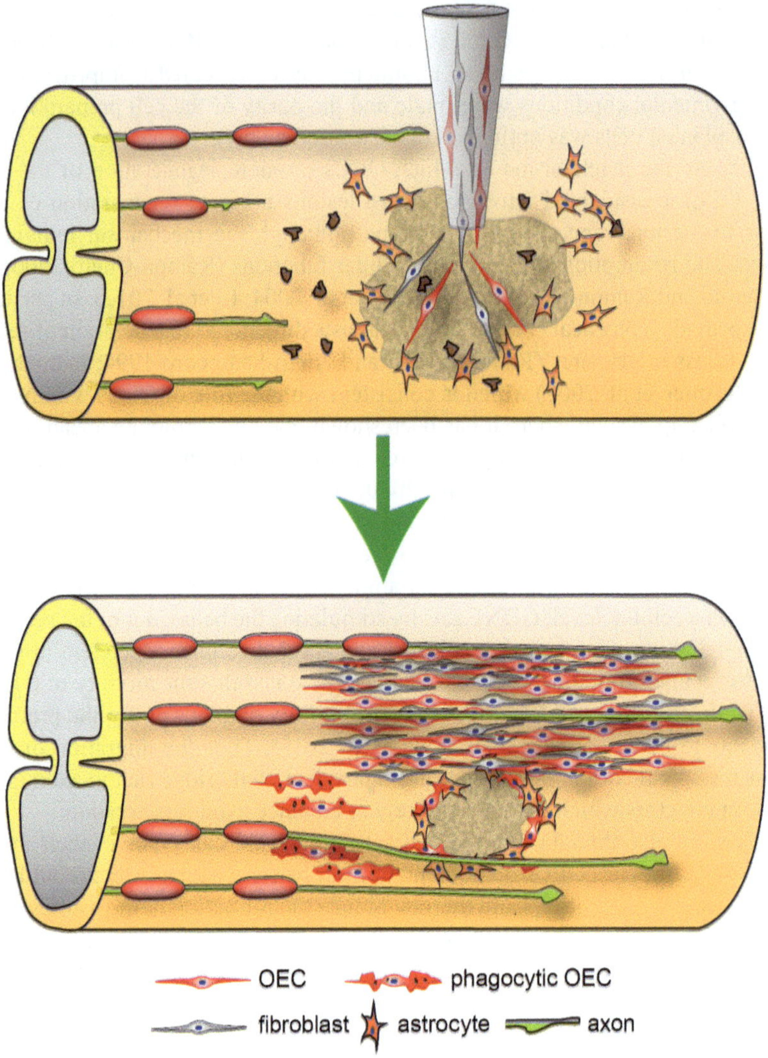

Fig. 13.4 The potential therapeutic use of OECs for spinal cord repair. OECs (*red*) can be transplanted with fibroblasts (*grey*) into the injury site. Strategies that enhance the integration, proliferation and migration of OECs are likely to result in improved axon growth. Targets for improvement can include increasing proliferation of transplanted cells; using fibroblasts to organise alignment and migration of OECs; enhancing how OECs phagocytose cellular debris; optimising the interaction between OECs and activated astrocytes to reduce glial scar formation

13.4.2 Human Trials of OEC Transplantation into the Injured Spinal Cord

Several human clinical trials have demonstrated that transplantation of OECs, of both autologous and of embryonic origin, is a safe and feasible treatment (Yan et al.

2001; Feron et al. 2005; Lima et al. 2006; Tabakow et al. 2013), even after 3 years post-implantation (Mackay-Sim et al. 2008; Huang et al. 2012). A Phase I clinical trial used autologous transplantation of OECs purified from nasal biopsies of spinal cord injury patients and showed that patients with chronic spinal cord injury did not show any adverse medical or surgical complications from the procedure, such as neuropathic pain or neurological deterioration (Feron et al. 2005; Mackay-Sim et al. 2008). In a more recent Phase I study, the transplantation of OECs together with fibroblasts was shown to be safe and feasible after 1 year (Tabakow et al. 2013). Another study showed partial recovery of bladder sensation and improvement of motor function scores in SCI patients transplanted with autologous non-purified OECs (whole olfactory mucosa grafts) (Lima et al. 2006) although this study was not conducted in accordance with Phase I or II standards and thus limited conclusions can be drawn from this study. Huang et al. (2003) reported improvement in both motor function scores and sensory perception in spinal cord injury patients aged 2–64 that received transplantation of human embryonic OECs cells, however, these results have been questioned for failing to meet safety standards for scientific protocols (Lim and Tow 2007). Encouraging results have indicated that OEC transplantation is a promising and safe option to improve recovery in human spinal cord injury patients. However, a more accurate, standardised experimental design, including appropriate controls as well as a larger number of patients, is needed in further clinical trials to confirm the feasibility and effectiveness of OEC therapy. The number of human trials conducted to date is not sufficient to conclude which degree of OEC purity is optimal. In the Phase I trial conducted by Feron et al. (2005) purified lamina propria-derived OECs from nasal biopsies were transplanted and shown to be safe, albeit no dramatic functional improvement was achieved (Feron et al. 2005; Mackay-Sim et al. 2008). Other groups (Lima et al. 2006; Chhabra et al. 2009; Aoki et al. 2010) used whole olfactory mucosa grafts (a mix of neural stem cells, OECs, fibroblasts and other associated cells) to fill the lesion cavity. In these studies some patients were reported to show improvement on motor functional and sensory neurological scores; however some adverse effects such as neuropathic pain were reported, and appropriate controls for strict comparisons were not included. The large expense and ethical concerns of conducting human trials, particularly those with large numbers of patients, confers the need to determine the optimal cell purity and cell mix to maximise the likelihood of successful outcomes. Therefore prior to conducting future human trials, it is important that the methods for purifying and identifying OECs and whether the inclusion of additional cell types is favourable should be addressed.

13.4.3 OECs are a Heterogeneous Population—Which Source Should be Used?

Within the olfactory system in vivo, OECs are not all the same, but have different characteristics and roles depending on their location in the olfactory nerve pathway. Are these different characteristics relevant for spinal repair therapies and are the differences in the OEC subpopulations maintained after culturing and after transplantation? If so, then the source of OECs that are selected for transplantation is likely to affect the outcomes of spinal repair therapies.

OECs express a range of different molecules depending on their location in the olfactory nerve. OECs within the lamina propria express $p75^{NTR}$ and S100β which are typically used to confirm the identity and purity of OECs that have been cultured (Table 13.2). OECs within the nerve fibre layer of the olfactory bulb are not homogeneous and two OEC subpopulations are present (Fig. 13.2) which each express distinct markers. The nerve fibre layer of the olfactory bulb can be further subdivided into outer and inner layers that contain different populations of OECs with different roles in vivo. OECs of the outer nerve fibre layer are responsible for the defasciculation and sorting of axons, and inner nerve fibre layer OECs, refasciculate and target axons to glomeruli (Fig. 13.2). Expression of OEC markers also differs between the subpopulations with OECs in the outer nerve fibre layer highly expressing S100β protein and $p75^{NTR}$ (Gong et al. 1994; Au et al. 2002), whereas, OECs in the inner nerve fibre layer express low levels of S100β and $p75^{NTR}$, but do express high levels of TROY (Hisaoka et al. 2004) and NPY (Ubink and Hokfelt 2000). Therefore, based simply upon the differential expression of these markers, OECs can be considered to be a heterogeneous population of cells each with distinct characteristics that may confer different benefits for regenerating axons.

What are the proposed roles of OECs for the guidance and targeting of olfactory axons? Within the lamina propria, OECs (green in Fig. 13.2) ensheathe fascicles of axons that arise from neurons expressing different odorant receptor types. When the axon fascicles reach the outer layer of the nerve fibre layer, the OECs (red in Fig. 13.2) contribute to the defasciculation and sorting of the axons as they search for their target glomeruli. Within the inner nerve fibre layer, the OECs (blue in Fig. 13.2) contribute to the refasciculation and targeting of axons according to their odorant receptor type. Thus the OECs play different roles in these three different regions: (i) fasciculation within the lamina propria; (ii) defasciculation and sorting within the outer layer of the nerve fibre layer; (iii) refasciculation and targeting within the inner layer of the nerve fibre layer.

Consistent with their in vivo roles, in vitro studies have shown that OECs that have been purified from different regions of the olfactory system have distinctly different behaviours during cell-cell contact (Windus et al. 2007, 2010; Roloff et al. 2013). Canine OECs from the olfactory bulb have been shown to migrate faster than OECs from the lamina propria (Roloff et al. 2013). In other studies, mouse OECs from the lamina propria tend to adhere to each other, consistent with their need to promote fasciculation (Windus et al. 2007). In contrast, OECs from the nerve fibre layer of the olfactory bulb exhibit various reactions to cell-cell contact including adhesion and repulsion consistent with their roles in mediating defasciculation of axons, refasciculation and targeting (Windus et al. 2010). When olfactory sensory neurons were co-cultured with different sources of OECs, axons formed fascicle-like structures with OECs from the lamina propria whereas they tended to disperse when grown with OECs from the olfactory bulb (Windus et al. 2010). In addition, OECs derived from different regions of the nerve fibre layer (rostral, dorsal, caudal, ventral) also exhibit varying responses (Windus et al. 2010). A lack of consistency in the purification of OECs from the olfactory bulb may account for the variation in results that are observed in the spinal transplant models. Some protocols purify OECs from the entire nerve fibre layer of the olfactory bulb (Li et al. 1997, 2007; Ramon-Cueto 2000; Lopez-Vales et al. 2006), while other protocols specify a more restricted region of the nerve fibre layer such as the rostral region (Lankford et al. 2008) or the ventral region (Guest et al. 2008). In other protocols, the region of the nerve fibre layer has not been specified (Teng et al. 2008; Torres-Espin et al. 2014) although it may be that the entire nerve fibre layer was used. Thus selecting the source of OECs could have crucial influences on the efficacy of these cells for repair of the injured spinal cord. Consistent with this, in vivo transplantation has demonstrated variable outcomes depending on the source of OECs that have been used. After transplantation into a crush injury model, OECs from the lamina propria migrated more extensively and reduced the size of the lesion cavity which resulted in more extensive axon sprouting compared to the results obtained after transplantation of OECs derived from the olfactory bulb (Richter et al. 2005). However, olfactory bulb derived OECs promoted greater angiogenesis and reduced autotomy (Richter et al. 2005). In other studies, however, the variation in outcomes depending on the source of OECs is less clear. High purity cultures of OECs from the lamina propria and from the olfactory bulb have both been shown to give similar results in reducing scar formation and promoting axon extension in comparison to control groups (Mayeur et al. 2013). If the OECs derived from the lamina propria give similar therapeutic outcomes compared to OECs derived from the olfactory bulb, then the use of the lamina propria OECs would be preferred for human therapies due to the ease of accessing the source of OECs from the olfactory mucosa.

OECs derived from the olfactory bulb offer other advantages that may make their use preferable. Typically, due to the amount of tissue that can be obtained from the nerve fibre layer, OECs from the olfactory bulb are easier to purify, result in large cell numbers, and have fewer contaminating cells such as fibroblasts (discussed below) which makes their use for animal trials attractive. Certainly, numerous studies have demonstrated improvements in anatomical or functional outcomes using OECs derived from the olfactory bulb (Table 13.2). In humans, the use of olfactory bulb derived OECs is feasible although much more invasive and primate studies have demonstrated that OECs can be successfully grown from olfactory bulbs (Rubio et al. 2008).

13.4.4 Purification of OECs and Contaminating Cells

One of the advantages of using OECs from the lamina propria for spinal repair therapies is that they can be easily harvested from the patient's own nasal mucosa. A small biopsy from the dorsal/caudal region of the nasal cavity can be obtained using local anaesthetic with minimal risk of complication. The OECs are then cultured in vitro to undergo population expansion and transplanted back into the same patient in the spinal injury site. This autologous transplantation avoids host recognition problems however it necessitates a window of several weeks while sufficient cells are expanded in culture. Therefore early intervention to treat the spinal injury cannot occur with this method. An alternative approach is to generate a bank of purified non-autologous OECs that is stored and ready for immediate transplantation for newly injured patients. To this end, extensive work has been conducted using rodent and canine models to identify protocols that result in purification of OECs and to enhance the proliferation of the cells in vitro.

In animal models, OECs can be purified from either the olfactory mucosa lining the nasal cavity or from the nerve fibre layer of the olfactory bulb. Both of these different sources have advantages and disadvantages. Within the olfactory mucosa, the axons of the olfactory sensory neurons form large fascicles surrounded by the OECs (Fig. 13.3a). By dissecting out the mucosa, or separating the lamina propria from the overlying olfactory epithelium, the OECs can be isolated from the explant tissue. However, other cell types are also present within the lamina propria which can potentially contaminate the cultures. The accessory olfactory nerve (part of the pheromone detection system in non-human mammals) projects along the septum and is often adjacent to the fascicles of the main olfactory system (Fig. 13.3b). To date, there has been no systemic comparison of the potential contamination of OEC cultures with accessory olfactory OECs and the possible differences in their contribution to spinal cord repair.

The branches of the trigeminal nerve in which the Schwann cells are present can be in close proximity to the olfactory nerve fascicles (Fig. 13.3c). In transgenic S100ß-DsRed reporter mice (Windus et al. 2007) the different morphologies and structures of the olfactory versus the trigeminal nerves are clearly distinguishable (Fig. 13.3d–e), but the close proximity of the trigeminal nerves means that cultures of cells obtained from olfactory mucosa are likely to contain both OECs and Schwann cells unless purification techniques to remove the Schwann cells are employed. For example, in dogs, HNK-1 is a marker of Schwann cells that is not expressed by OECs and the use of magnet-activated cell sorting based on HNK-1 is effective in removing the Schwann cells to obtain populations of OECs that are free of Schwann cells (Ziege et al. 2013). However, in mice there are no such suitable differential cell surface markers that have been identified to date that can be used to separate OECs and Schwann cells effectively. Comparisons have been made between the efficacy of Schwann cells versus OECs for repair of the injured spinal cord with the results typically showing that both OECs and Schwann cells can enhance the anatomical or functional outcomes although with differences between the cell types depending on the assay (Lankford et al. 2008; Li et al. 2012; Barbour et al. 2013). While the proportion of contaminating Schwann cells is likely to be low in OEC cultures, it should still be considered that Schwann cells could be contributing to the results of spinal repair transplantation studies and may be a source of the variation in outcomes that are observed.

Fibroblasts are also present in the olfactory nerve and often contaminate OEC cultures. To date, most research efforts have attempted to purify the OECs using sedimentation techniques and complement- or antibody-mediated purification techniques with the purpose to eliminate fibroblasts. However, in many cases, partially purified or unpurified mucosal tissues have been used with variable outcomes (Table 13.2). It has been proposed that the inclusion of fibroblasts with the OECs is essential for the functional improvements (Raisman and Li 2007) and transplantation of mixed olfactory cell populations (OECs + olfactory nerve fibroblasts) has resulted in enhanced recovery of motor function in rats when compared to OECs or fibroblasts alone, and suggested that the presence of the two cell types is necessary for optimal repair (Li et al. 2003). Recently, transplantation of semi-purified olfactory mucosa-derived cells (around 50 % OECs; the remainder being fibroblasts) in dogs with spinal cord injury demonstrated that functional recovery can be achieved (Granger et al. 2012). Dogs that received the autologous cells reported a significant gain of fore-hind limb coordination as well as an improvement in the communication across the damaged segments of the spinal cord. In contrast, it has been reported that the presence of fibroblasts on olfactory mucosa cultures used for transplantation results in higher numbers of activated astrocytes, as well as an increase in inhibitory extracellular molecules that reduce axonal regeneration when compared with transplantation of pure glial cells populations (Toft et al. 2013).

To date, the roles of fibroblasts in neuronal regeneration remain relatively uncharacterised, but emerging data suggest that these cells do have crucial and active roles significantly beyond those of inactive connective tissue cells. When fibroblasts are transplanted with neural progenitor cells, they provide a scaffold that facilitates axonal re-growth after spinal cord injury (Pfeifer et al. 2004). Furthermore, fibroblasts are able to populate the lesion cavity in damaged spinal cord (Vroemen et al. 2007) and cause neural progenitor cells to differentiate into glial lineage cells (Wu et al. 2013). Thus, it is evident that a characterisation of how OECs interact with fibroblasts and regenerating neurons in a multi-cellular environment is crucial, and the optimal ratio of cells needs to be determined. For example, in peripheral nerve repair it has been shown that fibroblasts organise and promote the migration of Schwann cells (Parrinello et al. 2010), but the effect of fibroblasts on the organization and growth of OECs has not been studied extensively. There is also a question of the method used to identify OECs and fibroblasts in vitro. The marker $p75^{NTR}$ is the principal one used to identify OECs in vitro, however cultured human fibroblasts have been shown to also express $p75^{NTR}$ (Garcia-Escudero et al. 2012). This raises the question of whether the cultures of reported high purity OECs in fact contain fibroblasts and whether the therapeutic outcomes in the spinal injury models are due solely to the OECs or rather to the combination of OECs and fibroblasts.

The inclusion of fibroblasts and other cells within the "purified" OEC preparations means that the purity of the cultures varies considerably. For those experiments which have sought to use pure OEC preparations, the majority have reported the percentage of cells that express $p75^{NTR}$ as being over 90 % (Table 13.2), while others have used preparations of OECs with less purity (e.g. Amemori et al. 2010; Table 13.2), but still achieve functional outcomes. Other studies have deliberately included other cells such as fibroblasts within preparations with reported purities such as 50 % $p75^{NTR}$ positive cells (OECs) and 50 % fibroblasts (Table 13.2). However, in a Phase I trial for humans in which the autologous purification strategies aimed to transplant high purity OECs, a lower purity was reached (76–88 %; Feron et al. 2005). In other autologous transplantation trials in which fibroblasts were deliberately included, the variation in purity of the separate cell preparations was high, with 20–50 % S100 positive cells (OECs) for a human trial (Tabakow et al. 2013) and 30–70 % for dogs (Granger et al. 2012). This indicates that the purification protocols do not result in predictable purities of cells or that the verification techniques (usually immunocytochemistry staining using antibodies against $p75^{NTR}$ and/or S100) are insufficiently accurate. For OEC transplantation to become a realistic therapy for treating the injured spinal cord, it is crucial that protocols are developed to improve the reliability of purifying, expanding and verifying the identity of cultures of OECs and/or fibroblasts.

13.5 Limitations of the Use of OECs

One of the disadvantages of using OECs for transplantation is the limited survival and low proliferation rates observed over time in culture (Doucette 2002), making large-scale expansion of OECs a main challenge. Moreover, changes in the local environment may induce changes in the morphology of OECs, affecting their behaviour (Vincent et al. 2005a) and causing cells to exhibit characteristics associated with specific roles in nerve regeneration (Doucette 1995; Woodhall et al. 2003).

OECs display a range of phenotypes in vitro and in vivo, and they are able to transform from one morphology type to another in response to modifications in the culture conditions (Doucette 1995; Windus et al. 2007, 2010; Huang et al. 2008a; Radtke et al. 2010). In culture, OECs display different morphologies, including a flattened sheet-like morphology (astrocyte-like), bipolar fusiform morphology (Schwann cell-like) and multipolar morphology (Chuah and Au 1994; Huang et al. 2008b). Morphology of OECs is known to be affected by variables as the method of isolation and purification, cell culture conditions as well as presence of extracellular and intracellular molecules such as cAMP, dBcAMP endothelin-1 and fibulin-3 (Vincent et al. 2003; Huang et al. 2008b; Vukovic et al. 2009). For example, expression of fibulin-3 in OECs results in a predominant bipolar morphology with extremely long processes (Vukovic et al. 2009), whereas local changes in cAMP levels lead to dynamic morphology alterations dramatically affecting OECs migration rate (Huang et al. 2008b).

Studies have shown that OEC plasticity is more evident after cells have been exposed to activated astrocytes and inflammatory mediators within a neuronal lesion site (Woodhall et al. 2003). This extreme plasticity in morphology and physiological changes in populations of OECs before and after transplantation may be a further cause for the discrepancy of results between studies. Thus, it is necessary to develop an optimal standard method by which OECs can be effectively propagated whilst maintaining a consistent or predictable phenotype. At the very least, a thorough characterisation of the mechanisms involved in regulating OEC morphology and behaviour is critical to guarantee better outcomes for OEC transplantation trials.

Another concern presented in transplantation of glial cells as a therapy for spinal cord injury is the inability of transplanted cells to integrate in the host spinal cord. This is due to the over-expression of proteins that inhibit migration, such as Nogo, known to alter the morphology of lamellipodial membrane protrusions in OECs to favor focal adhesion rather than migration (Xu et al. 2009). Some molecules such as proteoglycans, can block the production of these inhibitory proteins, thus promoting the migration of transplanted cells into injured spinal tracts (Huang et al. 2008a). In an olfactory injury model, it has been demonstrated that OECs migrate more extensively without the presence of axons, but that subsequent axon growth is then superior (Chehrehasa et al. 2010). After transplantation, OECs migrate ahead of regenerating axons through unfavorable growth environments increasing the potential of damaged axons to extend through the damaged site (Ramon-Cueto and Santos-Benito 2001; Chehrehasa et al. 2010; Ramon-Cueto 2011). Identification of factors or small molecules that enhance OEC migration properties could offer new alternatives for develop further strategies with emphasis in combinations of cell transplantation and drug therapy that could increase the efficacy of actual OECs transplantation treatments. For example, as described above and has been tested in vivo, OECs could be genetically engineered to express growth factors that would promote their function (Cao et al. 2004; Wu et al. 2008).

13.6 Future Outlook and Opportunities

The use of combination therapies involving OEC transplantation and neural growth factors or molecules that mimic their neurotrophic effect are likely to achieve better results than conventional therapies, and provide a more efficient solution for neural repair in humans (Doucette 1995). Neuregulins and other neurotrophic factors are involved in the survival, proliferation and migration of OECs, as well as Schwann cells, oligodendrocytes and astrocytes. All different types of glial cells present receptors for neuregulins, molecules that promote glial growth and regulate the role of glial cells on nerve regeneration. OECs express various neuregulins including Neu differentiation factor, glial growth factor and motor-neuron derived factor, as well as many receptors for these molecules, supporting the theory that OECs play an important role in the survival, growth promotion and regeneration of severed nerve cells (Woodhall et al. 2001, 2003; Vincent et al. 2005a). An increased understanding of how glial cells respond to neurotrophic molecules as well as the identification of novel compounds that enhance the migratory and integrative properties of OECs is therefore highly warranted in the field.

Another approach is the use of biomaterials and tissue engineering strategies that facilitate axon growth in the spinal cord, as well as to enhance the survival and integration of OECs or other cells transplanted to the injury site. Extracellular matrix motifs and bio-scaffolds can provide the transplanted cells with an enriched substrate that may increase cell proliferation and migration. For example, matrix proteins such as fibronectin and laminin, used on matrix-based scaffold have anti-apoptotic properties that may lead to an improvement on cell survival and migration (Friedman et al. 2002b, a).

To date, experimental and clinical studies suggest that OEC transplantation is safe and effective and may provide real benefits in the treatment of spinal cord injuries. However, it is important that a better understanding of the biology of OECs be pursued to enable new strategies to be developed, perhaps with a more multi-faceted approach, to achieve an effective use of OECs in the treatment of spinal cord injury in humans.

References

Amemori T, Jendelova P, Ruzickova K, Arboleda D, Sykova E (2010) Co-transplantation of olfactory ensheathing glia and mesenchymal stromal cells does not have synergistic effects after spinal cord injury in the rat. Cytotherapy 12:212–225

Antonic A, Sena ES, Lees JS, Wills TE, Skeers P, Batchelor PE, Macleod MR, Howells DW (2013) Stem cell transplantation in traumatic spinal cord injury: a systematic review and meta-analysis of animal studies. PLoS Biol 11:e1001738

Aoki M, Kishima H, Yoshimura K, Ishihara M, Ueno M, Hata K, Yamashita T, Iwatsuki K, Yoshimine T (2010) Limited functional recovery in rats with complete spinal cord injury after transplantation of whole-layer olfactory mucosa: laboratory investigation. J Neurosurg Spine 12:122–130

Arevalo MA, Santos-Galindo M, Bellini MJ, Azcoitia I, Garcia-Segura LM (2010) Actions of estrogens on glial cells: implications for neuroprotection. Biochim Biophys Acta 1800:1106–1112

Au WW, Treloar HB, Greer CA (2002) Sublaminar organization of the mouse olfactory bulb nerve layer. J Comp Neurol 446:68–80

Barbour HR, Plant CD, Harvey AR, Plant GW (2013) Tissue sparing, behavioral recovery, supraspinal axonal sparing/regeneration following sub-acute glial transplantation in a model of spinal cord contusion. BMC Neurosci 14:106

Barraud P, Seferiadis AA, Tyson LD, Zwart MF, Szabo-Rogers HL, Ruhrberg C, Liu KJ, Baker CV (2010) Neural crest origin of olfactory ensheathing glia. Proc Natl Acad Sci U S A 107:21040–21045

Barraud P, St John JA, Stolt CC, Wegner M, Baker CV (2013) Olfactory ensheathing glia are required for embryonic olfactory axon targeting and the migration of gonadotropin-releasing hormone neurons. Biol Open 2:750–759

Bartolomei JC, Greer CA (2000) Olfactory ensheathing cells: bridging the gap in spinal cord injury. Neurosurgery 47:1057–1069

Batzofin BM, Weiss YG, Ledot SF (2013) Do corticosteroids improve outcome for any critical illness? Curr Opin Anesthesio 26:164–170

Beattie MS, Bresnahan JC, Komon J, Tovar CA, Van Meter M, Anderson DK, Faden AI, Hsu CY, Noble LJ, Salzman S, Young W (1997) Endogenous repair after spinal cord contusion injuries in the rat. Exp Neurol 148:453–463

Block ML, Hong JS (2005) Microglia and inflammation-mediated neurodegeneration: multiple triggers with a common mechanism. Prog Neurobiol 76:77–98

Bolton AE (2005) Biologic effects and basic science of a novel immune-modulation therapy. Am J Cardiol 95:24C–29C; discussion 38C-40C

Boruch AV, Conners JJ, Pipitone M, Deadwyler G, Storer PD, Devries GH, Jones KJ (2001) Neurotrophic and migratory properties of an olfactory ensheathing cell line. Glia 33:225–229

Bunge RP, Bunge MB, Rish H (1960) Electron microscopic study of demyelination in an experimentally induced lesion in adult cat spinal cord. J Biophys Biochem Cytol 7:685–696

Byrnes KR, Stoica B, Riccio A, Pajoohesh-Ganji A, Loane DJ, Faden AI (2009) Activation of metabotropic glutamate receptor 5 improves recovery after spinal cord injury in rodents. Ann Neurol 66:63–74

Cabezas R, El-Bacha RS, Gonzalez J, Barreto GE (2012) Mitochondrial functions in astrocytes: neuroprotective implications from oxidative damage by rotenone. Neurosci Res 74:80–90

Cao L, Liu L, Chen ZY, Wang LM, Ye JL, Qiu HY, Lu CL, He C (2004) Olfactory ensheathing cells genetically modified to secrete GDNF to promote spinal cord repair. Brain 127:535–549

Cao L, Zhu YL, Su ZD, Lv BL, Huang ZH, Mu LF, He C (2007) Olfactory ensheathing cells promote migration of Schwann cells by secreted nerve growth factor. Glia 55:897–904

Centenaro LA, Jaeger MD, Ilha J, de Souza MA, Kalil-Gaspar PI, Cunha NB, Marcuzzo S, Achaval M (2011) Olfactory and respiratory lamina propria transplantation after spinal cord transection in rats: effects on functional recovery and axonal regeneration. Brain Res 1426:54–72

Chatzipanteli K, Garcia R, Marcillo AE, Loor KE, Kraydieh S, Dietrich WD (2002) Temporal and segmental distribution of constitutive and inducible nitric oxide synthases after traumatic spinal cord injury: effect of aminoguanidine treatment. J Neurotrauma 19:639–651

Chehrehasa F, Windus LC, Ekberg JA, Scott SE, Amaya D, Mackay-Sim A, St John JA (2010) Olfactory glia enhance neonatal axon regeneration. Mol Cell Neurosci 45:277–288

Chehrehasa F, Ekberg JA, Lineburg K, Amaya D, Mackay-Sim A, St John JA (2012) Two phases of replacement replenish the olfactory ensheathing cell population after injury in postnatal mice. Glia 60:322–332

Chhabra HS, Lima C, Sachdeva S, Mittal A, Nigam V, Chaturvedi D, Arora M, Aggarwal A, Kapur R, Khan TA (2009) Autologous olfactory [corrected] mucosal transplant in chronic spinal cord injury: an Indian Pilot Study. Spinal Cord 47:887–895

Chuah MI, Zheng DR (1992) The human primary olfactory pathway: fine structural and cytochemical aspects during development and in adults. Microsc Res Tech 23:76–85

Chuah MI, Au C (1994) Olfactory cell-cultures on ensheathing cell monolayers. Chemical Senses 19:25–34

Chuah MI, Teague R (1999) Basic fibroblast growth factor in the primary olfactory pathway: Mitogenic effect on ensheathing cells. Neuroscience 88:1043–1050

Chuah MI, Choi-Lundberg D, Weston S, Vincent AJ, Chung RS, Vickers JC, West AK (2004) Olfactory ensheathing cells promote collateral axonal branching in the injured adult rat spinal cord. Exp Neurol 185:15–25

Chuah MI, Hale DM, West AK (2011) Interaction of olfactory ensheathing cells with other cell types in vitro and after transplantation: glial scars and inflammation. Exp Neurol 229:46–53

Chugani DC, Kedersha NL, Rome LH (1991) Vault immunofluorescence in the brain: new insights regarding the origin of microglia. J Neurosci 11:256–268

Chung RS, Woodhouse A, Fung S, Dickson TC, West AK, Vickers JC, Chuah MI (2004) Olfactory ensheathing cells promote neurite sprouting of injured axons in vitro by direct cellular contact and secretion of soluble factors. Cell Mol Life Sci 61:1238–1245

Chung WS, Barres BA (2012) The role of glial cells in synapse elimination. Curr Opin Neurobiol 22:438–445

Collazos-Castro JE, Muneton-Gomez VC, Nieto-Sampedro M (2005) Olfactory glia transplantation into cervical spinal cord contusion injuries. J Neurosurg-Spine 3:308–317

Doucette JR (1984) The glial cells in the nerve fiber layer of the rat olfactory bulb. Anat Rec 210:385–391

Doucette R (1990) Glial influences on axonal growth in the primary olfactory system. Glia 3:433–449

Doucette R (1995) Olfactory ensheathing cells—potential for glial-cell transplantation into areas of CNS injury. Histol Histopathol 10:503–507

Doucette R (2002) Olfactory ensheathing cells: past, present and future. Faseb J 16:A751–A751

Doucette R, Devon R (1993) Olfactory ensheathing cells—factors influencing the phenotype of these glial-cells. Biol Pathol Astrocyte Neuron Interact 2:117–124

Ekberg JA, Amaya D, Chehrehasa F, Lineburg K, Claxton C, Windus LC, Key B, Mackay-Sim A, St John JA (2011) OMP-ZsGreen fluorescent protein transgenic mice for visualisation of olfactory sensory neurons in vivo and in vitro. J Neurosci Methods 196:88–98

Fairless R, Barnett SC (2005) Olfactory ensheathing cells: their role in central nervous system repair. Int J Biochem Cell Biol 37:693–699

Farbman AI, Squinto LM (1985) Early development of olfactory receptor cell axons. Dev Brain Res 19:205–213

Feron F, Perry C, Cochrane J, Licina P, Nowitzke A, Urquhart S, Geraghty T, Mackay-Sim A (2005) Autologous olfactory ensheathing cell transplantation in human spinal cord injury. Brain 128:2951–2960

Flora G, Joseph G, Patel S, Singh A, Bleicher D, Barakat DJ, Louro J, Fenton S, Garg M, Bunge MB, Pearse DD (2013) Combining neurotrophin-transduced schwann cells and rolipram to promote functional recovery from subacute spinal cord injury. Cell Transplant 22:2203–2217

Friedman JA, Windebank AJ, Yaszemski MJ, Moore MJ, Lewellyn EB (2002a) A Schwann cell-seeded, biodegradable polymer implant for promoting axon regeneration after spinal cord injury. Ann Neurol 52:S87–S87

Friedman JA, Windebank AJ, Moore MJ, Spinner RJ, Currier BL, Yaszemski MJ (2002b) Biodegradable polymer grafts for surgical repair of the injured spinal cord. Neurosurgery 51:742–751

Gao Z, Zhu Q, Zhang Y, Zhao Y, Cai L, Shields CB, Cai J (2013) Reciprocal modulation between microglia and astrocyte in reactive gliosis following the CNS injury. Mol Neurobiol 48:690–701

Garcia-Escudero V, Garcia-Gomez A, Langa E, Martin-Bermejo MJ, Ramirez-Camacho R, Garcia-Berrocal JR, Moreno-Flores MT, Avila J, Lim F (2012) Patient-derived olfactory mucosa cells but not lung or skin fibroblasts mediate axonal regeneration of retinal ganglion neurons. Neurosci Lett 509:27–32

Gong Q, Bailey MS, Pixley SK, Ennis M, Liu W, Shipley MT (1994) Localization and regulation of low affinity nerve growth factor receptor expression in the rat olfactory system during development and regeneration. J Comp Neurol 344:336–348

Granger N, Blamires H, Franklin RJ, Jeffery ND (2012) Autologous olfactory mucosal cell transplants in clinical spinal cord injury: a randomized double-blinded trial in a canine translational model. Brain 135:3227–3237

Graziadei PP, Levine RR, Graziadei GA (1978) Regeneration of olfactory axons and synapse formation in the forebrain after bulbectomy in neonatal mice. Proc Natl Acad Sci U S A 75:5230–5234

Gudino-Cabrera G, Pastor AM, de la Cruz RR, Delgado-Garcia JM, Nieto-Sampedro M (2000) Limits to the capacity of transplants of olfactory glia to promote axonal regrowth in the CNS. Neuroreport 11:467–471

Guest JD, Herrera L, Margitich I, Oliveria M, Marcillo A, Casas CE (2008) Xenografts of expanded primate olfactory ensheathing glia support transient behavioral recovery that is independent of serotonergic or corticospinal axonal regeneration in nude rats following spinal cord transection. Exp Neurol 212:261–274

Hermann GE, Rogers RC, Bresnahan JC, Beattie MS (2001) Tumor necrosis factor-alpha induces cFOS and strongly potentiates glutamate-mediated cell death in the rat spinal cord. Neurobiol Dis 8:590–599

Hisaoka T, Morikawa Y, Kitamura T, Senba E (2004) Expression of a member of tumor necrosis factor receptor superfamily, TROY, in the developing olfactory system. Glia 45:313–324

Huang H, Chen L, Wang H, Xiu B, Li B, Wang R, Zhang J, Zhang F, Gu Z, Li Y, Song Y, Hao W, Pang S, Sun J (2003) Influence of patients' age on functional recovery after transplantation of olfactory ensheathing cells into injured spinal cord injury. Chin Med J (Engl) 116:1488–1491

Huang H, Wang H, Chen L, Gu Z, Zhang J, Zhang F, Song Y, Li Y, Tan K, Liu Y, Xi H (2006) Influence factors for functional improvement after olfactory ensheathing cell transplantation for chronic spinal cord injury. Zhongguo Xiu Fu Chong Jian Wai Ke Za Zhi 20:434–438

Huang HY, Chen L, Xi HT, Wang HM, Zhang J, Zhang F, Liu YC (2008a) Fetal olfactory ensheathing cells transplantation in amyotrophic lateral sclerosis patients: a controlled pilot study. Clin Transplant 22:710–718

Huang ZH, Wang Y, Cao L, Su ZD, Zhu YL, Chen YZ, Yuan XB, He C (2008b) Migratory properties of cultured olfactory ensheathing cells by single-cell migration assay. Cell Res 18:479–490

Huang HY, Xi HT, Chen L, Zhang F, Liu YC (2012) Long-term outcome of olfactory ensheathing cell therapy for patients with complete chronic spinal cord injury. Cell Transplant 21:21S23–S31

Hulsebosch CE (2002) Recent advances in pathophysiology and treatment of spinal cord injury. Adv Physiol Educ 26:238–255

Hurlbert RJ, Hadley MN, Walters BC, Aarabi B, Dhall SS, Gelb DE, Rozzelle CJ, Ryken TC, Theodore N (2013) Pharmacological therapy for acute spinal cord injury. Neurosurgery 72:93–105

Jauregui-Huerta F, Ruvalcaba-Delgadillo Y, Gonzalez-Castaneda R, Garcia-Estrada J, Gonzalez-Perez O, Luquin S (2010) Responses of glial cells to stress and glucocorticoids. Curr Immunol Rev 6:195–204

Jessen KR (2004) Glial cells. Int J Biochem Cell Biol 36:1861–1867

Jessen KR (2006) A brief look at glial cells. Novartis Found Symp 276:5–14; discussion 54–17, 275–281

Jones DG, Anderson ER, Galvin KA (2003) Spinal cord regeneration: moving tentatively towards new perspectives. Neurorehabilitation 18:339–351

Kafitz KW, Greer CA (1999) Olfactory ensheathing cells promote neurite extension from embryonic olfactory receptor cells in vitro. Glia 25:99–110

Kigerl KA, Gensel JC, Ankeny DP, Alexander JK, Donnelly DJ, Popovich PG (2009) Identification of two distinct macrophage subsets with divergent effects causing either neurotoxicity or regeneration in the injured mouse spinal cord. J Neurosci 29:13435–13444

Kriegstein AR, Gotz M (2003) Radial glia diversity: a matter of cell fate. Glia 43:37–43

Lakatos A, Franklin RJM, Barnett SC (2000) Olfactory ensheathing cells and Schwann cells differ in their in vitro interactions with astrocytes. Glia 32:214–225

Lamond R, Barnett SC (2013) Schwann cells but not olfactory ensheathing cells inhibit CNS myelination via the secretion of connective tissue growth factor. J Neurosci 33:18686–18697

Lankford KL, Sasaki M, Radtke C, Kocsis JD (2008) Olfactory ensheathing cells exhibit unique migratory, phagocytic, and myelinating properties in the X-irradiated spinal cord not shared by Schwann cells. Glia 56:1664–1678

Lavdas AA, Chen J, Papastefanaki F, Chen S, Schachner M, Matsas R, Thomaidou D (2010) Schwann cells engineered to express the cell adhesion molecule L1 accelerate myelination and motor recovery after spinal cord injury. Exp Neurol 221:206–216

Leal-Filho MB (2011) Spinal cord injury: from inflammation to glial scar. Surg Neurol Int 2:112

Leung JY, Chapman JA, Harris JA, Hale D, Chung RS, West AK, Chuah MI (2008) Olfactory ensheathing cells are attracted to, and can endocytose, bacteria. Cell Mol Life Sci 65:2732–2739

Li BC, Xu C, Zhang JY, Li Y, Duan ZX (2012) Differing Schwann cells and olfactory ensheathing cells behaviors, from interacting with astrocyte, produce similar improvements in contused rat spinal cord's motor function. J Mol Neurosci 48:35–44

Li Y, Field PM, Raisman G (1997) Repair of adult rat corticospinal tract by transplants of olfactory ensheathing cells. Science 277:2000–2002

Li Y, Decherchi P, Raisman G (2003) Transplantation of olfactory ensheathing cells into spinal cord lesions restores breathing and climbing. J Neurosci 23:727–731

Li Y, Field PM, Raisman G (2005) Olfactory ensheathing cells and olfactory nerve fibroblasts maintain continuous open channels for regrowth of olfactory nerve fibres. Glia 52:245–251

Li Y, Yamamoto M, Raisman G, Choi D, Carlstedt T (2007) An experimental model of ventral root repair showing the beneficial effect of transplanting olfactory ensheathing cells. Neurosurgery 60:734–740; discussion 740–731

Lim PAC, Tow AM (2007) Recovery and regeneration after spinal cord injury: a review and summary of recent literature. Ann Acad Med Singapore 36:49–57

Lima C, Pratas-Vital J, Escada P, Hasse-Ferreira A, Capucho C, Peduzzi JD (2006) Olfactory mucosa autografts in human spinal cord injury: a pilot clinical study. J Spinal Cord Med 29:191–203

Lopez MV, Cuadrado MP, Ruiz-Poveda OM, Del Fresno AM, Accame ME (2007) Neuroprotective effect of individual ginsenosides on astrocytes primary culture. Biochim Biophys Acta 1770:1308–1316
Lopez-Vales R, Fores J, Verdu E, Navarro X (2006) Acute and delayed transplantation of olfactory ensheathing cells promote partial recovery after complete transection of the spinal cord. Neurobiol Dis 21:57–68
Lu P, Yang H, Culbertson M, Graham L, Roskams AJ, Tuszynski MH (2006) Olfactory ensheathing cells do not exhibit unique migratory or axonal growth-promoting properties after spinal cord injury. J Neurosci 26:11120–11130
Ma YH, Zhang Y, Cao L, Su JC, Wang ZW, Xu AB, Zhang SC (2010) Effect of neurotrophin-3 genetically modified olfactory ensheathing cells transplantation on spinal cord injury. Cell Transplant 19:167–177
Mackay-Sim A, Kittel P (1991a) Cell dynamics in the adult mouse olfactory epithelium: a quantitative autoradiographic study. J Neurosci 11:979–984
Mackay-Sim A, Kittel PW (1991b) On the life span of olfactory receptor neurons. Eur J Neurosci 3:209–215.
Mackay-Sim A, St John JA (2011) Olfactory ensheathing cells from the nose: clinical application in human spinal cord injuries. Exp Neurol 229:174–180
Mackay-Sim A, Feron F, Cochrane J, Bassingthwaighte L, Bayliss C, Davies W, Fronek P, Gray C, Kerr G, Licina P, Nowitzke A, Perry C, Silburn PAS, Urquhart S, Geraghty T (2008) Autologous olfactory ensheathing cell transplantation in human paraplegia: a 3-year clinical trial. Brain 131:2376–2386
Matthews MA, St Onge MF, Faciane CL, Gelderd JB (1979) Axon sprouting into segments of rat spinal cord adjacent to the site of a previous transection. Neuropathol Appl Neurobiol 5:181–196
Mayeur A, Duclos C, Honore A, Gauberti M, Drouot L, do Rego JC, Bon-Mardion N, Jean L, Verin E, Emery E, Lemarchant S, Vivien D, Boyer O, Marie JP, Guerout N (2013) Potential of olfactory ensheathing cells from different sources for spinal cord repair. PLoS ONE 8:e62860
Mombaerts P, Wang F, Dulac C, Chao SK, Nemes A, Mendelsohn M, Edmondson J, Axel R (1996) Visualizing an olfactory sensory map. Cell 87:675–686
Munoz-Quiles C, Santos-Benito FF, Liamusi MB, Ramon-Cueto A (2009) Chronic spinal injury repair by olfactory bulb ensheathing glia and feasibility for autologous therapy. J Neuropath Exp Neur 68:1294–1308
Ndubaku U, de Bellard ME (2008) Glial cells: old cells with new twists. Acta Histochem 110:182–195
Novikova LN, Lobov S, Wiberg M, Novikov LN (2011) Efficacy of olfactory ensheathing cells to support regeneration after spinal cord injury is influenced by method of culture preparation. Exp Neurol 229:132–142
Oudega M, Xu XM (2006) Schwann cell transplantation for repair of the adult spinal cord. J Neurotrauma 23:453–467
Panni P, Ferguson IA, Beacham I, Mackay-Sim A, Ekberg JA, St John JA (2013) Phagocytosis of bacteria by olfactory ensheathing cells and Schwann cells. Neurosci Lett 539:65–70
Park HW, Lim MJ, Jung H, Lee SP, Paik KS, Chang MS (2010) Human mesenchymal stem cell-derived Schwann cell-like cells exhibit neurotrophic effects, via distinct growth factor production, in a model of spinal cord injury. Glia 58:1118–1132
Parrinello S, Napoli I, Ribeiro S, Wingfield Digby P, Fedorova M, Parkinson DB, Doddrell RD, Nakayama M, Adams RH, Lloyd AC (2010) EphB signaling directs peripheral nerve regeneration through Sox2-dependent Schwann cell sorting. Cell 143:145–155
Pearse DD, Chatzipanteli K, Marcillo AE, Bunge MB, Dietrich WD (2003) Comparison of iNOS inhibition by antisense and pharmacological inhibitors after spinal cord injury. J Neuropathol Exp Neurol 62:1096–1107
Pfeifer K, Vroemen M, Blesch A, Weidner N (2004) Adult neural progenitor cells provide a permissive guiding substrate for corticospinal axon growth following spinal cord injury. Eur J Neurosci 20:1695–1704

Qiu J, Cai D, Dai H, McAtee M, Hoffman PN, Bregman BS, Filbin MT (2002) Spinal axon regeneration induced by elevation of cyclic AMP. Neuron 34:895–903
Radtke C, Lankford KL, Wewetzer K, Imaizumi T, Fodor WL, Kocsis JD (2010) Impaired spinal cord remyelination by long-term cultured adult porcine olfactory ensheathing cells correlates with altered in vitro phenotypic properties. Xenotransplantation 17:71–80
Raisman G, Li Y (2007) Repair of neural pathways by olfactory ensheathing cells. Nat Rev Neurosci 8:312–319
Ramer LM, Au E, Richter MW, Liu J, Tetzlaff W, Roskams AJ (2004) Peripheral olfactory ensheathing cells reduce scar and cavity formation and promote regeneration after spinal cord injury. J Comp Neurol 473:1–15
Ramon-Cueto A (2000) Olfactory ensheathing glia transplantation into the injured spinal cord. Neural Plast Regen 128:265–272
Ramon-Cueto A (2011) Olfactory ensheathing glia for nervous system repair. Exp Neurol 229:1–1
Ramon-Cueto A, Nieto-Sampedro M (1994) Regeneration into the spinal cord of transected dorsal root axons is promoted by ensheathing glia transplants. Exp Neurol 127:232–244
Ramon-Cueto A, Avila J (1998) Olfactory ensheathing glia: properties and function. Brain Res Bull 46:175–187
Ramon-Cueto A, Santos-Benito FF (2001) Cell therapy to repair injured spinal cords: olfactory ensheathing glia transplantation. Restor Neurol Neuros 19:149–156
Resnick DK, Cechvala CF, Yan Y, Witwer BP, Sun D, Zhang S (2003) Adult olfactory ensheathing cell transplantation for acute spinal cord injury. J Neurotrauma 20:279–285
Richter MW, Fletcher PA, Liu J, Tetzlaff W, Roskams AJ (2005) Lamina propria and olfactory bulb ensheathing cells exhibit differential integration and migration and promote differential axon sprouting in the lesioned spinal cord. J Neurosci 25:10700–10711
Roloff F, Ziege S, Baumgartner W, Wewetzer K, Bicker G (2013) Schwann cell-free adult canine olfactory ensheathing cell preparations from olfactory bulb and mucosa display differential migratory and neurite growth-promoting properties in vitro. BMC Neurosci 14:141
Rubio MP, Munoz-Quiles C, Ramon-Cueto A (2008) Adult olfactory bulbs from primates provide reliable ensheathing glia for cell therapy. Glia 56:539–551
Salehi M, Pasbakhsh P, Soleimani M, Abbasi M, Hasanzadeh G, Modaresi MH, Sobhani A (2009) Repair of spinal cord injury by co-transplantation of embryonic stem cell-derived motor neuron and olfactory ensheathing cell. Iran Biomed J 13:125–135
Santos-Benito FF, Ramon-Cueto A (2003) Olfactory ensheathing glia transplantation: a therapy to promote repair in the mammalian central nervous system. Anat Rec B New Anat 271:77–85
Santos-Silva A, Fairless R, Frame MC, Montague P, Smith GM, Toft A, Riddell JS, Barnett SC (2007) FGF/heparin differentially regulates Schwann cell and olfactory ensheathing cell interactions with astrocytes: a role in astrocytosis. J Neurosci 27:7154–7167
Singh S, Swarnkar S, Goswami P, Nath C (2011) Astrocytes and microglia: responses to neuropathological conditions. Int J Neurosci 121:589–597
St John JA, Ekberg JA, Dando S, Meedeniya ACB, Horton RE, Batzloff M, Owen SJ, Holt S, Peak IR, Ulett G, Mackay-Sim A, Beacham IR (2014) Burkholderia pseudomallei penetrates the brain via destruction of the olfactory and trigeminal nerves: implications for the pathogenesis of neurological melioidosis. mBIO In press
Stahel PF, VanderHeiden T, Finn MA (2012) Management strategies for acute spinal cord injury: current options and future perspectives. Curr Opin Crit Care 18:651–660
Stamegna JC, Felix MS, Roux-Peyronnet J, Rossi V, Feron F, Gauthier P, Matarazzo V (2011) Nasal OEC transplantation promotes respiratory recovery in a subchronic rat model of cervical spinal cord contusion. Exp Neurol 229:120–131
Su Z, He C (2010) Olfactory ensheathing cells: biology in neural development and regeneration. Prog Neurobiol 92:517–532
Su Z, Yuan Y, Chen J, Cao L, Zhu Y, Gao L, Qiu Y, He C (2009) Reactive astrocytes in glial scar attract olfactory ensheathing cells migration by secreted TNF-alpha in spinal cord lesion of rat. PLoS ONE 4:e8141

Su Z, Chen J, Qiu Y, Yuan Y, Zhu F, Zhu Y, Liu X, Pu Y, He C (2013) Olfactory ensheathing cells: the primary innate immunocytes in the olfactory pathway to engulf apoptotic olfactory nerve debris. Glia 61:490–503

Tabakow P, Jarmundowicz W, Czapiga B, Fortuna W, Miedzybrodzki R, Czyz M, Huber J, Szarek D, Okurowski S, Szewczyk P, Gorski A, Raisman G (2013) Transplantation of autologous olfactory ensheathing cells in complete human spinal cord injury. Cell Transplant 22:1591–1612

Teng X, Nagata I, Li HP, Kimura-Kuroda J, Sango K, Kawamura K, Raisman G, Kawano H (2008) Regeneration of nigrostriatal dopaminergic axons after transplantation of olfactory ensheathing cells and fibroblasts prevents fibrotic scar formation at the lesion site. J Neurosci Res 86:3140–3150

Tennent R, Chuah MI (1996) Ultrastructural study of ensheathing cells in early development of olfactory axons. Brain Res Dev Brain Res 95:135–139

Tharion G, Indirani K, Durai M, Meenakshi M, Devasahayam SR, Prabhav NR, Solomons C, Bhattacharji S (2011) Motor recovery following olfactory ensheathing cell transplantation in rats with spinal cord injury. Neurol India 59:77–83

Tisay KT, Key B (1999) The extracellular matrix modulates olfactory neurite outgrowth on ensheathing cells. J Neurosci 19:9890–9899

Toft A, Scott DT, Barnett SC, Riddell JS (2007) Electrophysiological evidence that olfactory cell transplants improve function after spinal cord injury. Brain 130:970–984

Toft A, Tome M, Barnett SC, Riddell JS (2013) A comparative study of glial and non-neural cell properties for transplant-mediated repair of the injured spinal cord. Glia 61:513–528

Torres-Espin A, Redondo-Castro E, Hernandez J, Navarro X (2014) Bone marrow mesenchymal stromal cells and olfactory ensheathing cells transplantation after spinal cord injury—a morphological and functional comparison in rats. Eur J Neurosci 39:1701–1717

Ubink R, Hokfelt T (2000) Expression of neuropeptide Y in olfactory ensheathing cells during prenatal development. J Comp Neurol 423:13–25

Valverde F, Santacana M, Heredia M (1992) Formation of an olfactory glomerulus: morphological aspects of development and organization. Neuroscience 49:255–275

Vassar R, Chao SK, Sitcheran R, Nunez JM, Vosshall LB, Axel R (1994) Topographic organization of sensory projections to the olfactory bulb. Cell 79:981–991

Vincent AJ, West AK, Chuah MI (2003) Morphological plasticity of olfactory ensheathing cells is regulated by cAMP and endothelin-1. Glia 41:393–403

Vincent AJ, West AK, Chuah MI (2005a) Morphological and functional plasticity of olfactory ensheathing cells. J Neurocytol 34:65–80

Vincent AJ, Taylor JM, Choi-Lundberg DL, West AK, Chuah MI (2005b) Genetic expression profile of olfactory ensheathing cells is distinct from that of Schwann cells and astrocytes. Glia 51:132–147

Vroemen M, Caioni M, Bogdahn U, Weidner N (2007) Failure of Schwann cells as supporting cells for adult neural progenitor cell grafts in the acutely injured spinal cord. Cell Tissue Res 327:1–13

Vukovic J, Ruitenberg MJ, Roet K, Franssen E, Arulpragasam A, Sasaki T, Verhaagen J, Harvey AR, Busfield SJ, Plant GW (2009) The glycoprotein Fibulin-3 regulates morphology and motility of olfactory ensheathing cells in vitro. Glia 57:424–443

Wewetzer K, Kern N, Ebel C, Radtke C, Brandes G (2005) Phagocytosis of O4+ axonal fragments in vitro by p75- neonatal rat olfactory ensheathing cells. Glia 49:577–587

Whitesides JG 3rd, LaMantia AS (1996) Differential adhesion and the initial assembly of the mammalian olfactory nerve. J Comp Neurol 373:240–254

Windus LC, Claxton C, Allen CL, Key B, St John JA (2007) Motile membrane protrusions regulate cell-cell adhesion and migration of olfactory ensheathing glia. Glia 55:1708–1719

Windus LC, Lineburg KE, Scott SE, Claxton C, Mackay-Sim A, Key B, St John JA (2010) Lamellipodia mediate the heterogeneity of central olfactory ensheathing cell interactions. Cell Mol Life Sci 67:1735–1750

Windus LC, Chehrehasa F, Lineburg KE, Claxton C, Mackay-Sim A, Key B, St John JA (2011) Stimulation of olfactory ensheathing cell motility enhances olfactory axon growth. Cell Mol Life Sci 68:3233–3247

Woodhall E, West AK, Chuah MI (2001) Cultured olfactory ensheathing cells express nerve growth factor, brain-derived neurotrophic factor, glia cell line-derived neurotrophic factor and their receptors. Mol Brain Res 88:203–213
Woodhall E, West AK, Vickers JC, Chuah MI (2003) Olfactory ensheathing cell phenotype following implantation in the lesioned spinal cord. Cell Mol Life Sci 60:2241–2253
Wu J, Sun TS, Ren JX, Wang XZ (2008) Ex vivo non-viral vector-mediated neurotrophin-3 gene transfer to olfactory ensheathing glia: effects on axonal regeneration and functional recovery after implantation in rats with spinal cord injury. Neurosci Bull 24:57–65
Wu XJ, Bolger WE, Anders JJ (2013) Fibroblasts isolated from human middle turbinate mucosa cause neural progenitor cells to differentiate into glial lineage cells. PLoS ONE 8 e76926
Xu XH, Zhou JF, Li TZ, Zhang ZH, Shan L, Xiang ZH, Yu ZW, Zhang WD, He C (2009) Polygalasaponin G promotes neurite outgrowth of cultured neuron on myelin. Neurosci Lett 460:41–46
Yamamoto M, Raisman G, Li DQ, Li Y (2009) Transplanted olfactory mucosal cells restore paw reaching function without regeneration of severed corticospinal tract fibres across the lesion. Brain Res 1303:26–31
Yan H, Bunge MB, Wood PM, Plant GW (2001) Mitogenic response of adult rat olfactory ensheathing glia to four growth factors. Glia 33:334–342
Zhang SX, Huang FF, Gates M, Holmberg EG (2011) Scar ablation combined with LP/OEC transplantation promotes anatomical recovery and P0-positive myelination in chronically contused spinal cord of rats. Brain Res 1399:1–14
Ziege S, Baumgartner W, Wewetzer K (2013) Toward defining the regenerative potential of olfactory mucosa: establishment of Schwann cell-free adult canine olfactory ensheathing cell preparations suitable for transplantation. Cell Transplant 22:355–367
Ziegler MD, Hsu D, Takeoka A, Zhong H, Ramon-Cueto A, Phelps PE, Roy RR, Edgerton VR (2011) Further evidence of olfactory ensheathing glia facilitating axonal regeneration after a complete spinal cord transection. Exp Neurol 229:109–119

Chapter 14
Characterization of the Phenotypic Features, Immuno-modulatory Properties and Therapeutic Potentials of Wharton's Jelly-Derived Mesenchymal Stromal Cells

Ike dela Peña, Marina Bastawrous, Diego Lozano, Daniela Aguirre, Diana Hernandez, Sandra Acosta, Mibel Pabon, Naoki Tajiri, Yuji Kaneko and Cesar V. Borlongan

Contents

C. V. Borlongan (✉) · I. dela Peña · M. Bastawrous · D. Lozano · D. Aguirre · D. Hernandez · S. Acosta · M. Pabon · N. Tajiri · Y. Kaneko
Center of Excellence for Aging and Brain Repair, Department of Neurosurgery and Brain Repair, University of South Florida College of Medicine, 12901 Bruce B. Downs Blvd. MDC 78, Tampa, FL 33612, USA
e-mail: cborlong@health.usf.edu

L.-R. Zhao, J. H. Zhang (eds.), *Cellular Therapy for Stroke and CNS Injuries*,
Springer Series in Translational Stroke Research, DOI 10.1007/978-3-319-11481-1_14

Abstract The Wharton's jelly (WJ) of the umbilical cord has been identified as a rich source of mesenchymal stromal cells (MSCs), which are considered as promising candidates for stem cell-based therapy to treat several diseases. In particular, MSCs harvested from the "young" WJ are believed to be more proliferative, immunosuppressive and therapeutically active stem cells than those derived from adult tissues, such as the bone marrow or adipose. MSCs derived from WJ also exhibit transplantable features such as ease of sourcing, in vitro expandability, differentiation capacities, immune-evasion and immune-regulation profiles. Indeed, the potentiality of WJ-derived stem cells to treat cancer, cardiovascular and liver diseases, and nerve and cartilage tendon injuries has been suggested. In this paper, we present an overview of the phenotypic characteristics, immune-modulatory properties and therapeutic potentials of WJ-derived stem cells, and suggest optimization protocols for successful advancement of WJ-derived stem cells into clinical use.

Abbreviations

AFP	Alpha-fetoprotein
BDNF	Brain-derived neurotrophic factor
bFGF	Basic fibroblast growth factor
BM-MSCs	Bone marrow-derived mesenchymal stromal/stem cells
cAMP	Cyclic adenosine monophosphate
CD	Cluster differentiation
CDKN	Cyclin-dependent kinase inhibitor
CFU-F	Colony-forming unit-fibroblast
CL-MSCs	Cord lining mesenchymal stem cells
COMP	Cartilage oligomeric protein
CXCR	C-X-C chemokine receptor
DCs	Dendritic precursors
ESCs	Embryonic stem cells
HIE	Hypoxic-ischemic encephalopathy
HIFs	hypoxia inducible factors
hiPSCs	human induced pluripotent stem cells
HLA	Human leukocyte antigen
HSCs	Hematopoietic stem cells
HUCPVCs	Human umbilical cord perivascular cells
HUVECs	Human umbilical vein endothelial cells
IFN-γ	Interferon gamma
IL	Interleukin
iPSC	Induced pluripotent stem cell
IV	Intravenous
IVF	In vitro fertilization
KDR	Kinase insert domain receptor
MHC	Major histocompatibility complex
MI	Myocardial infaction
MSCs	Mesenchymal stromal/stem cells

NGF	Nerve growth factor
NMII	Non-muscle myosin II
NP	Nucleus pulposus
PDGF	Platelet-derived growth factor
PHA	Phytohemagglutinin
PLLA	Poly-L-lactic-acid
SCID	Severe combined immunodeficiency
SMA	Smooth muscle actin
SPIO	Superparamagnetic iron oxide particles
SSEA4	Stage-specific embryonic antigen 4
TERT	Telomerase reverse transcriptase
TGF- β	Transforming growth factor- beta
UC	Umbilical cord
UCMSCs	Umbilical cord matrix stem cells
VEGF	Vascular endothelial growth factor
WJ	Wharton's jelly

14.1 Introduction

The area of stem cell research offers an avenue for an in-depth study of cell biology, as well as in the discovery of new strategies to treat diseases. There are various sources of stem cells and perhaps one of the most important is the Wharton's jelly (WJ), first described by Thomas Wharton in 1656 (Wharton 1996), which is a gelatinous layer in the umbilical cord that hosts different populations of stem cells (La Rocca 2011). Mesenchymal stromal cells (MSCs) constitute one of these populations, which display a specific group of surface markers including CD90, CD73 and CD105. Accordingly, these markers contribute to plastic adherence of WJ-MSCs, which is the prototypic feature of these cells. Aside from WJ, MSCs can also be derived from other sources such as adult bone marrow and adipose tissues. However, MSCs derived from different sources display different degrees of stemness and immune properties. For instance, WJ-MSCs have more potent proliferative, immunosuppressive and therapeutic properties than MSCs isolated from adult bone marrow or adipose tissue. McElreavey et al. (1991) first isolated and cultured WJ-MSCs. Thereafter, there have been several attempts to identify optimum isolation and differentiation of WJ-MSCs (Chacko and Reynolds 1954, Takechi et al. 1993; Naughton et al. 1997; Purchio et al. 1999; Kadner et al. 2002; Mitchell et al. 2003; Romanov et al. 2003; Kadner et al. 2004). The milestone discoveries on WJ-MSCs should encourage further research on stem cell biology and pave the way for advancing stem cells and stem cell-derived products to clinical applications.

14.2 Sources of Stem Cells for Transplantation

Stem cells can be classified into three major groups based on their biochemical and genomic markers: embryonic stem cells (ESCs), MSCs and hematopoietic stem cells (HSCs). Various types of stem cells have been isolated from human tissues including pre-implantation embryos, fetal sources such as aborted fetuses and birth-associated tissues, and adult organs. ESCs are pluripotent cells, which in theory, have the capacity to differentiate into any type of tissue in the human body. Nonetheless, there has been a moral debate about the use of ESCs in research. ESCs are isolated during the first 8 weeks of embryonic development from the inner cell mass of the blastocyst (Thomson et al. 1998) necessitating the destruction of the embryo. As there is no consensus as to whether a human life at the embryonic stage should be granted the moral status of a human being (Baldwin 2009), the use of ESCs remain controversial although blastocysts are usually obtained from discarded materials in in vitro fertilization (IVF) clinics. Moreover, the safety of ESCs for use in cell-based therapy has not been established, and there have been concerns with immunorejection and tumorigenic potentials of these cells. In 2006, a new method was introduced to derive pluripotent stem cells from adult tissues. Through induced pluripotent stem cell (iPSC) technology, autologous tissues from patients are transfected with pluripotency-inducing genetic factors to produce human induced pluripotent stem cells (hiPSCs) which when transplanted into the same patient will not cause an immune response. However, there have been reports of chromosomal duplications and deletions due to epigenetic changes in ensuing hiPSCs (Takahashi and Yamanaka 2006; Laurent et al. 2011). A previous study reported higher incidence of teratoma formation in immunocompromised mice transplanted with hiPSCs than in subjects that received ESCs (Gutierrez-Aranda et al. 2010). The risk of tumorigenesis is of particular importance when using pluripotent cells due to the propensity of these cells to induce teratomas in animal models (Thomson et al. 1998; Takahashi and Yamanaka 2006). Therefore, it is necessary to characterize exhaustively the differentiation status of transplanted cells to avoid delivery of residual pluripotent cells that may differentiate abruptly in vivo.

HSCs are isolated from the adult bone marrow and from the umbilical cord (UC). They are multipotent cells that give rise to a limited number of cell types namely, blood and blood related lineages. However, Ogawa et al. (2013) reported pluripotent characteristics of HSCs and their ability to differentiate into almost all cell types in the body. With regard to their clinical application in cell-based therapy, ex-vivo expansion prior to transplantation may be required due to the difficulty in harvesting sufficient amount of HSCs from the bone marrow or the UC.

MSCs can be derived from various sources, and the bone marrow has been considered as the primary source of MSCs (Pittenger et al. 1999). Similar to HSCs, however, there is an issue with obtaining ample amount of MSCs from the bone marrow such that in vitro expansion may be required to increase the number of cells for transplantation. However, in vitro expansion may promote some degree of cell differentiation, epigenetic modifications and contamination (Pittenger et al. 1999;

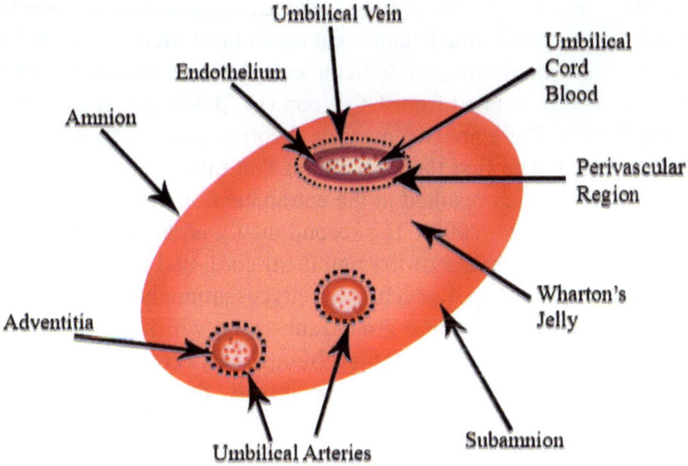

Fig. 14.1 Diagram of a cross-section of the human umbilical cord showing anatomical compartments including Wharton's jelly, a rich source of stem cells

Bongso and Fong 2013). As a favorable alternative source for MSCs, extraembryonic tissues have been considered, which include the placenta, fetal membrane (amnion and chorion), UC, UC blood, and amniotic fluid (Strakova et al. 2008; Ilancheran et al. 2009; Pappa and Anagnou 2009; Semenov et al. 2010). Since these tissues are normally discarded after birth, MSCs derived from this source raise no ethical controversies (Marcus and Woodbury 2008). Moreover, these cells exhibit partial pluripotent properties due to their ontogenic proximity from ESCs (Strakova et al. 2008; Ilancheran et al. 2009; Pappa and Anagnou 2009; Semenov et al. 2010). They have also been shown to display more preferable immunogenic and proliferative properties, as well as wider multipotent plasticity (Marcus and Woodbury 2008; Pappa and Anagnou 2009). In addition to the previously mentioned sources, MSCs can also be obtained from fetal tissues, which are controversial sources as they are derived from human abortuses, and from adipose tissues necessitating invasive procedures for isolation (Bunnell et al. 2008).

14.3 MSCs Isolated from the WJ and Other Anatomical Compartments of the UC

The umbilical cord functions to enclose two umbilical arteries and one umbilical vein that connect the fetus and the placenta enabling transfer of nutrients and waste products. These blood vessels are embedded in a mucous proteoglycan-rich matrix known as the Wharton's jelly (WJ), which is covered by the amniotic epithelium (Fig. 14.1). The WJ confers some elasticity to the UC, which helps prevent the vessels from bending and compressing. More than 10 years ago, the first fibroblast-like

MSC populations from the WJ have been isolated (Kobayashi et al. 1998). In literature, WJ-MSCs have been called "umbilical cord matrix stem cells (UCMSCs)" to make them distinguishable from endothelial cells isolated from the umbilical vein (HUVEC), and MSCs isolated from UC blood (UCB-MSCs) (Baudin et al. 2007; Markov et al. 2007). There are two possible theories accounting for the origins of stem cells in the WJ. The first theory assumes that there were two waves of fetal stem cell migration, which resulted in the entrapment of some of the MSCs in the WJ of the UC (Wang et al. 2008). The second theory argues that MSCs in the WJ evolved from the mesenchyme of the umbilical cord, and these cells function to secrete glycoproteins, mucopolysaccharides, glycosaminoglycans and extracellular matrix proteins that form gelatinous ground substance, which prevent the strangulation of UC blood vessels during gestation (Bongso and Fong 2013).

There are different compartments within the UC that are considered as hosts for stem cell populations. These compartments are described as distinct regions (Jeschke et al. 2011) and include the amniotic compartment (outer epithelial layer and inner subamniotic mesenchymal layer), the WJ compartment, the perivascular compartment surrounding the vessels, the media and adventitia compartment of the walls of UC blood vessels, the endothelial compartment (inner lining of the vein), and the vascular compartment (blood lying within the UC blood vessels) (Bongso and Fong 2013). There has been no standardized nomenclature for these structures although terms such as "subamnion", "cord lining (sub-amnion)", "intervascular", "perivascular" and "HUVEC" have been frequently used. Since there is no defined histological demarcation among these compartments, it is difficult to determine whether isolated populations from different compartments have the same identity. Also, the absence of a defined consensus on the region of interest, and differences in isolation methods and derivation protocols contribute to the difficulty in identifying cell population of interest (Bongso and Fong 2013). WJ-MSCs can be isolated from two regions: the perivascular zone and inter-vascular zone (Prasanna and Jahnavi 2011), although they can also be obtained from these three regions, namely the intervascular, subamnion, and the perivascular regions (Troyer and Weiss 2008). In vitro structural, immunohistochemical and functional analyses revealed significant differences in the number, nature and properties of cells isolated from the above-mentioned regions (Nanaev et al. 1997; Karahuseyinoglu et al. 2007). Moreover, it has been assumed that they originate from different pre-existing structures (Can and Karahuseyinoglu 2007). Human umbilical cord perivascular cells (HUCPVCs) are one of the cell populations isolated from around the umbilical cord vessels (Sarugaser et al. 2005; Baksh et al. 2007). Moreover, other groups isolated other equipotent stem-like cells from the subamnion (cord lining; CL) (Kita et al. 2010; Jeschke et al. 2011). Of note, previous studies have found that WJ-MSCs populations isolated in close proximity to the amniotic surface display higher proliferative and less differentiation properties than those isolated in proximity to umbilical vessels (Nanaev et al. 1997; Karahuseyinoglu et al. 2007).

14.4 Characteristic Properties of WJ-MSCs for Cell-Based Therapy

14.4.1 Phenotypic Characterization of WJ-MSCs

A groundwork on the characterization of WJ has been provided through an overview of the human UC (Conconi et al. 2011). In this paper, we review the phenotypic characteristics of human UC cells derived from various parts of the UC. There has been difficultly in determining the precise identity of UC stromal cells because of lack of standardized extraction, culture, and analysis methods. However, it was demonstrated that cells from WJ carry the mesenchymal phenotypic characteristics including the expression of specific lineage cytoskeletal markers, such as smooth muscle actin (SMA) and vimentin. Moreover, ESC markers Oct-4, stage-specific embryonic antigen 4 (SSEA4), nucleostemin, SOX-2 and Nanog are expressed by some populations, although HUCPV cells do not express Oct-4 and SSEA4. Moreover, HUCPV cells do not express cell surface molecules such as CD59, putatively involved in the regulation of the complement system and in preventing cell lysis, and CD146, an endothelial specific cell molecule also expressed on MSCs (Conconi et al. 2011). It has been hypothesized that HUCPV cells are more differentiated than WJ-MSCs. In a previous study, pan-cytokeratin staining was found to be stronger in HUCPV than in WJ-MSCs (Karahuseyinoglu et al. 2007), supporting the observation on the failure of HUCPV to differentiate into neuronal cells. Recently, another group characterized the role of the ESC marker SSEA4 in preserving the multipotency of WJ-MSCs. This group found that the stemness of WJ-MSC explants and their capacity to differentiate into osteocytes and adipocytes depended on the expression of SSEA4. Non-muscle myosin II (NMII), a protein that participates in cell division, migration and differentiation, is crucial in the promotion of the proliferative activity of WJ-MSCs. Accordingly, it has been found that inhibiting NMII leads to G_0/G_1 arrest of WJ-MSCs, thereby attenuating the proliferative property of WJ-MSCs. Consequentially, this leads to activation of cell cycle inhibitory genes, such as CDKN1A, CDKN2A and CDKN2B, and attenuation of the expression of genes that promote S/M phase transition (Sharma et al. 2013). A recent study compared WJ-MSCs transcriptomes after 4 and 12 passages to investigate the effect of prolonged cell culture on gene expression of WJ-MSCs (Gatta et al. 2013). Among the differentially expressed genes found after the 12th passage but not in the 4th passage are those involved in inflammatory and cell stress response, cell proliferation and maturation, and apoptosis. These findings suggest that WJ-MSCs develop characteristics common in aging cells such as decreased proliferative capacity and apoptosis resistance after multiple cycles of cell expansion (Gatta et al. 2013).

It has also been found that the macrophage marker CD14 is expressed only by cord lining mesenchymal stem cells (CL-MSCs) and not by WJ-MSCs (Kita et al. 2010). Furthermore, soluble CD14 has been shown to downregulate T-cell activation (Rey Nores et al. 1999). In addition, WJ-MSCs express HLA-G6 isoform, an immune-modulation protein, suggesting the suitability of WJ-MSCs for cell-based therapy. It was found that cell populations with different phenotypic profiles are not only present in various parts of the UC but also inside the same UC regions, indicating that UCMSCs are a unique cell family showing varying levels of stemness. Moreover, WJ CD105(+)/CD31(−)/KDR(−) cell populations have also been observed to differentiate in vivo into myogenic lineage and assist in muscle regeneration. This was also confirmed by in vitro assays which showed myogenic differentiation potential of CD105(+) cells (Conconi et al 2006).

Specific regions of the UC lining (sub-amnion) and WJ enriched with stem cell niches have been defined (Jeschke et al. 2011). A previous study reported MSCs isolation from sub-amnion (cord lining) of the UC and that sub-amnionic MSCs from ESCs do not show tumorigenicity in vitro (Kita et al. 2010).Moreover, isolated CL-MSCs expressed mesenchymal features in vitro but also showed several distinct features (Kita et al. 2010). As stated previously, UC-isolated multipotent cells show heterogeneity due to anatomically distinct zones in the UC from which they are obtained, and also differences in isolation techniques. However, although CL-MSCs have promising therapeutic potentials due to their multipotent and proliferative capacities, harvesting these cells is an extremely time-consuming process, and this limits their utility for cell-based therapy (Jeschke et al. 2011). Nevertheless, WJ is a good source for MSCs although WJ-MSCs show variation in cell quality. Therefore, it is important to consider both the quality and quantity of the stem cells for each specific application.

It is believed that the biological characteristics of MSCs can be affected by the perinatal environment. Studies have shown that the intrauterine metabolic disturbances occurring during pregnancy, such as those caused by maternal hyperglycemia, increase the odds of the offspring to develop diabetes and obesity (Dabelea and Pettitt 2001; Gillman et al. 2003; Clausen et al. 2008). Results from animal studies have shown that MSC commitment to pre-adipocytes and mature adipocytes begins during fetal development and the perinatal life (Tang et al. 2008). Further, it was found that obese subjects showed lower amounts of pre-adipocytes and mature adipocytes than normal subjects (Tchoukalova et al. 2007). However, more studies need to be conducted to examine the impact of the perinatal environment on differentiation of fetal MSCs, especially in unregulated gestational diabetes. The underlying mechanism causing metabolic diseases in offsprings of diabetic mothers has been investigated (Pierdomenico et al. 2001). One study found similarities in the types of markers expressed on isolated WJ-MSCs from the UC of healthy and diabetic mothers, but differences in the expression levels of these markers. This phenomenon can be explained by the differences in the functional features between the two groups (i.e., healthy and diabetic mothers). Accordingly, cells from the diabetic mother showed limited plasticity as exemplified by lower expression levels of CD90. However, these cells also showed proliferative features, attributable to high expressions of CD44, CD29, CD73, CD166, SSEA4 and telomerase reverse transcriptase (TERT) in isolated WJ-MSCs. Also, isolated WJ-MSCs from the diabetic group showed higher capacity to differentiate into adipocytes indicating a possible pre-commitment to adipocyte lineage. In summary, many studies suggest "pre-commitment" of MSCs to the adipocyte lineage due to a diabetic uterine environment. This causes higher adipocyte production and when coupled with an unhealthy diet lifestyle, will lead to diabetes and obesity.

14.4.2 Immunomodulatory Properties of WJ-MSCs

WJ-MSCs are characterized by low immunogenicity and high proliferative capacity, which make them favorable for allogenic and xenogenic transplants. In addition, WJ-MSCs possess immunosuppressive characteristics (Prasanna and Jahnavi 2011). Specifically, it was found that WJ-MSCs lack the cell surface expression of major histocompatibility complex (MHC) class II (HLA-DR) and the co-stimulatory antigens CD80 and CD86, which are involved in the activation of T and B cell responses (Rachakatla et al. 2008; Weiss et al. 2008; La Rocca et al. 2009; Prasanna and Jahnavi 2011). Moreover, MHC class I antigens (HLA-ABC) are less expressed in WJ-MSCs, which may protect them from natural killer cell-mediated lysis (Rachakatla et al. 2008; Weiss et al. 2008; La Rocca et al. 2009; Prasanna and Jahnavi 2011). Moreover, there appears to be differences in the induction of pro-inflammatory cytokines between bone marrow-derived MSCs (BM-MSCs) and WJ-MSCs, although both cell types display similarities in the expression of immune-stimulatory ligands. WJ-MSCs secrete immunosuppressive proteins including interleukin (IL)-6 and vascular endothelial growth factor (VEGF), which are key players in the immunosuppressive capability of MSCs (Djouad et al. 2007; Weiss et al. 2008). They also express HLA-G, a protein which suppresses the immune response against the fetus, induces the expansion of regulatory T cells (Djouad et al. 2007; Selmani et al. 2008; Weiss et al. 2008; La Rocca et al. 2009; Griffin et al. 2010; Prasanna et al. 2010; Deuse et al. 2011) and plays pivotal role in the activation of the regulatory T-cells during pregnancy. Some studies showed low rejection rate associated with expression of HLA-G and other antigens (e.g. IL-10, transforming growth factor [TGF]-β) in the blood, heart and liver/kidney grafts (Zarkhin et al. 2010). When compared to BM-MSCs, WJ-MSCs were found to exhibit higher immuno-modulating responses as exemplified by higher amounts of the anti-inflammatory molecules IL-10 and TGF-β (Selmani et al. 2008; Weiss et al. 2008; La Rocca et al. 2009; Prasanna et al. 2010). Also, another study compared the responses to the pro-inflammatory cytokine interferon (IFN)-γ between these two cell populations. Due to the absence of HLA-DR in WJ-MSCs, WJ-MSCs showed negligible induction by IFN-γ while BMSCs exerted substantial responses (Prasanna et al. 2010; Deuse et al. 2011). The immuno-modulatory capabilities of WJ-MSCs in diabetic rat models have also been suggested and some studies found suppressive effect of WJ-MSCs on the proliferation of allogenic T-cell response (Wang et al. 2014). Furthermore, these studies also demonstrated the absence of CD40, CD40L, CD80 and CD86 expression by WJ-MSCs. When co-cultured with peripheral blood lymphocytes, WJ-MSCs caused a remarkable decrease in the levels of IFN-γ secreted by these cells. Additionally, WJ-MSCs also exerted therapeutic benefits in type-1 diabetic rats by reducing pancreatic cell destruction and hyperglycemia, most likely by decreasing the inflammatory response against β-cells of the pancreas (Wang et al. 2014).

Nevertheless, UCMSCs can also elicit an immune response. Accordingly, although single injection of MHC-mismatched inactivated UCMSCs did not induce a detectable immune response, repeated injection in an inflamed region or upon stimulation with IFN-γ prior to injection unmasked immunogenic properties of UC-MSCs (Cho et al. 2008). These findings indicate that caution must be exercised when injecting UCMSCs repeatedly into inflamed or damaged regions. Generally, all MSCs have immunosuppressive properties as they inhibit CD4(+), CD8(+), CD2(+) and CD3(+) subsets (Aggarwal and Pittenger 2005). However, relative to BM-MSCs, WJ-MSCs have been found to display more robust immuno-modulating activity as evidenced by more remarkable mitogen-driven alloantigens and specific antigen-driven T cell responses. Some studies also revealed that WJ-MSCs caused dose-dependent attenuation of the T-cell immune response in vitro (Di Nicola et al. 2002). Moreover, WJ-MSCs also caused significant suppression of mitogen induced CD3(+) T cell activity even when administered at low doses, compared to their bone marrow-derived counterparts (Najar et al. 2009; Prasanna et al. 2010). Another study asserted that WJ-MSCs have more potent suppressive properties compared to BM-MSCs or adipose-derived MSCs when co-cultured with allogenic-stimulated T-cells (Prasanna and Jahnavi 2011). When compared to WJ-MSCs, the immuno-modulating activity of fetal liver-derived MSCs is limited only to the attenuation of mitogen-driven lympho-proliferative activity, although they have no effect on allo-proliferative responses (Gotherstrom et al. 2003). In this regard, perinatal MSCs not only seem to attenuate lymphoproliferation more robustly that BM-MSCs similar to WJ-MSCs, but also operate in a stimuli-independent fashion, unlike fetal MSCs (Prasanna and Jahnavi 2011). Another mechanism may be consequential to the influence of WJ-MSCs on the maturation and activation of dendritic precursors (DCs), as shown by experiments involving co-culture of WJ-MSCs with CD14(+) monocytes. Accordingly, WJ-MSCs inhibited differentiation of DCs in a contact-dependent manner when these cells were co-cultured with CD14(+) monocytes. Moreover, WJ-MSCs co-cultured monocytes appeared to be locked in an immature dendritic cell-stage and did not show upregulation of the co-stimulatory ligands essential for their maturation (Tipnis et al. 2010). Previous studies have focused on the kinetic patterns of pro-inflammatory cytokine secretion by phytohemagglutinin (PHA)-activated lymphocytes when co-cultured with WJ-MSCs and BM-MSCs (Prasanna et al. 2010). A change in the threshold and kinetics of the activating cytokine IL-2 was only observed in BM-MSC co-cultures. Moreover, results showed that activation of negative co-stimulatory ligands on peripheral blood lymphocytes was higher in WJ-MSCs co-cultures compared to BM-MSCs co-cultures (Prasanna et al. 2010). The secretion profiles of WJ-MSCs are similar to those of other MSC populations, except for the release of IL-12, IL-15 and platelet-derived growth factor (PDGF), which are secreted only by WJ-MSCs and cord blood MSCs. In summary, WJ-MSCs display robust immunomodulatory properties via several mechanisms including secretion of immunosuppressive soluble factors, upregulation of negative co-stimulatory ligands, production of memory cells, cell fusion to escape recognition, immune fetal-maternal specific immune avoidance mechanisms, attenuation of antigen-presenting cell functions, altered migration of immune cells, and T cell energy apoptosis tolerance (Prasanna and Jahnavi 2011).

14.5 Clinical Applications of WJ-Derived Stem Cells

14.5.1 Cancer Therapy

Stem cells can also be utilized for primary and metastatic cancers. Previous studies have demonstrated the capacity of un-engineered human and rat UCMSCs to attenuate amplification of multiple cancer cells in vivo and in vitro (Ayuzawa et al. 2009; Ganta et al. 2009). Tamura et al. (2011) also reported intrinsic stem cell-dependent regulation of cancer growth, probable mechanisms underlying the observed biological function, delivery of exogenous anti-cancer agents, and potential clinical applications. By showing intrinsic ability to secrete factors that can inhibit cancer cell growth and/or apoptosis in vitro and in vivo, UCMSC may have several advantages for cell-directed cancer therapy. Two potential mechanisms have been suggested to explain the manner by which naïve UCMSCs attenuate tumor growth. First, UCMSCs may produce multiple secretory proteins that induce death of cancer cells and cell cycle arrest. Caspases are activated by the secretion of many secretory proteins that lead to cell cycle arrest and eventually cell death (Deuse et al. 2011; Ganta et al. 2009; Nakamizo et al. 2005). Some evidence reported UCMSCs may stimulate caspase activities and arrest cell cyle even in the absence of direct contact with cancer cells. Microarray analysis of rat UCMSCs also revealed over-expression of multiple tumor suppressor genes (Tamura et al. 2011). The second mechanism may involve enhancement of an immune reaction to cancer cells. Immunohistochemistry analysis revealed CD8(+) T cell infiltration of the tumor tissues which is not in line with low immunogenicity profiles of UCMSCs (Ganta et al. 2009). Therefore, the immunogenicity of UCMSC in tumor-bearing animals may be dependent on the microenvironment of UCMSC and the tumor cells.

The interaction between cytokines/growth factors and their receptors mediate the homing ability of stem cells. Accordingly, tumor cells secrete large amount of cytokines and growth factors which are detected by their corresponding receptors expressed on the surface of the UCMSC and other MSCs (Tamura et al. 2011). Nevertheless, UCMSCs show greater migration capacity towards the tumor tissue than BM-MSCs due to overexpression of IL-8 receptor and C-X-C chemokine receptor (CXCR) in UMSCs. Furthermore, UMSCs can also be engineered to express cytotoxic cytokines before being delivered to the tumor, and can be preloaded with nanoparticle payloads which could attenuate tumors after they reach the tumor site (Rachakatla et al. 2007; Matsuzuka et al. 2010). Using these manipulations, human UCMSC expressing IFN–β produced sufficient amounts of IFN–β which induced death of human breast adenocarcinoma and bronchioalveolar carcinoma cells in vitro and in vivo (Rachakatla et al. 2008; Matsuzuka et al. 2010). Altogether, these studies suggest therapeutic potential of UCMSCs for the treatment of various cancers. Furthermore, UCMSCs are also suitable for allogenic transplantation due to their abundance, low immunogenicity, lack of CD34 and CD45 expression, and less complex methods involved in harvesting these cells and during in vitro cell expansion. Importantly, the homing ability of UCMSCs to inflammatory tissues including cancer tissues, and the tumoricidal ability of these cells make them ideal stem cell-based intervention for cancer therapy.

14.5.2 Liver Disease

Based on the potentiality of UCMSCs to differentiate into an endodermal lineage including hepatocyte-like cells, WJ-MSCs can also be considered for the treatment of liver disease, furthermore, as an alternative to orthotopic liver transplantations. The in vitro and in vivo use of UCMSCs for rescuing injured hepatocytes has been described previously (Scheers et al 2011). UCMSCs have also been shown to display several hepatic markers that characterize the sequential steps of liver development. In vivo studies showed the capacity of transplanted undifferentiated UCMSCs to express human hepatic markers such as albumin and alpha- fetoprotein (AFP) in the liver of severe combined immunodeficiency (SCID) mice with partial hepatectomy at 2, 4, and 6 weeks after in vivo transplantation (Campard et al. 2008). Furthermore, there is some evidence which showed decreased rate of hepatic cell fibrosis due to compensatory effects of transplanted undifferentiated UCMSCs. This finding thus indicates supporting role of UCMSCs in enhancing functional recovery of recipient livers, probably via stimulation of the differentiation of endogenous parenchymal cells and promoting degradation of fibrous matrix, even in the absence of a transdifferentiation process (Anzalone et al. 2010). Meanwhile, the capacity of UCMSCs to differentiate into a hepatic lineage can be enhanced in vitro and in vitro via co-culture with hepatogenic factors. A previous study also found relationship between anti-inflammatory and anti-fibrotic actions of UCMSCs due to secreted metalloproteinases (Anzalone et al. 2010). Nevertheless, further research is required to clarify the ability of UCMSCs to differentiate into hepatocyte-like cells and their capacity to repopulate and rescue liver functions.

14.5.3 Cardiovascular Diseases

The therapeutic value of WJ may also be explored for cardiovascular diseases (Semenov and Breymann 2011). Due to disadvantages of surgical treatment using non-autologous valves or conduits (Mayer 1995; Schoen and Levy 1999), cardiovascular tissue engineering focused on in vitro fabrication of autologous, living tissues with the potential for regenerating heart muscles (Shinoka et al. 1996). A number of animal studies have validated WJ-MSC-based cardiovascular tissue engineering. In one study, autologous trileaflet heart valves derived from human WJ-MSCs were successfully implanted in sheep models for up to 20 weeks. The functional, biochemical and structural features of these valves were comparable with those of the native semilunar heart valves. The feasibility of obtaining stem cells from UC for cardiovascular tissue engineering in contrast with obtaining cells from other sources has also been demonstrated (Kadner et al. 2004). Another study was conducted to develop a myocardial patch to repair myocardial infarction tissue or slow down tissue damage, and enhance long-term function of the heart (Kenar et al. 2011). This study used a 3D aligned microfibrous myocardial tissue construct cultured under transient perfusion. Accordingly, 3D constructs from static and perfused cultures increased cell viability, provided uniform cell distribution and alignment due to supply of nutrients from the 3D structure.

Indeed, WJ-MSCs have great potential in cardiovascular tissue engineering. Nevertheless, we need to understand deeply the clinical limitations of these cells and establish realistic clinical protocols for their application. For instance, the use of natural, synthetic or hybrid polymers as scaffolds resulted in reduced growth and remodeling of cells, and increased risk for infection and formation of thromboembolus. Thus, it is necessary to develop materials that are biocompatible with cardiovascular tissue for them not to negatively affect the regenerative and immune-modulatory characteristics of WJ-MSCs in engineered scaffolds (Semenov and Breymann 2011). Furthermore, despite advanced knowledge on the characteristics of WJ-MSCs and their successful preclinical and clinical applications, standardized criteria for isolation, characterization, long-term cultivation and maintenance of human MSCs are required, as long-term survival of stem cells in the host tissue and establishment of treatment regimen are still critical issues which hamper broad clinical application of stem cells.

14.5.4 Cardiac Differentiation of Human WJ-Derived Stem Cells

Due to the capacity of undifferentiated MSCs to differentiate spontaneously into multiple lineages when transplanted in vivo, BM-MSCs may differentiate into multiple tissue types after transplantation, and its developmental fate may not be restricted by surrounding tissues after myocardial infarction. Undifferentiated cells are known to undergo maldifferentiation within the infarcted myocardium posing life-threatening consequences (Breitbach et al. 2007). Thus, cardiac differentiation must be initiated prior to transplantation of cells, which could be achieved by culturing the cells in defined culture conditions. It has been hypothesized that cardiac differentiation prior to transplantation would result in enhanced myocardial differentiation and recovery of heart functions (Bittira et al. 2002; Tomita et al. 2002). Accordingly, WJ-MSCs can be induced to differentiate into heart cells by treatment with 5-azacytidine for 3 weeks; thereafter these cells would express cardiocyte markers such as cardiac troponin I, connexin 43, and desmin, and display cardiac myocyte morphology (Wang et al. 2004). Additionally, cells may also be co-cultured with oxytocin, whose levels are higher in the fetal but not adult cardiac tissues indicating their involvement in cadiomyocyte differentiation (Fathi et al. 2009), or with embryo-like aggregates and other growth factors such as TGF-β1, PGDF and basic fibroblast growth factor (bFGF) to transform cells into myocytes (Maltsev et al. 1993; Matsuura et al. 2004; Hollweck et al. 2011). It has been observed that the treatment of UCMSC with oxytocin and 5-azacytidine caused cardiac differentiation and also the formation of "embryoid bodies" (Hollweck et al. 2011). A detailed morphological and immunocytochemical analyses showed that oxytocin is a more potent inducer of cardiac differentiation than 5-azacytindine in the formation of embryoid bodies.

The long-term therapeutic effects of MSCs derived from the BM-MSCs and WJ-MSCs were compared (Lopez et al. 2013) in a rat model of myocardial infaction (MI). Results showed significant improvement in groups which received MSCs at 25–31 weeks after treatment with MSCs. Moreover, cardiac differentiation potential of WJ-MSC was enhanced when these cells were co-cultured with fetal MSCs, compared with those co-cultured with adult MSCs. Furthermore, combination treatment with WJ-MSC and fetal MSCs induced myotube formation in 2–3 days and spontaneous contractions (beating) after 5–7 days. Collectively, the above-mentioned studies indicate that MSCs administered 24–48 h after MI exerted strong beneficial effects which lasted longer than 25 weeks after MI, and that WJCs may be useful sources of off-the-shelf cellular therapy for MI. Due to the accessibility and differentiation capability of USMCs into cardiomyocyte-like cells, UCMSC are a favorable treatment intervention for cell-based therapies and cardiac engineering. However, in order to use UCMSCs, as well WJ-derived stem cells in cardiac repair, we need to determine whether these cells possess functional properties of cardiomyocytes.

14.5.5 Cartilage Regeneration

Exposure to traumatic injury or autoimmune diseases damages the cartilage, a tissue which displays poor in vivo regeneration and self-repair capacity. Reports on the capacity of stem cells to differentiate into chondrocyte-like cells raised enthusiasm for the use of cell therapy to repair cartilage damage. In particular, WJ-MSCs have been shown to differentiate into chondrocyte-like cells in vivo and in vitro (Lo Iacono et al. 2011). Analysis of the chondrogenic potential of WJ-MSCs revealed their multipotency and chondrogenic capacity (Arufe, et al. 2011), which are essential properties for cell therapy in articular diseases (Arufe et al. 2011). Moreover, WJ-MSCs are also capable of increasing the production of hyaluronic acid, glycosaminoglycans and key genes such as SOX9, cartilage oligomeric protein (COMP), collagen type II and fibromodulin (Fong et al. 2012). The beneficial effects of seeding density of WJ-MSCs in poly-glycolic acid (PGA) scaffolds, in the presence of a chondrogenic medium, on chondrogenic potentials of WJ-MSCs have also been described (Wang et al. 2009). This study revealed the potential for chondrogenic differentiation of WJ-MSCs in three-dimensional tissue engineering; higher seeding densities promoted better biosynthesis and mechanical integrity, and that a seeding density of at least 25 million cells/mL was sufficient for fibrocartilage tissue engineering with umbilical cord mesencyhmal stromal cells (Wang et al. 2009). Culturing WJ-MSCs on nanofibrous substrates with sequential two culture medium environments also promoted chondrogenic differentiation of WJ-MSCs (Fong et al. 2012). For successful osteochondral regeneration, cell sources and tissue integration between cartilage and bone regions must be considered because osteochondral tissue consists of cartilage and bone. A supportive structure, which mimics native osteochondral tissues, has been recently developed (Wang et al. 2011). In their study, investigators showed that embedding WJ-MSCs into two poly-L-lactic-acid (PLLA) scaffolds with chondrogenic and osteogenic media for 3 weeks, followed by suturing of the chondrogenic and osteogenic constructs to form four different osteochondral assemblies resulted in enhanced integration and transition of the matrices between two layers in the composite group compared to other control composites (Wang et al. 2011). WJ-MSCs were also found to differentiate into intervertebral disc-like tissues exhibiting immature nucleus pulposus (NP) phenotype in a pseudo-three-dimensional culture system (Chon et al. 2013). Moreover, it has been shown that these NP-derived cells possess specific laminin isoforms and laminin binding receptors, which may lead to the formation of NP-like cells. As a result, WJ-MSCs are not only promising candidates in osteochondral regeneration but also in intervertebral disc repair. Also, their low immunogenicity suggests their ideal use in autoimmune disorders such as osteoarthritis and rheumatoid arthritis. In view of the high variability of cell sources, the need for scaffold and matrices, and the necessity of co-culture with growth factors, more research is required to optimize this cellular therapy approach and translate the results obtained from bench to clinic for cartilage repair.

14.5.6 Peripheral Nerve Repair

Tissue engineering has been considered as another strategy to restore neural function of the peripheral nervous system after injury. Bioartificial nerve conduits have been developed and placed between neural gaps to guide axonal regrowth (Hall 2001; Ishikawa et al. 2007). However, a limitation of these structures is their restricted growth-promoting action when the nerve gap is wide. In order to yield better nerve growth, Schwan cells, which are essential in myelin formation, were added to the microenvironment of damaged nerves (Ohta et al. 2004). To secure a proper source of Schwan cells, MSCs from different tissues were utilized instead of isolating cells from other peripheral nerve tissues. The first report of induction system for differentiating Schwann cells from BM-MSCs appeared in 2001 (Dezawa et al. 2001). Recent studies showed that UCMSCs differentiated into

Schwann cells capable of supporting neural regeneration and in constructing myelin (Matsuse et al. 2010; Kuroda et al. 2011). Some studies suggest that UC-Schwan cells are viable alternatives to native Schwann cells and may be used for cell-based therapy for nerve injuries. Accordingly, when human UC-Schwan cells were transplanted into a rat with transected sciatic nerve, these cells maintained their differentiated phenotype after transplantation, and contributed to axonal regeneration and functional recovery. UC-Schwann cells differentiated from WJ also produced neurotrophic factors such as NGF and BDNF (Peng et al. 2011; Xu et al. 2011). Schwan cells play important roles in axonal regeneration and myelin construction in spinal cord injuries. As Schwann cells can be differentiated from various sources of MSCs and not only from WJ, a vis-à-vis comparison of various sources of MSCs will reveal the potential of WJ-derived MSCs for therapeutic application in spinal cord injury (Kuroda et al. 2011). Furthermore, human umbilical cord-derived MSCs (hUCMSCs) have been shown to ameliorate neonatal hypoxic-ischemic encephalopathy (HIE) in rat models (Zhang et al. 2014). Differentiation of the UCMSCs into neurons, and better locomotor functions in 24 h-transplanted HIE rats compared to rats transplanted after 72 h were also observed (Zhang et al. 2014). In addition, more efficient homing of UCMSCs to the ischemic frontal cortex after intravenous (IV) administration of cells compared to intraperitoneal administration were also demonstrated. These findings suggest that IV administration of UCMSCs at an early stage after HIE could be used as a therapeutic approach for hypoxic-ischemic encephalopathy. Due to the ability of WJ-MSCs to promote neural generation and neuroprotection, the potential of these cells for stroke treatment has been suggested (Hsieh et al. 2013). Secretome analysis of WJ-MSCs cultured in an oxygen-glucose deprived environment revealed secretion of proteins that promote neural differentiation and cell migration, and a reduced rate of apoptosis in primary cortical cells. More recently, the capability of WJ-MSCs to differentiate into dopaminergic neural-like cells in the presence of forskolin, which influences intracellular levels of cAMP, has been reported (Paldino et al. 2014). Microarray analysis reported modulation of 1,532 genes, with most of these differentially expressed genes involved in signaling pathways of neurons, and also in neuronal dopaminergic induction. Immunohistochemistry and western blot analysis revealed up-regulation of Nurr1, NeuroD1, and tyrosine hydroxylase proteins, which are specific dopaminergic phenotypic markers.

Effort has been made to maximize isolation and differentiation of stem cells derived from WJ, such as designing optimized cell harvest protocols via controlling oxygen concentration and plating density (Lopez et al. 2011), which has been known to allow the expansion and maintenance of colony-forming unit-fibroblast (CFU-F). Accordingly, adjusting oxygen concentration from 21 to 5% during expansion increased cell yield and maintained CFU-F without affecting expression of surface markers or differentiation capacity of WJ-MSCs, whereas reducing plating density from 100 to 10 cells/cm^2 increased CFU-F frequency. Reducing oxygen concentration in cultures may upregulate hypoxia inducible factors (HIFs) and induce increased cell proliferation and maintenance of CFU-C via downstream effects of HIF activation and also by affecting telomerase activity (Lopez et al. 2011).

Thus, oxygen concentration and plating density are key factors to successfully expand the rate and frequency of CFU-F or WJ-MSCs. Moreover, these factors need to be considered to produce different input populations for tissue engineering or cell-based therapy.

14.6 New Directions for WJ Research

14.6.1 Clonal MSCs

Non-hematopoietic (CD45−, CD34−, SH2+, Thy-1, CD44+) HUCPVC cell populations were isolated and found to display non-hematopoietic myofibroblastic MSC phenotypic markers (CD45−, CD34−, CD105+, CD73+, CD90+, CD44+, CD106+, 3G5+, CD146+) with similar immunological properties to BM-MSCs. Additionally, HUCPVCs have robust quinti-potential differentiation capacity in vitro, and can contribute to musculo-skeletal and dermal wound healing in vivo (Sarugaser et al. 2009). Identification of the perivascular region of human UC as a rich source of HUCPVs led to the first single cell clonal confirmation of a theoretical hierarchy of MSC differentiation (Sarugaser et al. 2005; Schugar et al. 2009; Farias et al. 2011; Sarugaser et al. 2009). High MHC−/− phenotype, high frequencies of CFU-F and CFU-osteogenic subpopulation, and their rapid doubling time are properties of HUCPVCs that allow them to be used for allogenic cell-based therapies. More research on the clonal expansion of stem cells derived from WJ is required in order to maximize their therapeutic potentials.

14.6.2 Magnetic Resonance Imaging as a Labeling Method for UC Stem Cells

Studies have been conducted to examine marker expression patterns and the proliferation and differentiation capacities of stem cell populations isolated from the interventricular and periventricular regions of the UCM (Lange-Consiglio et al. 2011). Compared to cells isolated from the perivascular regions, intervascular regions-isolated cells are less differentiated and they exhibit faster doubling times. However, both isolates exhibited expression of MSC mRNA markers (CD29, CD105, CD44, CD166), and were negative for CD34 and MHC-II. Data from staining and gene expression profiling analysis confirmed osteogenic, adipogenic, chondrogenic and neurogenic differentiation of the isolated cells from both regions. Studies conducted to assess the in vitro labeling efficiency of MSCs with magnetic resonance agents, mainly superparamagnetic iron oxide particles (SPIO) and magnetic chloride, showed greater sensitivity for SPIO, although both agents have been shown to be simple, effective and safe labeling methods. Thus, magnetic resonance labeling by SPIO can be used to study migration of stem cells to injured and non-injured tissues, and the mechanism of actions of cell therapy.

Conclusions

While we have provided insights on the phenotypic markers and therapeutic potentials of stem cells derived from the Wharton's Jelly, we have also discussed knowledge gap on the biological properties of these cells and the important issues that need to be addressed with regard to their translational application. In order to render WJ applicable for cell-based therapy, further investigations are warranted to determine their advantages and limitations, and also to design optimal transplantation regimens required for specific diseases.

References

Aggarwal S, Pittenger MF (2005) Human mesenchymal stem cells modulate allogeneic immune cell responses. Blood 105:1815–1822

Anzalone R, Lo Iacono M, Corrao S, Magno F, Loria T, Cappello F, Zummo G, Farina F, La Rocca G (2010) New emerging potentials for human Wharton's jelly mesenchymal stem cells: immunological features and hepatocyte-like differentiative capacity. Stem Cells Dev 19:423–438

Arufe MC, De la Fuente A, Mateos J, Fuentes I, De Toro FJ, Blanco FJ (2011) Analysis of the chondrogenic potential and secretome of mesenchymal stem cells derived from human umbilical cord stroma. Stem Cells Dev 20:1199–1212

Ayuzawa R, Doi C, Rachakatla RS, Pyle MM, Maurya DK, Troyer D, Tamura M (2009) Naive human umbilical cord matrix derived stem cells significantly attenuate growth of human breast cancer cells in vitro and in vivo. Cancer Lett 280:31–37

Baksh D, Yao R, Tuan RS (2007) Comparison of proliferative and multilineage differentiation potential of human mesenchymal stem cells derived from umbilical cord and bone marrow. Stem Cells 25:1384–1392

Baldwin T (2009) Morality and human embryo research. Introduction to the talking point on morality and human embryo research. EMBO Rep 10:299–300

Baudin B, Bruneel A, Bosselut N, Vaubourdolle M (2007) A protocol for isolation and culture of human umbilical vein endothelial cells. Nat Protoc 2:481–485

Bittira B, Kuang JQ, Al-Khaldi A, Shum-Tim D, Chiu RC (2002) In vitro preprogramming of marrow stromal cells for myocardial regeneration. Ann Thorac Surg 74:1154–1160

Bongso A, Fong CY (2013) The therapeutic potential, challenges and future clinical directions of stem cells from the Wharton's jelly of the human umbilical cord. Stem Cell Rev 9:226–240

Breitbach M, Bostani T, Roell W, Xia Y, Dewald O, Nygren JM, Fries JW, Tiemann K, Bohlen H, Hescheler J, Welz A, Bloch W, Jacobsen SE, Fleischmann BK (2007) Potential risks of bone marrow cell transplantation into infarcted hearts. Blood 110:1362–1369

Bunnell BA, Flaat M, Gagliardi C, Patel B, Ripoll C (2008) Adipose-derived stem cells: isolation, expansion and differentiation. Methods 45:115–120

Campard D, Lysy PA, Najimi M, Sokal EM (2008) Native umbilical cord matrix stem cells express hepatic markers and differentiate into hepatocyte-like cells. Gastroenterology 134:833–848

Can A, Karahuseyinoglu S (2007) Concise review: human umbilical cord stroma with regard to the source of fetus-derived stem cells. Stem Cells 25:2886–2895

Chacko AW, Reynolds SRM (1954) Architecture of distended and nondistended human umbilical cord tissues, with special reference to the arteries and veins. Contrib Embryol 35:135–150

Cho PS, Messina DJ, Hirsh EL, Chi N, Goldman SN, Lo DP, Harris IR, Popma SH, Sachs DH, Huang CA (2008) Immunogenicity of umbilical cord tissue derived cells. Blood 111:430–438

Chon BH, Lee EJ, Jing L, Setton LA, Chen J (2013) Human umbilical cord mesenchymal stromal cells exhibit immature nucleus pulposus cell phenotype in a laminin-rich pseudo-three-dimensional culture system. Stem Cell Res Ther 4(5):120

Clausen TD, Mathiesen ER, Hansen T, Pedersen O, Jensen DM, Lauenborg J, Damm P (2008) High prevalence of type 2 diabetes and pre-diabetes in adult offspring of women with gestational diabetes mellitus or type 1 diabetes: the role of intrauterine hyperglycemia. Diabetes Care 31:340–346

Conconi MT, Burra P, Di Liddo R, Calore C, Turetta M, Bellini S, Bo P, Nussdorfer GG, Parnigotto PP (2006) CD105(+) cells from Wharton's jelly show in vitro and in vivo myogenic differentiative potential. Int J Mol Med 18:1089–1096

Conconi MT, Di Liddo R, Tommasini M, Calore C, Parnigotto PP (2011) Phenotype and differentiation potential of stromal populations obtained from various zones of human umbilical cord: an overview. The Open Tissue Eng Regen Med J 4:6–20

Dabelea D, Pettitt DJ (2001) Intrauterine diabetic environment confers risks for type 2 diabetes mellitus and obesity in the offspring, in addition to genetic susceptibility. J Pediatr Endocrinol Metab JPEM 14:1085–1091

Deuse T, Stubbendorff M, Tang-Quan K, Phillips N, Kay MA, Eiermann T, Phan TT, Volk HD, Reichenspurner H, Robbins RC, Schrepfer S (2011) Immunogenicity and immunomodulatory properties of umbilical cord lining mesenchymal stem cells. Cell Transplant 20:655–667

Dezawa M, Takahashi I, Esaki M, Takano M, Sawada H (2001) Sciatic nerve regeneration in rats induced by transplantation of in vitro differentiated bone-marrow stromal cells. Eur J Neurosci 14:1771–1776

Di Nicola M, Carlo-Stella C, Magni M, Milanesi M, Longoni PD, Matteucci P, Grisanti S, Gianni AM (2002) Human bone marrow stromal cells suppress t-lymphocyte proliferation induced by cellular or nonspecific mitogenic stimuli. Blood 99:3838–3843

Djouad F, Charbonnier LM, Bouffi C, Louis-Plence P, Bony C, Apparailly F, Cantos C, Jorgensen C, Noel D (2007) Mesenchymal stem cells inhibit the differentiation of dendritic cells through an interleukin-6-dependent mechanism. Stem Cells 25:2025–2032

Farias VA, Linares-Fernandez JL, Penalver JL, Pava Colmenero JA, Ferron GO, Duran EL, Fernandez RM, Olivares EG, O'Valle F, Puertas A, Oliver FJ, Ruiz de Almodovar JM (2011) Human umbilical cord stromal stem cell express CD10 and exert contractile properties. Placenta 32:86–95

Fathi F, Murasawa S, Hasegawa S, Asahara T, Kermani AJ, Mowla SJ (2009) Cardiac differentiation of P19CL6 cells by oxytocin. Int J Cardiol 134:75–81

Fong CY, Subramanian A, Gauthaman K, Venugopal J, Biswas A, Ramakrishna S, Bongso A (2012) Human umbilical cord Wharton's jelly stem cells undergo enhanced chondrogenic differentiation when grown on nanofibrous scaffolds and in a sequential two-stage culture medium environment. Stem Cell Rev 8:195–209

Ganta C, Chiyo D, Ayuzawa R, Rachakatla R, Pyle M, Andrews G, Weiss M, Tamura M, Troyer D (2009) Rat umbilical cord stem cells completely abolish rat mammary carcinomas with no evidence of metastasis or recurrence 100 days post-tumor cell inoculation. Cancer Res 69:1815–1820

Gatta V, D'Aurora M, Lanuti P, Pierdomenico L, Sperduti S, Palka G, Gesi M, Marchisio M, Miscia S, Stuppia L (2013) Gene expression modifications in Wharton's Jelly mesenchymal stem cells promoted by prolonged in vitro culturing. BMC Genomics 14:635

Gillman MW, Rifas-Shiman S, Berkey CS, Field AE, Colditz GA (2003) Maternal gestational diabetes, birth weight, and adolescent obesity. Pediatrics 111:221–226

Gotherstrom C, Ringden O, Westgren M, Tammik C, Le Blanc K (2003) Immunomodulatory effects of human foetal liver-derived mesenchymal stem cells. Bone Marrow Transplant 32:265–272

Griffin MD, Ritter T, Mahon BP (2010) Immunological aspects of allogeneic mesenchymal stem cell therapies. Hum Gene Ther 21:1641–1655

Gutierrez-Aranda I, Ramos-Mejia V, Bueno C, Munoz-Lopez M, Real PJ, Macia A, Sanchez L, Ligero G, Garcia-Parez JL, Menendez P (2010) Human induced pluripotent stem cells develop teratoma more efficiently and faster than human embryonic stem cells regardless the site of injection. Stem Cells 28:1568–1570

Hall S (2001) Nerve repair: a neurobiologist's view. J Hand Surg Am 26:129–136

He H, Nagamura-Inoue T, Tsunoda H, Yuzawa M, Yamamoto Y, Yorozu P, Agata H, Tojo A (2013) Stage-specific embryonic antigen 4 in Wharton's jelly-derived mesenchymal stem cells is not a marker for proliferation and multipotency. Tissue Eng Part A 20(7–8):1314–1324

Hollweck T, Hartmann I, Eblenkamp M, Wintermantel E, Reichart B, ï¾berfuhr P, Eissner G (2011) Cardiac differentiation of human Wharton's jelly stem cells—experimental comparison of protocols. Open Tissue Eng Regen Med J 4:95–102

Hsieh JY, Wang HW, Chang SJ, Liao KH, Lee IH, Lin WS, Wu CH, Lin WY, Cheng SM (2013) Mesenchymal stem cells from human umbilical cord express preferentially secreted factors related to neuroprotection, neurogenesis, and angiogenesis. PLoS ONE 8(8):e72604

Ishikawa N, Suzuki Y, Ohta M, Cho H, Suzuki S, Dezawa M, Ide C (2007) Peripheral nerve regeneration through the space formed by a chitosan gel sponge. J Biomed Mater Res Part A 83:33–40

Ilancheran S, Moodley Y, Manuelpillai U (2009) Human fetal membranes: a source of stem cells for tissue regeneration and repair? Placenta 30:2–10

Jeschke MG, Gauglitz GG, Phan TT, Herndon DN, Kita K (2011) Umbilical cord lining membrane and Wharton's jelly-derived mesenchymal stem cells: the similarities and differences. Open Tissue Eng Regen Med J 4:21–27

Kadner A, Hoerstrup SP, Tracy J, Breymann C, Maurus C, Melnitchouk S, Kadner G, Zund G, Turina M (2002) Human umbilical cord cells: a new cell source for cardiovascular tissue engineering. The Ann Thorac Surg 74:S1422–1428

Kadner A, Zund G, Maurus C, Breymann C, Yakarisik S, Kadner G, Turina M, Hoerstrup SP (2004) Human umbilical cord cells for cardiovascular tissue engineering: a comparative study. Eur J Cardiothorac Surg 25:635–641

Karahuseyinoglu S, Cinar O, Kilic E, Kara F, Akay GG, Demiralp DO, Tukun A, Uckan D, Can A (2007) Biology of stem cells in human umbilical cord stroma: in situ and in vitro surveys. Stem Cells 25:319–331

Kenar H, Kose GT, Toner M, Kaplan DL, Hasirci V (2011) A 3d aligned microfibrous myocardial tissue construct cultured under transient perfusion. Biomaterials 32:5320–5329

Kita K, Gauglitz GG, Phan TT, Herndon DN, Jeschke MG (2010) Isolation and characterization of mesenchymal stem cells from the sub-amniotic human umbilical cord lining membrane. Stem Cells Dev 19:491–502

Kobayashi K, Kubota T, Aso T (1998) Study on myofibroblast differentiation in the stromal cells of Wharton's jelly: expression and localization of alpha-smooth muscle actin. Early Hum Dev 51:223–233

Kuroda Y, Kitada M, Wakao S, Dezawa M (2011) Mesenchymal stem cells and umbilical cord as sources for schwann cell differentiation: their potential in peripheral nerve repair. Open Tissue Eng Regen Med J 4:54–63

Lange-Consiglio A, Corradetti B, Rutigliano L, Cremonesi F, Bizzarro D (2011) In vitro studies of horse umbilical cord matrix-derived cells: from characterization to labeling for magnetic resonance imaging. Open Tissue Eng Regen Med J 4:120–133

La Rocca G (2011) Connecting the dots: The promises of Wharton's jelly stem cells for tissue repair and regeneration. Open Tissue Eng Regen Med J 4:3–5

La Rocca G, Anzalone R, Corrao S, Magno F, Loria T, Lo Iacono M, Di Stefano A, Giannuzzi P, Marasã L, Cappello F, Zummo G, Farina F (2009) Isolation and characterization of Oct-4 + / HLA-G + mesenchymal stem cells from human umbilical cord matrix: differentiation potential and detection of new markers. Histochem Cell Biol 131:267–282

Laurent LC et al (2011) Dynamic changes in the copy number of pluripotency and cell proliferation genes in human ESCs and iPSCs during reprogramming and time in culture. Cell Stem Cell 8:106–118

Lo Iacono M, Anzalone R, Corrao S, Giuffrè M, Di Stefano A, Giannuzzi P, Cappello F, Farina F, La Rocca G (2011) Perinatal and wharton's jelly-derived mesenchymal stem cells in cartilage regenerative medicine and tissue engineering strategies. Open Tissue Eng Regen Med J 4:72–81

Lopez Y, Seshareddy K, Trevino E, Cox J, Weiss ML (2011) Evaluating the impact of oxygen concentration and plating density on human Wharton's jelly-derived mesenchymal stromal cells. Open Tissue Eng Regen Med J 4:82–94

Lopez Y, Lutjemeier B, Seshareddy K, Trevino EM, Hageman KS, Musch TI, Borgarelli M, Weiss ML (2013) Wharton's jelly or bone marrow mesenchymal stromal cells improve cardiac function following myocardial infarction for more than 32 weeks in a rat model: a preliminary report. Curr Stem Cell Res Ther 8:46–59

Maltsev VA, Rohwedel J, Hescheler J, Wobus AM (1993) Embryonic stem cells differentiate in vitro into cardiomyocytes representing sinusnodal, atrial and ventricular cell types. Mech Dev 44:41–50

Marcus AJ, Woodbury D (2008) Fetal stem cells from extra-embryonic tissues: do not discard. J Cell Mol Med 12:730–742

Markov V, Kusumi K, Tadesse MG, William DA, Hall DM, Lounev V, Carlton A, Leonard J, Cohen RI, Rappaport EF, Saitta B (2007) Identification of cord blood-derived mesenchymal stem/stromal cell populations with distinct growth kinetics, differentiation potentials, and gene expression profiles. Stem Cells Dev 16:53–73

Matsuse D, Kitada M, Kohama M, Nishikawa K, Makinoshima H, Wakao S, Fujiyoshi Y, Heike T, Nakahata T, Akutsu H et al (2010) Human umbilical cord-derived mesenchymal stromal cells differentiate into functional schwann cells that sustain peripheral nerve regeneration. J Neuropathol Exp Neurol 69:973–985

Matsuura K, Nagai T, Nishigaki N, Oyama T, Nishi J, Wada H, Sano M, Toko H, Akazawa H, Sato T, Nakaya H, Kasanuki H, Komuro (2004) I Adult cardiac Sca-1-positive cells differentiate into beating cardiomyocytes. J Biol Chem 279:11384–11391

Matsuzuka T, Rachakatla RS, Doi C, Maurya DK, Ohta N, Kawabata A, Pyle MM, Pickel L, Reischman J, Marini F, Troyer D, Tamura M (2010) Human umbilical cord matrix-derived stem cells expressing interferon-beta gene significantly attenuate bronchioloalveolar carcinoma xenografts in scid mice. Lung Cancer 70:28–36

Mayer JE Jr (1995) Uses of homograft conduits for right ventricle to pulmonary artery connections in the neonatal period. Semin Thorac Cardiovasc Surg (Annu) 7:130–132

McElreavey KD, Irvine AI, Ennis KT, McLean WH (1991) Isolation, culture and characterisation of fibroblast-like cells derived from the Wharton's jelly portion of human umbilical cord. Biochem Soc Trans 19:29S

Mitchell KE, Weiss ML, Mitchell BM, Martin P, Davis D, Morales L, Helwig B, Beerenstrauch M, Abou-Easa K, Hildreth T, Troyer D, Medicetty S (2003) Matrix cells from Wharton's jelly form neurons and glia. Stem Cells 21:50–60

Najar M, Rouas R, Raicevic G, Boufker HI, Lewalle P, Meuleman N, Bron D, Toungouz M, Martiat P, Lagneaux L (2009) Mesenchymal stromal cells promote or suppress the proliferation of t lymphocytes from cord blood and peripheral blood: the importance of low cell ratio and role of interleukin-6. Cytotherapy 11:570–583

Nakamizo A, Marini F, Amano T, Khan A, Studeny M, Gumin J, Chen J, Hentschel S, Vecil G, Dembinski J, Andreeff M, Lang FF (2005) Human bone marrow-derived mesenchymal stem cells in the treatment of gliomas. Cancer Res 65:3307–3318

Nanaev AK, Kohnen G, Milovanov AP, Domogatsky SP, Kaufmann P (1997) Stromal differentiation and architecture of the human umbilical cord. Placenta 18:53–64

Naughton BA, San Roman J, Liu K, Purchio A, Pavelec R, Rekettye L (1997) Cells isolated from Wharton's jelly of the human umbilical cord develop a cartilage phenotype when treated with TGF-β in vitro. FASEB J 11:A19

Ogawa M, Larue AC, Mehrotra M (2013) Hematopoietic stem cells are pluripotent and not just "hematopoietic". Blood Cells Mol Dis 51:3–8

Ohta M, Suzuki Y, Chou H, Ishikawa N, Suzuki S, Tanihara M, Suzuki Y, Mizushima Y, Dezawa M, Ide C (2004) Novel heparin/alginate gel combined with basic fibroblast growth factor promotes nerve regeneration in rat sciatic nerve. J Biomed Mater Res Part A 71:661–668

Paldino E, Cenciarelli C, Giampaolo A, Milazzo L, Pescatori M, Hassan HJ, Casalbore P (2014) Induction of dopaminergic neurons from human Wharton's jelly mesenchymal stem cell by forskolin. J Cell Physiol 229(2):232–244

Pappa KI, Anagnou NP (2009) Novel sources of fetal stem cells: where do they fit on the developmental continuum? Regen Med 4:423–433

Peng J, Wang Y, Zhang L, Zhao B, Zhao Z, Chen J, Guo Q, Liu S, Sui X, Xu W et al (2011) Human umbilical cord Wharton's jelly-derived mesenchymal stem cells differentiate into a schwann-cell phenotype and promote neurite outgrowth in vitro. Brain Res Bull 84:235–243

Pierdomenico L, Lanuti P, Lachmann R, Grifone G, Cianci E, Gialã L, Pacella S, Romano M, Vitacolonna E, Miscia S (2001) Diabetes mellitus during pregnancy interferes with the biological characteristics of Wharton's jelly mesenchymal stem cells. Open Tissue Eng Regen Med J 4:103–111

Pittenger MF, Mackay AM, Beck SC, Jaiswal RK, Douglas R, Mosca JD, Moorman MA, Simonetti DW, Craig S, Marshak DR (1999) Multilineage potential of adult human mesenchymal stem cells. Science 284:143–147

Prasanna SJ, Jahnavi VS (2011) Wharton's jelly mesenchymal stem cells as off-the -shelf cellular therapeutics: a closer look into their regenerative and immunomodulatory properties. Open Tissue Eng Regen Med J 4:28–38

Prasanna SJ, Gopalakrishnan D, Shankar SR, Vasandan AB (2010) Pro-inflammatory cytokines, IFNgamma and TNFalpha, influence immune properties of human bone marrow and Wharton jelly mesenchymal stem cells differentially. PloS One 5:e9016

Purchio AF, Naughton BA, Roman JS (1999) Production of cartilage tissue using cells isolated from Wharton's jelly. US Patent 5(919):702

Rachakatla RS, Marini F, Weiss ML, Tamura M, Troyer D (2007) Development of human umbilical cord matrix stem cell-based gene therapy for experimental lung tumors. Cancer Gene Ther 14:828–835

Rachakatla RS, Pyle MM, Ayuzawa R, Edwards SM, Marini FC, Weiss ML, Tamura M, Troyer D (2008) Combination treatment of human umbilical cord matrix stem cell-based interferon-beta gene therapy and 5-fluorouracil significantly reduces growth of metastatic human breast cancer in scid mouse lungs. Cancer Invest 26:662–670

Rey Nores JE, Bensussan A, Vita N, Stelter F, Arias MA, Jones M, Lefort S, Borysiewicz LK, Ferrara P, Labeta MO (1999) Soluble cd14 acts as a negative regulator of human t cell activation and function. Eur J Immunol 29:265–276

Romanov YA, Svintsitskaya VA, Smirnov VN (2003) Searching for alternative sources of postnatal human mesenchymal stem cells: candidate MSC-like cells from umbilical cord. Stem Cells 21:105–110

Sarugaser R, Lickorish D, Baksh D, Hosseini MM, Davies JE (2005) Human umbilical cord perivascular (HUCPV) cells: a source of mesenchymal progenitors. Stem Cells 23:220–229

Sarugaser R, Hanoun L, Keating A, Stanford WL, Davies JE (2009) Human mesenchymal stem cells self-renew and differentiate according to a deterministic hierarchy. PloS ONE 4:e6498

Sarugaser R, Ennis J, Stanford WL, Davies JE (2009) Isolation propagation, and characterization of human umbilical cord perivascular cells (HUCPVCs). Methods Mol Biol 482:269–279

Scheers I, Lombard C, Najimi M, Sokal E (2011) Cell therapy for the treatment of metabolic liver disease: an update on the umbilical cord derived stem cells candidates. Open Tissue Eng Reg Med J 4:48–53

Schoen FJ, Levy RJ (1999) Tissue heart valves: current challenges and future research perspectives. J Biomed Mater Res 47:439–465

Schugar RC, Chirieleison SM, Wescoe KE, Schmidt BT, Askew Y, Nance JJ, Evron JM, Peault B, Deasy BM (2009) High harvest yield, high expansion, and phenotype stability of CD146 mesenchymal stromal cells from whole primitive human umbilical cord tissue. J Biomed Biotechnol 2009:789526

Selmani Z, Naji A, Zidi I, Favier B, Gaiffe E, Obert L, Borg C, Saas P, Tiberghien P, Rouas-Freiss N, Carosella ED, Deschaseaux F (2008) Human leukocyte antigen-g5 secretion by human mesenchymal stem cells is required to suppress t lymphocyte and natural killer function and to induce cd4 + cd25highfoxp3 + regulatory t cells. Stem Cells 26:212–222

Semenov O, Breymann C (2011) Mesenchymal stem cells derived from Wharton's jelly and their potential for cardio-vascular tissue engineering. Open Tissue Eng Regen Med J 4:64–71

Semenov OV, Koestenbauer S, Riegel M, Zech N, Zimmermann R, Zisch AH, Malek A (2010) Multipotent mesenchymal stem cells from human placenta: critical parameters for isolation and maintenance of stemness after isolation. Am J Obstet Gynecol 202:e191–e193

Sharma T, Kumari P, Pincha N, Mutukula N, Saha S, Jana SS, Ta M (2013) Inhibition of non-muscle myosin II leads to G_0/G_1 arrest of Wharton's jelly-derived mesenchymal stromal cells. Cytotherapy 0:1–13

Shinoka T, Ma PX, Shum-Tim D, Breuer CK, Cusick RA, Zund G, Langer R, Vacanti JP, Mayer JE Jr (1996) Tissue-engineered heart valves. Autologous valve leaflet replacement study in a lamb model. Circulation 94:64–168

Strakova Z, Livak M, Krezalek M, Ihnatovych I (2008) Multipotent properties of myofibroblast cells derived from human placenta. Cell Tissue Res 332:479–488

Takahashi K, Yamanaka S (2006) Induction of pluripotent stem cells from mouse embryonic and adult fibroblast cultures by defined factors. Cell 126:663–676

Takechi K, Kuwabara Y, Mizuno M (1993) Ultrastructural and immunohistochemical studies of Wharton's jelly umbilical cord cells. Placenta 14:235–245

Tamura M, Kawabata A, Ohta N, Uppalapati L, Becker KG, Troyer D (2011) Wharton's jelly stem cells as agents for cancer therapy. Open Tissue Eng Regen Med J 4:39–47

Tang W, Zeve D, Suh JM, Bosnakovski D, Kyba M, Hammer RE, Tallquist MD, Graff JM (2008) White fat progenitor cells reside in the adipose vasculature. Science 322:583–586

Tchoukalova Y, Koutsari C, Jensen M (2007) Committed subcutaneous preadipocytes are reduced in human obesity. Diabetologia 50:151–157

Thomson JA, Itskovitz-Eldor J, Shapiro SS, Waknitz MA, Swiergiel JJ, Marshall VS, Jones JM (1998) Embryonic stem cell lines derived from human blastocysts. Science 282:1145–1147

Tipnis S, Viswanathan C, Majumdar AS (2010) Immunosuppressive properties of human umbilical cord-derived mesenchymal stem cells: role of b7-h1 and ido. Immunol Cell Biol 88:795–806

Tomita S, Mickle DA, Weisel RD, Jia ZQ, Tumiati LC, Allidina Y, Liu P, Li RK (2002) Improved heart function with myogenesis and angiogenesis after autologous porcine bone marrow stromal cell transplantation. J Thorac Cardiovasc Surg 123:1132–1140

Troyer DL, Weiss ML (2008) Wharton's jelly-derived cells are a primitive stromal cell population. Stem Cells 26:591–599

Wang HS, Hung SC, Peng ST, Huang CC, Wei HM, Guo YJ, Fu YS, Lai MC, Chen CC (2004) Mesenchymal stem cells in the Wharton's jelly of the human umbilical cord. Stem Cells 22:1330–1337

Wang XY, Lan Y, He WY, Zhang L, Yao HY, Hou CM, Tong Y, Liu YL, Yang G, Liu XD, Yang X, Liu B, Mao N (2008) Identification of mesenchymal stem cells in aorta-gonad-mesonephros and yolk sac of human embryos. Blood 111:2436–2443

Wang L, Seshareddy K, Weiss ML, Detamore MS (2009) Effect of initial seeding density on human umbilical cord mesenchymal stromal cells for fibrocartilage tissue engineering. Tissue Eng Part A 15:1009–1017

Wang L, Zhao L, Detamore MS (2011) Human umbilical cord mesenchymal stromal cells in a sandwich approach for osteochondral tissue engineering. J Tissue Eng Reg Med 5:712–721

Wang H, Qiu X, Ni P, Qiu X, Lin X, Wu W, Xie L, Lin L, Min J, Lai X, Chen Y, Ho G, Ma L (2014) Immunological characteristics of human umbilical cord mesenchymal stem cells and the therapeutic effects of their transplantion on hyperglycemia in diabetic rats. Int J Mol Med 33(2):263–270

Weiss ML, Anderson C, Medicetty S, Seshareddy KB, Weiss RJ, VanderWerff I, Troyer D, McIntosh KR (2008) Immune properties of human umbilical cord Wharton's jelly-derived cells. Stem Cells 26:2865–2874

Wharton T (1996) Adenographia [trans. S. Freer]. Oxford University, Oxford

Xu Q, Zhang HT, Liu K, Rao JH, Liu XM, Wu L, Xu BN (2011) In vitro and in vivo magnetic resonance tracking of sinerem-labeled human umbilical mesenchymal stromal cell-derived schwann cells. Cell Mol Neurobiol 31:365–375

Zarkhin V, Talisetti A, Li L, Wozniak LJ, McDiarmid SV, Cox K, Esquivel C, Sarwal MM (2010) Expression of soluble HLA-G identifies favorable outcomes in liver transplant recipients. Transplantation 90:1000–1005

Zhang X, Zhang Q, Li W, Nie D, Chen W, Xu C, Yi X, Shi J, Tian M, Qin J, Jin G, Tu W (2014) Therapeutic effect of human umbilical cord mesenchymal stem cells on neonatal rat hypoxic-ischemic encephalopathy. J Neurosci Res 92(1):35–45

Index

L.-R. Zhao, J. H. Zhang (eds.), *Cellular Therapy for Stroke and CNS Injuries*,
Springer Series in Translational Stroke Research, DOI 10.1007/978-3-319-11481-1